The Quaternary History of the Irish Sea

Geological Journal Special Issue No. 7

GEOLOGICAL JOURNAL SPECIAL ISSUES

ISSN 0435 — 3951
ISBN 0 902354 07 8

The Quaternary History of the Irish Sea

Edited by

C. Kidson

Professor of Physical Geography and Head of the Department of Geography,
University College of Wales, Aberystwyth.

and

M. J. Tooley

Lecturer in Geography, University of Durham.

Series Editor

G. Newall

International Union for Quaternary Research.
Commission on Shore-Lines.
Subcommission on Shorelines of Northwestern Europe.

Seel House Press, Liverpool

SEEL HOUSE PRESS
Seel Street, Liverpool L1 4AY

First Edition, March, 1977

Printed in Great Britain by
SEEL HOUSE PRESS
Seel Street, Liverpool L1 4AY

Contents

Contributors

C. Kidson is Professor of physical geography and head of the department of geography at the University College of Wales, Aberystwyth. Formerly he was lecturer at the University of Exeter and head of the Coastal Unit of the Nature Conservancy. His research interests include coastal geomorphology and the 'inheritance' of coasts during the Quaternary in South West England, East Anglia and Cardigan Bay.

M. J. Tooley is lecturer in geography at the University of Durham, and tutor at St. Aidan's College. He is the United Kingdom representative of the UNESCO-IGCP Sea-level project. His research interests include sea-level changes in northern England, north Wales, Lincolnshire and the north Fenlands.

D. Q. Bowen is Reader in the University College of Wales, Aberystwyth and is a member of the Royal Society INQUA Committee. His research interest is Quaternary stratigraphy, particularly in south and west Wales.

P. A. Carter is a teacher at Whinney Hill School, Durham City and formerly held an NERC research studentship at the University of Durham. His research interest is coastal palaeoenvironments and contemporary and fossil diatom assemblages in Cumbria and the Isle of Man.

M. R. Dobson is Senior Lecturer in the department of geology, University College of Wales, Aberystwyth, with special responsibility for the marine research programme. His research interests include the stratigraphy and structural evolution of the western margin of the British Isles, especially in the Irish, Celtic and Malin Seas.

R. A. Garrard is a member of the exploration department of Burmah Oil (North Sea Ltd.), now a subsidiary of the British National Oil Corporation. Formerly, he held a research post in the department of geology, University College of Wales, Aberystwyth.

D. Huddart is lecturer in Quaternary Studies at Trinity College, Dublin. After doctoral research at the University of Reading he was a post-doctoral research fellow at the University of Newcastle. His research interests include the interpretation of Pleistocene sediments in Cumbria, eastern Ireland and northern Europe.

W. G. Jardine is Senior Lecturer in the department of geology, University of Glasgow. Formerly, he was a member of the Soil Survey of Scotland. He is Secretary-General of the tenth INQUA Congress. His research interests are Quaternary stratigraphy and shorelines with particular reference to south west Scotland.

A. M. McCabe is lecturer in the department of biology and earth sciences at the Northern Ireland Polytechnic. His research interests include late glacial raised beach features and ice limits in north and east Ireland.

D. N. Mottershead is Senior Lecturer in geography at Portsmouth Polytechnic. His research interests include periglacial features in south west England, north west Scotland and the Canadian Arctic.

H. M. Pantin is a member of the Institute of Geological Sciences (Continental Shelf Unit South) where he has worked on Quaternary sediments in the Irish and Celtic Seas. Formerly, he was a member of the New Zealand Oceanographic Institute at Wellington.

N. Stephens is Professor of geography at the University of Aberdeen. Formerly, he was lecturer in geography at the Queen's University of Belfast. His research interests include coasts, mass movement and Pleistocene geomorphology, mainly in Ireland and south west England.

F. M. Synge is a member of the Geological Survey of Ireland. His research interest lies in the mapping of Quaternary deposits and in shoreline studies in Ireland, Finland, Norway and Arctic Canada.

G. S. P. Thomas is lecturer in geography at the University of Liverpool. His research interests include Quaternary stratigraphy of the Isle of Man, glacio-tectonic structures and glacio-marine sediments.

Preface

1977 is an important year for Quaternary Studies in the British Isles; it is the year in which the International Union for Quaternary Researcn (INQUA) holds its tenth congress at Birmingham University. In addition, it marks the end of a fifty year period during which the tempo of research into vegetational history, ice limits, stratigraphy of unconsolidated sediments, climatic change and sea-level movements has increased. Although there is a long history of research in Quaternary sediments in the British Isles, visits by the late Professor G. Erdtman in 1922 and the late Professor L. von Post in 1930 and their subsequent publications in 1924, 1926, 1928 and 1933 showed the potential of pollen and stratigraphic analysis for the elucidation of problems such as vegetational history and sea-level changes. At about the same time, Professor Sir Harry Godwin developed a sound foundation for the successful and imaginative development of Quaternary Studies in the British Isles based on pollen, macro-fossil and stratigraphic analyses and later radiocarbon dating at Cambridge. Both Erdtman, Godwin and von Post made contributions to the history of the Irish Sea, applying techniques that have been used with effect subsequently.

This collection of essays is a fitting tribute to their pioneer work. The volume is one of three prepared under the auspices of the Subcommission on Shorelines of Northwestern Europe of the Commission on Shore-Lines of the International Union for Quaternary Research. Together with its companion volumes on the Baltic Sea and North Sea it is presented as a contribution to the Tenth International Congress of INQUA to be held in Birmingham, England from 16th to 24th August 1977. It marks also the symposium on The Quaternary history of the Baltic, the North Sea and the Irish Sea held in March 1977 to celebrate the 500th anniversary of Uppsala University.

The essays, that constitute this volume do not offer a consensus view, but present each author's research material, interpretation, synthesis and correlations. No attempt has been made to standardise nomenclature, although most authors use the recommended stratigraphical nomenclature of the British Quaternary proposed by Shotton (1973). The exception is the use of Holocene by some authors in place of the stage name Flandrian, applied to the last 10,000 radiocarbon years. Correlation of Irish regional stages with British regional stages is not explicit, and reference to the table in Mitchell *et al.* (1973) provides a correlation of these regional stages.

MITCHELL, G. F., COLHOUN, E. A., STEPHENS, N. and SYNGE, F. M. 1973. Ireland. *In*, MITCHELL, G. F. *et al.* 1973 A correlation of Quaternary deposits in the British Isles. *Geol. Soc. Lond.* Special Report No. **4,** 67–80.
SHOTTON, F. W. 1973. General principles governing the subdivision of the Quaternary system. *In*, MITCHELL, G. F. *et al.* 1973. A correlation of Quaternary deposits in the British Isles. *Geol. Soc. Lond.* Special Report No. **4,** 1–7.

January 1977   *C. Kidson*
                                                                               M. J. Tooley

Some problems of the Quaternary of the Irish Sea

C. Kidson

---

1. Introduction

The Irish Sea is less easy to define than either the Baltic or North Seas to which the companion volumes in this series are devoted. A strict interpretation would confine it to the area of sea lying between England and Ireland. The sea between Wales and Ireland is named St. George's Channel. It is, however, more practicable to regard all the seas lying between the whole of Great Britain and Ireland as the Irish Sea. That part lying north of a line from Holyhead to Dublin may be described as the northern Irish Sea Basin with the area to the south as the southern Basin. For the purposes of this series of essays, the boundaries are cast even wider. Dr. Jardine's area includes not only the coast of southwest Scotland which can, *sensu stricto*, be defined as part of the coast of the Irish Sea, but also the whole of the Hebridean region. Similarly, while southwest England and the English Channel coasts (C. Kidson and D. N. Mottershead, this volume) would by any definition be regarded as outside the limits of the Irish Sea and part of the "Western Approaches" they have been included here as a link with the North Sea volume. To avoid undue overlap, the eastern boundary of the Channel coast of England has been taken as Dungeness. Increasingly the term "Celtic Sea" has come to be applied to the area between the south coasts of Ireland and Wales and the Atlantic coast of Cornwall. This terminology, arising largely as the result of exploration for oil and gas on the continental shelf, has been adopted by some of the contributors to this volume.

A primary purpose of this first short essay is to introduce the succeeding 11 which constitute essays on those areas of western Britain which can conveniently be grouped under the title, "The Irish Sea". In addition, attention will be drawn to a number of problems of special significance along the western seaboard of Britain.

Dobson, Pantin, Whittington and Garrard (this volume) are concerned with the structure, sediments and geological history of the "basin" itself while the remaining eight attempt to analyse the development of the surrounding coasts during the Quaternary. Each of the 11 contributions has been written as a separate essay. No attempt has been made to arrive at a common view of the Quaternary development of the area. It is not surprising, therefore, that differing approaches and interpretations are to be found within the following pages. Not all the authors would regard themselves as specialists in the Quaternary. Some degree of overlap has been regarded as inevitable and, in some respects, desirable.

It must be said at once that there is little general agreement on many aspects of Quaternary history in western Britain. Mitchell (1972) entitled his second review, "The Pleistocene History of the Irish Sea: second approximation". This choice of words indicates that much of the evidence is unsatisfactory and equivocal and many uncertainties remain. Even his "approximation" does not command general support. This is not surprising since many sections lack dateable organic material and this, together with the absence of radiometric dating techniques covering the Middle and Upper Pleistocene, means that they are capable of widely differing interpretations. Until very recently, opinions, particularly on Quaternary events in the southern parts of the Irish Sea Basin, could be grouped into a so-called "Irish School" and a so-called "Anglo-Welsh School". The former view, summarised in Mitchell (1972) revolved around the argument that "the shores of the Irish Sea are . . . reluctant to reveal deposits of the Ipswichian warm stage". The interpretations of Mitchell (1972), Mitchell *et al.* (1973), Stephens (1966, 1970 and 1973) and Synge (1970), were all based on the argument that what may be termed the "main raised beach" of the area as exemplified at Courtmacsherry in southern Ireland, on the Gower coast of South Wales and around Barnstaple Bay in southwest England, could be dated as Hoxnian. The deposits overlying this beach then had all to be viewed as post-Hoxnian in age. It is therefore not surprising in this context, that raised beaches from the last interglacial, the Ipswichian, were not easily recognised.

The alternative view that the main raised beach is Ipswichian in age was put forward by Bowen (1970 and 1973), Kidson (1971) and Kidson and Wood (1974). The essential arguments in this debate are given in detail by Bowen and Kidson (this volume). However, at least one member of the so-called "Irish School" has radically changed his views. Dr. Synge (this volume) now suggests that much of the Irish succession, including the Courtmacsherry beach, is "much younger than previously thought". Indeed, he seriously argues for a Middle Devensian age for the beach formerly ascribed to the Hoxnian! Since he also suggests a mid-Devensian age for the Fremington Till (regarded as Wolstonian by Kidson, this volume), it is unlikely that Dr. Synge's new ideas will have greater success in converting the "Anglo-Welsh School" to the new Irish chronology than was achieved with the earlier, more extensive version!

If differences of view between the two sides of the Irish Sea seem to have been confined in the above discussion to the south, this does not mean that agreement is to be found in the north. Here, isostatic adjustment has had a more significant role and differences of view have revolved around this aspect. These differences are noted in section 5b below.

2. Glacial limits in the Irish Sea

Mitchell's two review papers (1960, 1972) illustrated a degree of uncertainty about the maximum limits of ice advances in the Irish Sea which has persisted to the present day. Indeed, in this volume the uncertainty is considerably extended.

Mitchell's Devensian ice limits varied between (1960) a line running from the Lleyn Peninsula in North Wales to Wexford on the Irish side and (1972) a line between Mathry on the Pembrokeshire coast of Wales and just north of Carnsore Point on the Irish Coast. It is perhaps not too broad a generalisation to suggest that in the past discussion (see for example Bowen, this volume) on these Devensian limits has been concerned with relatively local "detail"—whether or not ice reached the coastal area of central Cardigan Bay and the relationship of Devensian tills in the eastern part of the Gower Peninsula to the main advances of Devensian ice in the Irish Sea. In the present volume much more fundamental debate centres on whether Devensian Irish Sea ice, which Dr. Thomas (this volume) claims left the mountain core of the Isle of Man unglaciated, could have extended to Pembroke-shire, as argued by Bowen (1973) and questioned by Thomas, or even, as suggested by Dr. Synge (this volume) extended as far south as north Devon. Dr. Thomas suggests that the different views might be reconciled if the idea of a single ice mass thrusting down the Irish Sea Basin were replaced by two separate masses "bisecting about the pivot of the Isle of Man". He suggests that different views of the "geo-metry" of Devensian Irish Sea ice can be reconciled in terms of the varying advance and retreat stages. How difficult such reconciliations may be is indicated by the discussion in this volume on Devensian stages in northwest England and Ulster.

The evidence from the sea floor presented by Dr. Garrard (this volume) supports neither the views of Dr. Thomas in his arguments for a more northerly Devensian limit nor the views of Dr. Synge, when he suggests that the tills at Ballycroneen and Fremington are related to a more southerly maximum extension of the ice at around 22000 B.P.! The existence of a sheet of Irish Sea till of the last glaciation in St. George's Channel between Pembrokeshire and Leinster and some little way to the south, as demonstrated by Dr. Garrard, destroys Dr. Thomas's arguments on ice gradients. Similarly, his evidence that the Devensian till did not formerly extend much further south indicates that Dr. Synge's arguments for a late Devensian age for the Fremington till are not well founded. The evidence from the floor of the southern Irish Sea Basin supports the broadly "orthodox" view of Devensian limits as put forward, for example, by Mitchell (1972). Garrard's interpretations of these limits at many coastal sites, particularly in Wales, may be regarded by some as confusing and complicating the debate in these areas.

It cannot be claimed that knowledge of pre-Devensian glacial limits is greatly advanced by the present volume.

In his 1960 paper Mitchell suggested an Anglian limit from Courtmacsherry on the coast of Southern Ireland to Porthleven in Cornwall and a Wolstonian ice front stretching from Ballycroneen in County Cork to Fremington in Devon. His 1972 paper revised this view. He argued that deposits of an Anglian ice sheet had not been identified even though some of the "giant" erratics of Cornwall and Devon, and even erratics in the raised beach at Slindon in Sussex, "may be of Anglian or still earlier age". His revised Wolstonian limit ran from Ballycroneen through Fremington to Pencoed in South Wales in a wide arc which touched the northern shores of the Isles of Scilly and skirted the Atlantic Coast of Cornwall. Dr. Garrard describes in this volume the distribution of pre-Devensian till in the Southern Irish

Sea Basin and in the Celtic Sea. His "pre-Devensian Glacial Limit" (his figure 9) is virtually the same as Mitchell's (1972) Wolstonian limit, except at the head of the Bristol Channel, and suggests that he has accepted Mitchell's interpretations of coastal sites in Devon and Cornwall. The papers by Whittington and Garrard cannot, therefore, be regarded as providing additional evidence in support of this limit. The distribution of till on the sea floor could equally be interpreted as supporting a more northerly limit. It certainly does not provide support for any extension of pre-Devensian ice into the English Channel as suggested by Kellaway *et al.* (1975).

It is perhaps also important to note that the term pre-Devensian is used and not Wolstonian or Anglian, although Garrard suggests that the evidence does not preclude two pre-Devensian glaciations. Bowen (this volume) also remains agnostic on the age of the glacial event preceding the Ipswichian for which evidence is to be found on the Welsh coast. In this volume Kidson firmly assigns the Fremington Till, regarded as part of the same pre-Ipswichian glacial event, to a Wolstonian glaciation and in this follows Stephens (1970, 1973) and Mitchell (1960, 1972). It has to be admitted that there is little solid evidence to justify a Wolstonian age. The depth of weathering and the degree of decalcification in the Fremington Till would support a Wolstonian age but the caution shown by Garrard and Bowen is justified.

3. Devensian high sea-levels

Evidence for sea-levels in the last glaciation only a little below that of the present has been accumulating since Curry's (1961) pioneering paper. Thus, Donovan (1962) lent support to Curry's ideas from the Bristol Channel. Despite problems of the validity of some 14C dates on shell and other materials (Milliman and Emery 1968) and difficulties arising from tectonic and isostatic movements, the weight of evidence from widely scattered parts of the world has steadily become convincing. Three papers, in particular, in the present volume add to this evidence and contribute to the debate on the relationships between such high sea-levels and the various local advance and retreat stages of Devensian ice. Dr. Jardine reviews the evidence for the Scottish west coast. Dr. Thomas adds further data from the Isle of Man and Drs Stephens and McCabe present some particularly interesting and convincing information from Ulster. Of special interest is the evidence of high sea-levels, presumably in response to world wide eustatic movements, while ice remained in the Northern Irish Sea Basin. Such confirmation of the possibility of high sea-levels close to the limits of ice advance in the Irish Sea may help to resolve difficulties relating to earlier glaciations. One of the intriguing unsolved problems (Kidson and Bowen 1976) of the southern part of the area covered by this volume (Kidson and Mottershead) is the method of emplacement of the many giant erratics well outside any generally accepted glacial limits. Ice rafting has long been suggested as the agency for the distribution of these giant blocks along the English Channel coast from Cornwall to Sussex (Reid 1892). The major difficulty in accepting such an explanation has, of course, been that of reconciling the idea of a sea-level close to that of the present, necessary for the emplacement of those erratics found high up on the contemporary shore platform, with glacial maxima. The evidence for Devensian high sea-levels in the Northern Irish Sea Basin lends support to the view that local ice advances could have coincided with eustatic rises of sea-level responding to major ice retreats in other parts of the world. If similar evidence were to be found for Wolstonian or Anglian high sea-levels, one unsolved problem, at least, relating to glacial maxima in the area in the Middle Pleistocene might be solved.

4. The nature of the Holocene rise of sea-level

Differences of view, referred to in section 6c below, on whether or not sea-level in the recent past has exceeded that of the present, are only part of the debate on the nature of the recovery of the sea from low Late Devensian levels. Some, including many who accept broadly the ideas initiated by Fairbridge (1961) on higher Late Holocene sea-levels, regard the Flandrian transgression as composed, particularly in its later phases, of a series of transgressions separated by regressive phases. While they accept that some of the alternations of peats and marine clays in Holocene coastal sections may be explained in purely local terms, by for example the breaching of beach barriers, they relate many of the intercalating layers of terrestrial and marine sediments to actual changes of sea-level. They correlate them with palaeo-temperature and other data on climatic change. Dr. Tooley, who contributes to this volume and who presented his ideas on Holocene sea-level changes in the Northern Irish Sea Basin in more detail in a recent paper (Tooley 1974) belongs to this "school". Others, including many who argue with Shepard (1963) that Holocene sea-levels have not exceeded that of the present, regard the Flandrian transgression as resulting from a smooth rise in sea-level, initially rapid but decelerating progressively in the later Holocene. They explain alternations of marine and terrestrial sediments in Holocene coastal sections partly in terms of local geomorphic changes but also, more importantly, as consequences of differences between rates of sea-level rise, of sedimentation and, in places, of isostatic rebound. Kidson, whose work with Heyworth (1973) in the southern part of the "Irish Sea" is discussed briefly in this volume, belongs to this school of thought. These differences of approach cannot, as yet, be reconciled but it is probable that further work in the "Irish Sea" area will contribute to this end.

5. Problems relating to tectonic history and isostatic rebound

5a. Tectonics

The northern and western regions of the British Isles have a complex geological and structural history. Trend lines are dominantly Caledonian in the north and Hercynian in the south. Folding and faulting of all ages from Pre-Cambrian to Tertiary have affected the region and vertical movements of tectonic origin can be regarded as the rule rather than the exception. In assessing the events of the Quaternary, the most significant movements are those of Cretaceous and Tertiary age. Dr. Dobson draws attention in this volume to a period of intense tectonic activity throughout virtually the whole of western Britain, particularly in the late Palaeogene and early Neogene, producing uplift and eastward tilting of basement blocks. Oligocene deposits of great thickness from Skye in the north to Bovey Tracey, Devon, in the south, give evidence of warping and faulting on a massive scale. However, he concludes that, "Major vertical movements on the coastal faults had largely ceased by the early Pliocene, as evidenced by the Neogene overstep across the south of Ireland fracture and the structure west of St. Bride's Bay".

Despite the fact that the Pliocene-Pleistocene appears to have been a tectonically quiet period along the western seaboard of Britain. Dobson draws attention to the probability of movement in the Quaternary along major basin faults in the southern Irish Sea Basin. In assessing Quaternary events in the region, tectonic activity cannot be lightly dismissed. In south and southwestern Britain the possibility of movement along faults like the Sticklepath in Devon (see Dobson, this volume) is complicated by possible involvement in the known down warping in the Flemish Bight. Despite this, it is argued by Kidson and Mottershead here that the south has been tectonically stable at least since the Middle Pleistocene.

5b. Isostatic rebound

One of the main arguments for stability in the south lies in the presence over wide areas of a series of shore platforms up to a height above O.D. of the order of 12 m. They appear to indicate a return of the sea at intervals, at least during the Middle and Upper Pleistocene, to a level only marginally different from that of today. In many localities the overlying sequence of deposits, raised beach, old dune sands and "head", leaves little doubt that individual fragments are part of the same sequence along the length of the English Channel coast and into south Wales. The appearance of stability is reinforced by higher, older benches, possibly of Early Pleistocene age, which have been traced over the same region of southern and western Britain. It is not possible to follow these erosional features with the same confidence into north-west England. In Scotland the debate now centres on whether the low level shore platform(s), which was formerly regarded as Interglacial in age, could have been cut in the Late Devensian. The equivalents of the main shore platform(s) of south and west Britain are, in Scotland, to be found many metres above the heights recorded in southern and southwestern Britain. The so-called "Pre-glacial rock platform" described by Wright (1911) in the Inner Hebrides at heights of 27 and 41 m above high water of Spring Tides is probably contemporaneous with the platforms at and below 10 to 12 m above O.D. in southwest Britain. (In the Inner Hebrides, High Water of Ordinary Spring Tides is from 2 to 4 m above Ordnance Datum (O.D.), *i.e.* the mean level of the sea as determined at Newlyn in Cornwall.) Wright (1937) suggested that the "high level" shore platform of the Hebrides might be contemporaneous with shoreline features now close to sea-level in northwestern England. Many of the workers who succeeded Wright have identified this "high level" platform in much of western Scotland at heights ranging from 21 to 43 m above high tide level. Many of them, as recorded by Dr. Jardine in this volume, have regarded the platform(s) as warped and tilted. The suggestion now is that, like its southern analogue, it is a composite feature of possibly "multiple" age. If the "high level" shore platform(s) of western Scotland is indeed analogous to the "low level" platform(s) of southern England, differential isostatic adjustment of the order or 50 m between north and south is involved. One of the major problems of the Quaternary of western Britain concerns the nature and distribution of this adjustment.

Account has also to be taken of the known major down-warping in the geosynclinal area of the Flemish Bight which certainly affects southeastern England, and the effects of which may extend into eastern Scotland. If this is so, it complicates and reinforces isostatic adjustments to ice loading and unloading centres (Sissons 1974) in northwestern Scotland. Little precise work has been possible so far along the Western seaboard to demonstrate either in time or space, the isostatic effects of successive ice advances and retreats or the associated withdrawal and return of waters in the shelf seas. There is no hard evidence that isostatic rebound is, or is not, still affecting western Britain. In the south it has been argued (Donovan 1962; Kidson 1971) that no isostatic effects are now discernible in the Bristol Channel and southwest England. As yet unpublished work suggests that at least as far north as the Dovey Estuary, in the centre of the Cardigan Bay coast, any isostatic movements which may have taken place in the late Pleistocene and the early Holocene have cancelled themselves out. The sea-level curve, produced for Bridgwater Bay in the Bristol Channel (Kidson and Heyworth 1973) where Holocene sediments do not reveal any evidence of isostatic rebound, appears to have equal validity in Cardigan Bay.

Dr. Synge has discussed in this volume possible isostatic depression, in the Lower Boyn Valley, of the order of 140 m but is still able to refer to southern Leinster as lying outside the area affected by isostatic movements. Stephens and McCabe argue here that in Ulster, "recovery could be in excess of 80 m since Pollen Zone I about 12000 years B.P.". They quote Sissons (1976) however, who suggests that the "Main Post-Glacial Raised Shoreline" is downtilted from, "about 14 m near Glasgow to 8 m on the Antrim Coast". The high values of isostatic uplift along the coast of western Scotland taken together with the suggested zero values for south and southwestern Britain, and the evidence of tilt in Scotland and Ulster, could be reconciled in terms of a hinge line through southern Leinster and northwest England. A number of papers (Wright 1928; Churchill 1966) have suggested a tilt in Britain from northwest to southeast along axes broadly compatible with this. Some (Valentin 1953; Lennon 1975) have suggested that a tilting movement is discernible at the present time. While the data on which some of the earlier papers were based would now be regarded as inadequate to support their conclusions, the sophistication of the work described by Lennon (1975) merits very serious consideration. He argues for secular variations for the Baltic and North Sea areas along a nodal line from Latvia to northeast England, with the land rising to the north and east of this line and falling to the south and west. He gives a rate of 1 mm per year, "in the whole of southern England". One of the stations which supports this secular variation is Newlyn in Cornwall, where the Mean Sea Level now in use in Britain was calculated. Lennon does, however, carefully draw attention to the many complications which inhibit unquestioning acceptance of these secular movements, within the context of the "growing awareness of the hazardous nature of the coastal environment in general for precise measurements" (see section 6 below). One of the major complications is the response of the earth's surface not only to marine tides but also to Earth body tides. Lennon observes, "ironically, it is in the southwest peninsula, where the fundamental Newlyn bench mark is located, that these terrestrial tidal phenomena reach their maximum development".

Even if secular changes of the order discussed by Lennon, on the basis of observations over a 50 year period, are taking place, they are not of themselves sufficient to negate a conclusion of "long term" stability in southwest England as suggested above from morphological evidence. It is much too simplistic to extrapolate over, say, five millenia and arrive at a displacement of 5 m. The story is, neither simple nor straightforward. There is quite clearly need for more detailed evidence and for a review of the available data. There is also a need for caution in correlating Quaternary sites until possible vertical movements are more fully understood. Some of the complexities become apparent in the papers in the Special Number of the Transactions of the Institute of British Geographers (1966) on Vertical Displacement of Shorelines in Highland Britain.

B

6. Problems relating to the levels of operation of marine processes

6a. Tidal range

Any reader of the British literature on Quaternary coastal sections will immediately be aware that some confusion exists in relation to the datum levels to which features such as raised beaches and shore platforms are referred. The current tendency is to use Ordnance Datum (O.D.). However, many references in the literature relate to High Water Mark, sometimes qualified by "of ordinary tides" or "of ordinary spring tides". The height of any high tide varies enormously from place to place around the British coasts. Thus, for example, High Water of Ordinary Spring Tides at Saunton in Devon ranges from 4·9 to 5·5 m above O.D., while at the southern end of Orfordness in Suffolk, on the opposite coast, it ranges from 0·9 to 1·5 m above O.D. (It was noted in section 2 above, that in the Hebrides the comparable figures are approximately 2 to 4 m above O.D.) These large variations are a function of very variable tidal ranges. Dr. Jardine draws attention in this volume to variations on the Scottish west coast from 1 m to 9 m. At their maximum, tidal ranges can vary from as high as 12·3 to 14·4 m (Equinoctial Spring Range and Extreme Spring Range in the Bristol Channel) to 2·0 to 2·8 m (the comparable figures for Lowestoft on the coast of East Anglia). Since variation in tidal range results from the combined effects of position in relation to amphidromic systems and position on a crenulate coastline, it is not regular in operation. Thus, while Ayr has a Neap Range of 1·5 m and a Spring Range of 2·6 m, Glasgow, a short distance away in the Clyde, has ranges of 2·4 and 4·1 m. Similarly, while Avonmouth, close to the head of the Bristol Channel, has a Neap Range of 6·5 m and a Spring Range of 12·3 m, Ilfracombe, close to the mouth, has figures of 4·3 and 8·5 m and Padstow, on the Atlantic coast of Cornwall, of 3·9 and 6·6 m.

The significance of these height variations for the correlation of coastal features is obviously considerable. Altitude in such a situation is a most misleading parameter. When it is appreciated that changes of sea-level during the Quaternary varied the position of any site in terms of the shape of the adjacent coast and in relation to the amphidromic systems affecting it, the sources of potential error appear considerable. They are probably more significant in the Irish Sea area than anywhere else in Europe. It has to be said that where the factors discussed above are not referred to in the literature, it is probably because authors were unaware of their importance. Much evidence on Quaternary coastal sections and sea-level change will need, therefore to be re-evaluated. A particular significance attaches to evidence of isostatic rebound discussed in the previous section. It is by no means clear, in this regard, that present tidal ranges and possible variations in them throughout the Quaternary have been given adequate attention by workers on coastal problems. These words of caution probably apply equally to the essays in the present volume as to earlier literature. They are of special significance in comparing results in an area like the "Irish Sea" where very high tidal ranges occur, and locations in seas, like the Baltic, where tidal ranges are negligible.

6b. Wave action

Exposure to the action of both storm and swell waves is even more variable in the area covered by the present volume than tidal range. In parts of southwest England and northwest Scotland, fetches in excess of 12000 km are to be found. In the more sheltered parts of the Irish Sea, maximum fetches are measured in hundreds and, in places, in tens of kilometres. Similarly, wave refraction and the distribution of wave energy on such a crenulate coastline is almost infinitely variable. In these circumstances the heights of coastal features resulting from the operations of contemporary marine processes cover a very wide range. Even over short distances heights on a single shore platform (Kidson 1971; Kidson and Manton 1973) can differ by several metres and the crests of beach ridges show similar variation. Thus Chesil Beach in Dorset reaches under 7 m above O.D. at its northwestern end but attains almost 15 m at its southwestern extremity. These responses to contemporary processes differ most between headlands and bays and are to be expected in a storm wave environment such as that of western Britain. If this is true of the contemporary coastline, it was also true during the Quaternary. Many differences in altitude of features of similar age can be explained in terms not of variations in tectonic or isostatic activity, but in terms of differences of wave climate or tidal range. Undoubtedly many mis-interpretations have arisen because of a failure to appreciate these simple relationships. The essential point is that the range of altitudinal variation arising from wave and tidal differences can and frequently does exceed differences caused by isostatic or tectonic activity. In assessing the varying interpretations in the present volume the reader should understand that it is set in a macro-tidal storm wave environment.

The careful reader of this volume will also observe differences between authors in describing the height ranges of features which they regard as different parts of a single whole. Thus, in southwest England, Kidson refers to the main (composite) shore platform in terms of a normal height range of 6 to 9 m above O.D. while referring to particular fragments reaching as high as 12 m or more. In South Wales, Bowen extends that range to from 3·6 to 15·24 m for the same composite feature. These variations are to be expected not necessarily for the reasons given above but because the chance of survival of this or that fragment of old beach or shore platform in such variable stormy seas is itself infinitely variable.

6c. Limits of marine penetration

In the light of the discussion above on tidal ranges and wave activity in the Irish Sea area, some comments on the upper limits of marine processes and landward penetrations are appropriate. They may help in assessing the evidence put forward in the present volume for high sea-levels during the Quaternary. The discussion is particularly relevant when considering Flandrian marine limits, and sea-levels during the later Flandrian said to be higher than that of today. Most authors in this volume refer to these Holocene problems. In particular, Dr. Jardine, whose paper (1973) on this aspect of the Quaternary is of particular relevance, discusses here Flandrian marine sedimentation well landward of the present coastline in western Scotland where isostatic adjustments are, as yet, imperfectly understood. Similarly, in Ulster, Stephens and McCabe survey here in detail post-glacial strandlines above present sea-level. Here the discussion again is largely concerned with Flandrian isostatic rebound, but the authors refer to Mörner's (1971) arguments for a world wide eustatic transgression which, "may have carried sea-level to some 4 m above present" sometime between 6000 and 3500 years ago. It is the belief of the present writer that evidence for such higher Holocene sea-levels should be looked at most critically in the context of the height range of contemporary marine processes. Certainly in areas such as southwest England (Kidson, this volume) where isostatic and tectonic activity during the Holocene are believed to be of little significance, there is no evidence for sea-levels in the later Flandrian higher than the present. Much of the evidence which has in the past been used to postulate a so called "Romano-British transgression" (Godwin 1943, 1956) can be explained in terms of interactions between rate of eustatic rise and rates of sedimentation (Kidson and Heyworth 1973).

Some earlier interpretations show a lack of appreciation of the capacity of the sea to operate at heights well above its normal still water levels and to penetrate well landward of the present strandline which is, in many places, highly artificial. Again it must be emphasised that the area of this Irish Sea Volume is a macro-tidal, storm wave environment where the factors discussed below are of special significance.

It is universally recognised that waves can move sediment above the height of the highest tide. "Storm" beaches, often of pebbles and cobbles, are clear evidence of this capacity. Less well understood is the frequency with which even "storm" beaches are overtopped by the sea even in calm conditions. For example, at the mouth of the River Parrett in Somerset, where the highest part of the storm beach lies at 7·3 m above O.D., well above the level of high water of ordinary spring tides, the lower lying land behind is regularly flooded by the over-topping of the shingle ridges. This can happen during equinoctial spring tides or it can occur when onshore winds blowing for sustained periods raise ordinary spring tides above the levels resulting from astronomical causes. At much wider time intervals, storm surges raise sea-levels several metres above normal. Thus in the storm surge of January 1953 in the North Sea, Steers (1953) records tide levels 2·5 m higher than predicted. Had the surge coincided with high water, the resultant coastal flooding which extended many tens of kilometres inland, would have been much worse and would have reached even greater heights above "normal". The significance of storm surges has not been appreciated in many studies of Quaternary high sea-levels. Frequently they are ignored in consideration of, for example, Flandrian marine sedimentation landward of present shorelines. Recent studies (Gottschalk 1971, 1975) have, however, clearly indicated that in some areas at least they occur much more frequently than has hitherto been appreciated. In the southern North Sea, events of the same order of magnitude as the surge of 31st January 1953 have occurred on average twice a century since the seventh century A.D. There is no reason to assume that their frequency was any lower during the Holocene or for that matter during the Ipswichian or earlier interglacials. The present writer believes that the cumulative effects of exceptional storms, exceptional high tides and storm surges can account for much of the "evidence" for Flandrian sea-levels above that of the present and for Flandrian marine sedimentation landward of the present coast in areas unaffected by isostatic or tectonic activity. This is particularly the case on low coasts where, not infrequently, the land lies below the level of even ordinary spring tides. Thus in the Somerset Levels fronted by storm beaches reaching heights in excess of 7 m above O.D., levels fall, several miles inland, to below 4, and in some places to below 3 m above O.D. In the days before artificial embanking and pumping, such areas were regularly flooded by the present sea. Yet Carbon-14 dates from these localities have been used to suggest fluctuations in Flandrian sea-level curves which cannot possibly be sustained. All evidence for former sea-levels and strandlines must be examined therefore, in the context of how contemporary marine processes operate, something which is, unfortunately, only rarely done.

7. Conclusion

It is clear from this introduction to the contributions which follow that there are differences of view and approach between their authors. It is equally clear that many unsolved problems remain which must await new techniques and particularly the development of new dating methods. It is not surprising that this should be so since the area lay astride the limits of ice advance in each of the major phases at least of the Later Pleistocene. The pace of investigation is quickening and new advances will undoubtedly result from the impending acceleration of the exploration of these western parts of the shelf seas for oil and gas. It is hoped that the essays which follow present a reasonable picture of the debate concerning Quaternary events in the "Irish Sea".

References

BOWEN, D. Q. 1970. South-east and central south Wales. *In*, Lewis, C. A. (Editor). *The Glaciations of Wales and adjacent regions*. Longmans, London 378 pp.
—— 1973. The Pleistocene Successions of the Irish Sea. *Proc. Geol. Assoc.* **84**, 249–273.
CHURCHILL, D. M. 1966. The Displacement of deposits formed at sea-level 6,500 years ago in southern Britain. *Quaternaria* 7, 239–257.
CURRY, J. R. 1961. Late Quaternary sea-levels: a discussion. *Geol. Soc. Am. Bull.* **72**, 159–162.
DONOVAN, D. T. 1962. Sea-levels of the last glaciation. *Geol. Soc. Am. Bull.* **73**, 1297–1298.
FAIRBRIDGE, R. W. 1961.Eustatic Changes in Sea-Level. *In*, Ahrens, L. H. *et al.* (Editors). *Physics and Chemistry of the Earth.* **4.** Pergamon Press, London. 98–185.
GODWIN, H. 1943. Coastal peat beds of the British Isles and North Sea. *J. Ecol.* **31**, 199–217.
—— 1956. *The History of the British flora*. Cambridge University Press 384 pp.
GOTTSCHALK, M. K. E. 1971. *Stormvloeden en Rivieroverstromingen in Nederland. I—de periode voor 1400*. Assen, Von Gorcum. 581 pp.
—— 1975. *Stormvloeden en Rivieroverstromingen in Nederland. II—de periode 1400-1600*. Assen, Von Gorcum. 896 pp.
INSTITUTE OF BRITISH GEOGRAPHERS 1966. Special Number on the Vertical Displacement of shorelines in Highland Britain. *Trans. Inst. Brit. Geogr.* **39**, 145 pp.
JARDINE, W. G. 1973. The Determination of Former Sea-Levels in Areas of Large Tidal Range. *R. Soc. N.Z. Bulletin* **13**, 163–168.
KELLAWAY, G. A., REDDING, J. H., SHEPARD-THORN, E. R. and DESTOMBES, J.-P. 1975. The Quaternary History of the English Channel. *Phil. Trans. R. Soc. A.* **279**, 189–218.
KIDSON, C. 1971. The Quaternary History of the coasts of southwest England with special reference to the Bristol Channel coast. *In*, Gregory, K. J. and Ravenhill, W. L. D. (Editors). *Exeter Essays in Geography*. University of Exeter Press. 1–22.
—— and BOWEN, D. Q. 1976. Some Comments on the History of the English Channel. *Quaternary Newsletter* **18**, 8–10.
——and HEYWORTH, A. 1973. The Flandrian sea-level rise in the Bristol Channel. *Proc. Ussher. Soc.* **2**, 565–584.
—— and MANTON, M. M. M. 1973. Assessment of Coastal Change with the Aid of Photogrammetric and Computer-Aided Techniques. *Est. and Mar. Coastal Sci.* **1**, 271–283.
—— and WOOD, R. 1974. The Pleistocene stratigraphy of Barnstaple Bay. *Proc. Geol. Assoc.* **85**, 223–237.
LENNON, G. W. 1975. Coastal Geodesy and the Relative Movements of Land and Sea-Levels. In, *Geodynamics Today*. The Royal Society, London. 97–104.
MILLIMAN, J. D. and EMERY K. O. 1968. Sea-Levels during the past 35,000 years. *Science* **162**, 1121–1123.
MITCHELL, G. F. 1960. The Pleistocene History of the Irish Sea. *Advmt. Sci.* **17**, 313–325.
—— 1972. The Pleistocene History of the Irish Sea: a second approximation. *Sci. Proc. R. Dubl. Soc.* **4**, 181–199.
—— COLHOUN, E. A., STEPHENS, N. and SYNGE, F. M. 1973. Ireland. *In*, Mitchell *et al*. A Correlation of Quaternary Deposits in the British Isles. *Geol. Soc. Lond.* Special Report No. 4, 99 pp.

MORNER, N.-A. 1971. The Holocene eustatic sea-level problem. *Geologie Mijnb.* **50,** 699–702.

REID, C. 1892. The Pleistocene Deposits of the Sussex Coast and their equivalents in other districts. *Q. J. geol. Soc. Lond.* **48,** 344–363.

SHEPARD, F. P. 1963. Thirty-five thousand years of Sea-Level. *In*, Clements, T. (Editor). *Essays in Marine Geology in Honour of K.O. Emery*. University of South California Press, Los Angeles 1-10.

SISSONS, J. B. 1974. The Quaternary in Scotland: a review. *Scott. J. Geol.* **10,** 311–337.

—— 1976. *The Geomorphology of the British Isles: Scotland*. Methuen, London. 150 pp.

STEERS, J. A. 1953. The East Coast Floods. *Geogr. J.* **119,** 280–298.

STEPHENS, N. 1966. Some Pleistocene Deposits in North Devon. *Biul. periglac.* **15,** 103–114.

—— 1970. The West Country and Southern Ireland. *In*, Lewis, C. A. (Editor). *The Glaciations of Wales and adjoining regions*. Longmans, London 378 pp.

——1973. South West England. *In*, Mitchell *et al.* A Correlation of Quaternary Deposits in the British Isles. *Geol. Soc. Lond.* Special Report No. 4 99 pp.

SYNGE, F. M. 1970. The Pleistocene Period in Wales. *In*, Lewis, C. A. (Editor). *The Glaciations of Wales and adjoining regions*. Longmans, London 378 pp.

TOOLEY, M. J. 1974. Sea-level changes during the last 9,000 years in North-West England. *Geogr. J.* **140,** 18–42.

VALENTIN, H. 1953. Present vertical movements of the British Isles. *Geogr. J.* **119,** 299–305.

WRIGHT, W. B. 1911. On a pre-glacial shoreline in the western isles of Scotland. *Geol. Mag.* **48,** 97–109.

—— 1928. The raised beaches of the British Isles. In, *First Report of the Commission on Pliocene and Pleistocene Terraces*. International Geographical Union 99–106.

—— 1937. *The Quaternary Ice Age*. 2nd edition. Macmillan, London 464 pp.

C. Kidson, The University College of Wales, Department of Geography, Llandinam Building, Penglais, Aberystwyth, Dyfed, Wales. SY23 3DB.

The geological structure of the Irish Sea

M. R. Dobson

The Irish Sea consists of Caledonoid trending basement ridges separated by elongated down-warps or grabens. In the north these downwarps contain Carboniferous and Permo-Triassic sediments which although thick are only remnants. Towards the south progressively more complete sequences are preserved such that the St George's Channel Basin in the South Irish Sea records a history of virtually continuous sedimentation from Carboniferous and Permo-Triassic to later Mesozoic and Tertiary. Although the basement ridges have been affected by faulting it is the downwarps and particularly their marginal zones that have been complicated by phases of intense movement. The most prominent result of this activity include block faulting, causing both uplift and erosion and downwarping with rapid sedimentation. In addition salt tectonics, which are widely recorded in the south, complicate the structural picture further. Seismic data calibrated by coastal boreholes together with shallow offshore coring has allowed the thickness age and character of these basin infills to be understood in broad terms.

1. The North Irish Sea

1a. Introduction

The area here referred to as the Irish Sea is that shown in Figure 1. It is virtually an inland sea being enclosed on all sides with the exception of the narrow North Channel in the northwest and the restricted channel part of the Irish Sea between Dublin and Holyhead. The region covers an area of 900 sq. km.

All the available geological and geophysical data have been used to construct the map. The 49 boreholes and 14 core stations collected up to the end of 1973 form the basis for interpreting the shallow sparker evidence (to about 300 m). The coastal evidence, which includes deep borehole results, coalfield exploration information as well as the fully documented and detailed outcrop data, has been heavily relied upon particularly for the offshore contouring of the two major Permo-Triassic basins. The extension of these contours offshore has been attempted using, in addition, refraction data (Bacon and McQuillin 1972) and the gravity model work of Bott (1964) and the Institute of Geological Sciences (Wright *et al.* 1971).

The general character of the outcrop geology is at the moment uncertain in two important areas, the approaches to the North Channel and southwest of the Isle of Man. In the latter case thick deposits of presumed Quaternary prevent the subcrop geology from being recorded on shallow seismic equipment. In addition because there is very little impedence between Quaternary and Tertiary material (see Dobson *et al.* 1973) it is not known whether Neogene deposits are present in this area.

The regional fold and fault patterns established for the coastal areas have been included in Figure 1 whilst the named features may be located on Figure 2.

1b. The structural framework

Essentially, the region of the North Irish Sea is developed within a Caledonoid trending structural setting, but as emphasised by Moseley (1973), no particular time period is responsible for the evolutionary structural sequence which probably originated in the Pre-Cambrian basement; however, each successive phase is initially dependent on earlier ones.

For example, in the Lake District, Caledonoid structures trending NE/SW to ENE/WSW are developed, yet towards the south and into the region of thick Carboniferous cover the basement exerted a strong influence on the orientation of the Variscan structures which tend to be N/S monoclines, these gradually give way to NE/SW trending folds known collectively as the Ribblesdale folds. It was the post-Carboniferous (Saalian) movements which were responsible for the major tectonics in west Lancashire.

Further west in North Wales the Carboniferous structures as far as the Flintshire coalfield trend NNW/SSE, and near the Great Orme and throughout Anglesey they trend NE/SW. Similarly in central Ireland, the Variscan fold axes turn gradually NE or even NNE. The degree of intensity which is less in the north is attributed to the thickness (relative thinness) of the Upper Palaeozoic for the Carboniferous is really only gently rippled. Thus with the possible exception of parts of North Wales the regional trend of the Variscan structures likely to be found in the North Irish Sea will be the same as that found in central Ireland and in west Lancashire as far as the Craven faults, namely NE/SW to ENE/WSW.

Large scale folding is geophysically recorded only in the Carboniferous and Permo-Triassic in the Irish Sea, the older formations register as acoustic basement. Post-Triassic earth movements were essentially confined to the reactivation of existing faults as normal faults especially along the Cumberland/Lancashire coast. Numerous NNW faults have downthrows towards the southwest. As Moseley (1973) demonstrates the uplifted Askrigg Block is separated from the Irish Sea basin by a series of step faults some of which like the Boundary fault having throws in excess

Fig. 1. Pre-Pleistocene sub-crop map of the North Irish Sea, based in part, on data supplied by the Institute of Geological Sciences.

of 1000 m. Most of these step faults have been included on the map as have the important faults in Anglesey and Ireland.

Within the region three major basement elements are present: the Southern Uplands–Longford Down Massif, the Balbriggan–Manx Ramsey–Whitehaven Arch, and the North Wales Massif; the latter a part of the more extensive Irish Sea geanticline. Geophysical investigations in northern England (Bott 1967) and more recently the Isle of Man (Cornwall 1971) have served to explain in large measure the disposition of the basement ridges. These features prove to be intruded by low density granitic masses which through isostatic adjustment have remained positive elements. The Southern Uplands and their Irish extensions are largely cored by granite batholiths. The long rather narrow Manx-Whitehaven Arch is similarly dominated by a series of granites which may be traced to Ireland. Marine geophysical surveys in the western part of the Irish Sea favour a granite at depth interpretation (Wright *et al.* 1971) rather than the earlier proposals of a deep sedimentary basin (Bott 1964). Masses like the Leinster granite may represent the western extension of this positive zone. Further south the extensive Anglesey region is a part of the weakly positive ridge (Dobson *et al.* 1973) that extends southwest to Rosslare. The present elevation of the Anglesey Precambrian may owe its position more to the regional block tilting than isostatic adjustment. Hinge belts (Johnson 1967) controlled by normal faults, bound the intervening troughs, and whilst mass deficiency arguments (Bott and Masson Smith 1960) may explain the granite cored Massifs on the one hand the compensation mantle flow theory is difficult to apply to these restricted, but persistent elongated basins on the other.

1c. The nature of the regional lows

(i) *The Carboniferous.* Carboniferous rocks are present throughout almost the whole of the North Irish Sea, for they are considered to underlie the Permo-Triassic almost everywhere, and only fail to sub-crop where the Pre-Cambrian and Lower Palaeozoic extend a few kilometres offshore.

Turning first to the western side of the Irish Sea reference must be made to the tectonic frame established by the Caledonian orogeny for northeast Ireland and may be directly compared with the Lancashire region. Like the Balbriggan Massif, which has already been linked to the Manx-Whitehaven Arch, links may also be made with negative regions like the Dublin trough, a marked structural feature that extends into central Ireland. Eastwards it persists as a deep synclinal feature that is developed between Anglesey and the Isle of Man. It may continue and ultimately join with the Carboniferous rocks that outcrop in the Ribblesdale fold belt.

Regional links with the Solway Firth trough are less obvious for it appears to be replaced towards the west by a thin veneer of Carboniferous. Along the east Irish Coast the relationships and subcrop configuration offshore are speculative. For example, the Longford Down Massif was submerged receiving calcareous sediments throughout most of Visean time yet the strata resulting are little more than a 300 m veneer. Consequently on the geological map in the absence of reliable data the southeast margin of the Massif has been extended offshore, with a slight offset to accommodate the Carlingford fault zone, to just south of Clogher Head, leaving Balbriggan as an inlier. The Carboniferous around Drogheda is thus extended to the Irish Sea sub-crop and the Carboniferous around Carlingford Lough is mapped as an outlier although the effects of the several major faults that trend NNW are not known.

Large scale open folding on a general Caledonoid trend is a feature of the

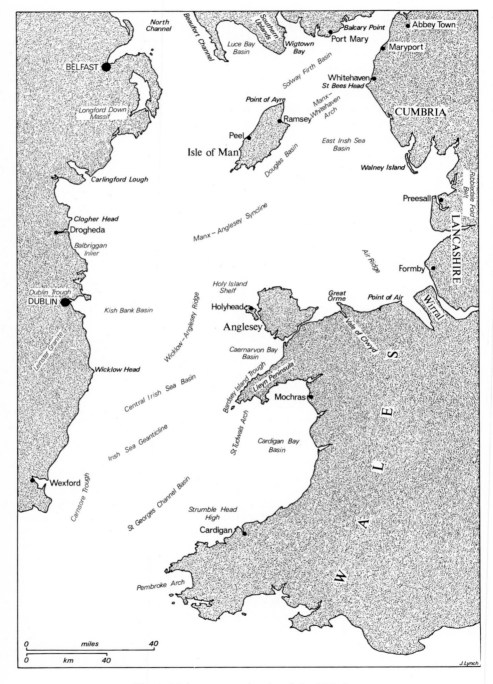

Fig. 2. Major structural units of the Irish Sea.

Carboniferous along the east Irish coast. Distortion of this trend by the presence of the Leinster granite is restricted to the region of the Dublin trough. Offshore these trends persist and extend as far as one can judge, right across the Irish Sea. However, to the northeast, off Strangford Lough, the open folding is abruptly replaced by steep northeasterly dips. The line of separation between the two structural regimes coincides with the projected southern margin disturbance along the edge of the Southern Uplands. This possible line of structural weakness is included on the geological map. One might speculate that separate down faulting in the North Channel along the eastern side may be contributing to the steep dips recorded there.

The tectonic frame, established earlier in the analysis of the west side of the Irish Sea, broadly divides the eastern side into the Solway basin and the East Irish Sea basin. The former is margined on the north side by five outliers of Lower Carboniferous from Port Mary to Balcary Point throughout the Carboniferous and is in fault contact with the Lower Palaeozoic of the Southern Uplands. This fault has a maximum downthrow to the southeast of 600 m (Craig and Nairn 1956).

On the south side of the Solway Firth, the limestone group of West Cumberland resembles in thickness and facies the strata at the north end of the Isle of Man. The gravity models proposed by Bott (1964) for the Solway "Low" envisage a 2300 to 5300 m fill of Carboniferous plus Permo-Triassic. It is unlikely, however, that the Permo-Triassic is less than 1000 m in the west area of Carlisle, which might suggest a Carboniferous fill figure of about 3300 m.

Coastal exposures of Lower Carboniferous in North Wales occur as a series of isolated outcrops, a feature due to the original embayed Carboniferous coastline and emphasised by subsequent structural adjustments. Nevertheless, more than 1000 m of clear water limestone are present. To the north at the southern end of the Isle of Man, 200 m of thin reef facies of comparable age are developed.

The Upper Carboniferous in North Wales is located chiefly in Flintshire and Denbighshire. Workings associated with the Flintshire Coalfields extend under the Dee estuary, close to the Point of Air, and it is this evidence, coupled with the regional structural data, that makes it probable that the Productive Coal Measures occur under the Irish Sea, certainly the pre-Permo-Triassic sub-crop map of Colter and Barr (1975) shows this as a clear possibility.

Direct evidence for the offshore Carboniferous in the East Irish Sea basin area has come from boreholes and four gas wells, two of which penetrated to Carboniferous of Namurian to Westphalian age (Colter and Barr 1975).

Although a broad rather regional syncline is recognised between Anglesey and the Isle of Man, the essential character consists of a series of open shallow folds orientated in the line of the Ribblesdale fold belt. There is considerable seismic evidence, which indicates that this offshore structure compares with the sub-cropping Carboniferous along the Lancashire coast, in that pre-Permian erosion will have removed much of the Westphalian from the anticlines. Any preserved Upper Carboniferous will most probably occur as narrow strips, broken not only by north-northwest normal faulting, but also by east-west dislocation.

A still unresolved area concerns the offshore configuration of the Carboniferous associated with the inshore outcrops in Anglesey, Arvon and around the Great Orme. Although all are of the same zonal age, their possible offshore connections are difficult to determine because of the presence of the Berw and Dinorwic faults. The throw of the faults is of the order of 700 m. Figure 1 shows therefore the likely arrangement, bearing in mind the fault control.

(ii) *The Permo-Triassic*. Rocks of this age are exposed over a wide area of central and northern England, whilst local developments are recorded in southern Scotland and along the North Wales coast. Of most importance in terms of the North Irish Sea, however, are the outcrops along the Cumbria and Lancashire coasts and in the Carlisle Basin. In this area, Audley (1970) used six boreholes for the stratigraphic correlation of the Triassic deposits around the North Irish Sea; these are the onshore boreholes on Figure 1 and correspond to 37–42 on plate 2 in his paper. Evidence of these boreholes and many others along the northwest coast and the associated outcrop data, the thickness of the underlying Permian deposits has been determined. From the total data available contours of the Pre-Permian floor have been included on the map. Both Kent (1949) and the Tectonic map of Great Britain and Northern Ireland include coastal contours and these have been incorporated into Figure 1. The two deep gas wells which reached the Carboniferous (Coulter and Barr *op. cit.*) indicated that thicknesses of Permo-Triassic approaching 3000 m occur towards the centre of the basin.

These boreholes, together with an extensive shallow seismic cover have confirmed that two major Permo-Triassic basins are present (the East Irish Sea Basin and the Solway Firth Basin) and two minor developments, one in Wigtown Bay and the other in Luce Bay, the latter is divided into two subcrops on the map.

Structurally the seismic evidence indicates shallow NE trending folds which cannot be traced laterally any great distance. Recent reappraisal of the sparker evidence suggests that, particularly in the East Irish Sea Basin, salt activity down to 300 m/sec. has been minimal. This inferred salt quiescence compares with the low level of movement noted for the Cheshire Basin.

In the Solway Firth Basin disturbed bedding is a feature of the sparker records, and salt with anhydrite, which is well developed at the Point of Ayre, Isle of Man, may have responded to the more pronounced basin folding recognised here.

As previously indicated, an attempt has been made in Figure 1 to extend the Pre-Permian basement contours developed along the Lancashire and Cumbria coasts, but clearly at this stage of our knowledge they can only be very approximate. This is particularly true of the extension of the structural hinges northwards from the Point of Air, Flintshire, which is also termed the Neston Fault. If this Air Ridge is a well developed feature, than a pronounced graben, as Audley (1970) implied in his cross-section, is a reality.

In the Solway Firth Basin, the proposed contours are extensions of the Carlisle basin evidence, and controlled by the gravity data and refraction data which suggests a thickness for the 3·64 km/sec. velocity layer of 1·2 km.

North of the extensive sub-crop of Permo-Triassic in the Solway Firth Basin, six boreholes have been put down, of these only two—70/2 and 71/63—penetrated to sandstones, neither of which provided an age. Sparker surveys in the bays have allowed their lateral extent to be mapped. An eastern fault margin has been recognised in Loch Ryan by Bott (1964) and Kelling and Welsh (1970). The latter authors estimated the westward downthrow to be 1525 m; a fault margin can only be inferred for Wigtown Bay. On the coast at Loch Ryan, Permian rocks have been recognised but offshore correlations are difficult. Audley (*op. cit.*) restricts Triassic deposition to south of the structural break margining the south Scottish coast.

Finally, although there is very little information available, the North Channel is shown as an asymmetric graben with a pronounced fault along the eastern side. The gravity data collected by I.G.S. is the only evidence available that can support this structural contention. Permo-Triassic rocks are extensively developed in the area

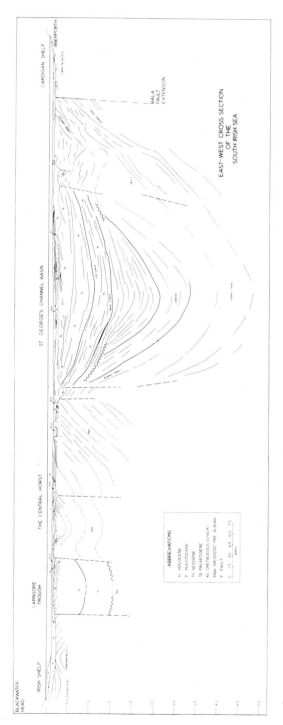

Fig. 3. East-west cross-section of the South Irish Sea, based on both deep and shallow seismic reflection data.

of Belfast Lough and may be reasonably expected to extend towards the east, whilst the use of the Beaufort Channel for ammunition dumping has effectively prevented gravity cores being taken to confirm this.

2. The South Irish Sea

2a. Introduction and general structural framework

The region displayed in Figure 1 consists in physiographic terms of a Channel (St George's Channel) and two associated Welsh embayments (Cardigan Bay and Caernarvon Bay), which together cover an area of 13000 sq. km.

In terms of sub-Quaternary out-crop, however, the region is dominated structurally by the Pre-Cambrian and Lower Palaeozoic rocks which are prominent as three sub-parallel positive features and possessing a strong Caledonoid trend. The most northerly two are in reality one being separated by a narrow mainly fault controlled Upper Palaeozoic basin. Thus the Anglesey–Wicklow ridge and the Irish Sea geanticline, which includes on land the Lleyn Peninsula, dominate the geology of the northern half of this region.

The southern ridge is granite cored and termed the Pembroke Arch now broken towards the west by north-south faults associated with a Tertiary rift, yet it is still a major structural element with a Caledonoid trend which has had a long history as a positive feature. These positive features in the north and south of the region will as a result of their geological history have only a thin to absent Lower Palaeozoic cover along their crests such that much of the sub-cropping Geanticline will consist of Precambrian as is the case with the most westerly part of Lleyn.

Separating the northern ridges from the Pembroke Arch is a major rift filled with 6000 m of Mesozoic and Tertiary sediments and considered to be floored by a thick Lower Palaeozoic succession; smaller horsts and rifts are associated with this feature which are distorted towards the northeast by a basement arch termed St. Tudwal's, that extends into Cardigan Bay and which may have a granitic core.

Bordering this rift, termed the St George's Channel and Cardigan Bay Basin throughout the whole area, but notably on the east, is a wide basement shelf. This shelf of sub-cropping Lower Palaeozoic rocks isolates the rift from the adjacent coastline with the exception of northeast Cardigan Bay where it narrows to allow the downfaulted basin to impinge on the coast at Mochras.

2b. The basins north of the Geanticline

The bedrock of the region (Fig. 1) records that the ENE trending basinal region separating the northern ridges extends from Wexford to Anglesey. The principal evidence for this consists of gravity anomalies, which are interpreted as due to light sediment fill (Al-Shaikh 1970) and the deep seismic data, which indicates that two distinct types of reflecting horizons appear to be present. In the central part of this basin where a graben structure is developed the sedimentary fill is considered to include both Carboniferous and Permo-Triassic. The calculated thickness of Permo-Triassic is 1300 m, this rests unconformably on eroded and folded rocks possibly of Carboniferous age. Towards the southwest, only the folded strata of the Carboniferous are adjudged to persist and ultimately link with the Irish coastal exposures. In Caernarvon Bay, both Carboniferous and Permo-Triassic strata appear to be present, although towards the Malldreath syncline in Anglesey only the Carboniferous persist. Thus in this complex basinal region between the two positive basement horsts, Carboniferous is thought to be present throughout and providing a link between the Irish and Welsh coastal exposures. It is only in the fault-bounded central trough and adjacent to the Dinorwic fault that Permo-Triassic sediments occur. The fault, an extension of the Dinorwic fracture, progressively truncates the synclinal axis when traced northeast towards the Menai Straits. In this case there is discernible axial thickening of the upper sequence, here interpreted as Permo-Triassic. This basin remnant is linked to the structurally complete and extensive Central Irish Sea basin which incorporates the half graben between the Holy Island shelf and the Wicklow Head shelf. The nature and full extent of the supposed Permo-Triassic is poorly known at present although it is possible, indeed probable, that these rocks extend closer to the Irish Coast than the map (Fig. 1) indicates, at least to the north end of the Carnsore Trough. Finally, in the Kish Bank basin immediately east of Dublin it now seems likely that a Permo-Triassic sequence is present at depth and overlain by a thick shaly Liassic sequence and Middle Jurassic limestones. This likelihood is based, in the absence of deep seismic data, on the widespread occurrence of the Permo-Triassic throughout the Irish Sea in general and in the basinal developments in particular, and on the observable fact that the gravity contours display a distinct Caledonoid trend. Major faults along the northern and southwestern margins serve to ensure a marked asymmetry for the sedimentary fill. Gravity evidence supports the view that there is a structurally controlled link between this basin and the Central Irish Sea basin. Two additional points emerge. Firstly Carboniferous strata have been sort for but not recorded on the St. Tudwal's Arch; this may be due to Permo-Triassic overstep; certainly the eastern fault margin of the Cardigan Bay Basin offers no support as to its possible presence in the area, for in most places the strata dip into the fault plane. Secondly there is some rather poor evidence on the deep seismic records to suggest that strata with marked reflecting horizons come to sub-crop along the southern margin of the Irish Sea geanticline. Long line correlation from Permo-Triassic boreholes does not discount a Carboniferous interpretation for these horizons.

Aside from geophysical evidence it may be noted that the stratigraphy of southern Ireland and south Wales has been linked by George (1958, fig. 14), although exact correlation is uncertain particularly of the Tournasian and Visean bases. Again the northern limit of Tournasian seas and the subsequent Visean overstep in the St George's Channel area can only be speculated upon. George concluded that a trough could have been formed repeatedly if not continuously in the area of St George's Channel to link the southwest and central Carboniferous provinces.

The Pembrokeshire positive axis, now broken by faults, but still recognisable on the Celtic Sea gravity map (Day and Williams 1970) and extending WSW towards the present Continental margin, may have been a considerable feature during the Lower Carboniferous as was the Irish Sea geanticline. If so it would help to explain the difficulty in correlating the Carboniferous a distance of 90 km across St George's Channel. It is tempting to suggest that a narrow NE orientated Gulf between the Pembroke Arch, and the geanticline presisted during the Lower Carboniferous. Although there was a Caledonoid trending "down-warp" from Shannon to Dublin in Namurian times which was more extensive later, there is no indication that Upper Carboniferous is present immediately north of the geanticline. The seismic sections indicate thick sequences but the recording characteristics most closely compare with rocks of Dinantian ages. The only other indication is the presence of Upper Carboniferous at Malldreath, Anglesey. One might infer from this that

Upper Carboniferous rocks are present in the Caernarvon Bay basin.

Complicating the Carboniferous structural picture further is the tectonic overprint which is severe up to the Killarney–Dungarvan line, but north of the Hercynian front the folding, in particular, is much more gentle and orientated to the Caledonian trend. The period of marked erosion occasioned by the post Westphalian phase of the Hercynian orogeny may well have removed the Upper Carboniferous and the evidence afforded by the South Wales sequence especially around the Gower would not favour preservation. Interpreted deep seismic sections for St George's Channel, suggests that a graben structure floundered early in the Mesozoic, unfortunately, the complexity of the Upper Carboniferous surface notably, south the the Dungarvan thrust line, and well displayed along the north Devon coast makes it most unlikely that a reflective interface could be recorded on the seismic that would assist in a complete analysis.

2c. The St George's Channel and Cardigan Bay basins

Unlike the North Irish Sea the geology of the major rift is largely unrepresented in the coastal sections. No post-Palaeozoic sediments occur either along the Irish east coast or the Welsh west coast. Consequently the identification of that great thickness of sediment filling the St George's Channel Basin and the associated structure of Cardigan Bay Basin must rely on offshore boreholes and deep seismic reflection and refraction techniques. However, the single most important source of information has been the coastal borehole at Mochras (Woodland 1971). This borehole penetrated to a depth of 1938 m and provided the essential control for the deep reflection survey which was a direct result of the dramatic rock sequences recovered. The offshore boreholes, although concentrated for technical reasons in the shallow waters of Cardigan Bay, have recovered Mesozoic from the flanks of the main St George's Channel Basin, thus confirming the geophysically inferred geology of much of the main trough.

The stratigraphic sequence recognised for the main basin and displayed on the cross-section (Fig. 3) has been determined using the offshore boreholes. The peculiar recording characteristics of the Chalk and the recognition of the two major unconformities are widespread and well documented throughout southern England namely the pre-Albian and the Upper Palaeogene. In addition the horizon ascribed to the Chalk may be traced to the area of southeast Ireland where it was directly cored. (Curry *et al.* 1967). Further support for the interpretation proposed may be derived from the refraction data (Blundell *et al.* 1971) which readily distinguishes in velocity terms between layer two (2·1 km/sec.) and considered to be Palaeogene and layer three (3·5 km/sec.) here regarded as Chalk.

The central rift is determined structurally by a combination of east-northeasterly (Caledonoid) and northerly (Tertiary) directions (for structural details of this basin see Dobson *et al.* 1973 p. 24). The western margin is dominated by normal faults whose adjustments have created a number of elongated blocks which are stepped down towards the basin axis. The exception is the Carnsore Trough. The eastern flank of the basin may be resolved as presenting a combination of different tectonic effects including salt tectonics, mainly broad pillows at depth, upthrown blocks and regional normal faults. The basin itself is a downwarp of major proportions and although distorted throughout by salt tectonics, is generally a simple structure. The constriction of the basin is due to positive elements of Caledonoid trend which plunge towards the basin from the flanks. These include the St Tudwal's Arch, the Strumble Head High and faulted elements of the Geanticline.

As Liassic and younger Jurassic sequences subcrop extensively in Cardigan Bay in water and overburden depths accessible to the drill ship employed by the Institute of Geological Sciences more than 24 holes have been drilled in the area since 1970. By westward extrapolation from Cardigan Bay it is possible to recognise the major subdivisions of the Mesozoic within the St George's Channel Basin, and appreciate the degree of basinal thickening. Seismic sections across the main basin reveal a Jurassic sequence that is markedly lensed, although a clear distinction can be drawn between structural adjustments that generated basinal growth and those faults that truncate the basin.

On the deep seismic record it is clear that sedimentation has been continuous throughout the whole of the Mesozoic in the central part of the main basin but the thicknesses of the Middle and Upper Jurassic and Lower Cretaceous cannot be determined separately. A possible exception to this is the Lower Cretaceous as on the basin flanks a clear and major unconformity is recorded (later Kimmerian/Pre-Albian unconformity). Extrapolation of this flank feature recognized only on the western side is not easy, however an approximate figure of 1200 m is proposed assuming a velocity of 3–5 km/sec. This compares with 1500 m in the north Celtic Sea proposed by Ziegler (1975).

The Chalk, the top Mesozoic facies, has a characteristic seismic signature and may be readily recognised. It is entirely confined to the main basin being faulted down on the east, and terminated to the west by a combination of thinning onlap and fracturing. This gives a maximum present subcrop width of 18 km. Towards the north the distinctive reflectors are absent and it is concluded that only marginal facies are developed. The Chalk is uniformly thick (600 m) with no indication of thinning towards the east, consequently a palaeogeographic reconstruction must extend the Chalk Sea at least to the present Welsh coast if not beyond.

Although very small isolated pockets of Tertiary have been recorded on the Welsh mainland notably at Flimstone in Pembrokeshire and in the Halkyn Hills the only extensive Tertiary legacy is the landform. The offshore position is completely different as was evidenced by the early geophysical surveys which indicated the existence of low velocity material of considerable thickness both in Tremadoc Bay and St George's Channel. The Mochras borehole provides the first direct evidence in support of the geophysical findings; and subsequent offshore drilling by the I.G.S. further defined the extent of Tertiary.

All the Tertiary confirmed to date and particularly the subcrop drawn on Figure 1 is of Palaeocene to Miocene in age, the younger Neogene although considered present along the axis of St George's Channel basin, is difficult to recognise and map with confidence. Interpreted sparker sections included in the paper by Dobson *et al.* (1973) indicate the probable development of the Neogene (see also the cross section Fig. 3).

The Tertiary of the Mochras borehole is reckoned to extend downwards from −77·47 m O.D. to −601·83 m where it rests abruptly on weathered Liassic of the topmost Toarcian. This great thickness of low velocity material (2000 m/sec.) which was recognised by Griffiths *et al.* (1961) using refraction techniques may be traced offshore into Tremadoc Bay using the results of the deep reflection seismic surveys.

The 600 m contour (Fig. 1) extends away from the fault line into Tremadoc Bay a distance of only 5 km this is essentially a fault localised doubling of the average thickness (300 m) to the south the picture is confused by the effect of the Mawddach estuary fault. A tongue of Tertiary extends beyond this fault recently confirmed by the Tonfannau borehole. The deposits rapidly terminate to the north, although further

C

deposits of low velocity material occur on the Glaslyn Valley. The deepest point of this basin or asymmetric graben is offset to the east by several kilometers compared to the Bouguer anomaly contours. This implies that the underlying Mesozoic which has a significant gravity effect, is thickest some 5 km west of the Mochras fault.

The subcrop of the thick post-Cretaceous sediments in the South Irish Sea is clearly indicated on Figure 1. For the main development in St George's Channel Basin the presented isopachs are calculated from the deep seismic sections an average velocity of 2000 m/sec is assumed. For the Carnsore Trough to the west the thicknesses are based on shallow seismic analysis and are only tentative. In St George's Channel Basin the isopachs are drawn with some confidence as the presence of Upper Cretaceous Chalk facies at depth allows the Tertiary above to be readily identified. The marginal facies of the Upper Cretaceous developed adjacent to the flanks of the geanticline lacking a strong upper reflecting interface makes the Tertiary contouring in the north less reliable, but it may be traced as far as the Bardsey Island trough. The cross-section (see Fig. 3), drawn using both sparker and deep reflection data, shows the complex nature of the upper surface of the Lower Tertiary. Above the folded and bevelled Palaeogene lies ill-defined, frequently poorly-bedded Neogene sequences and a blanket of Quaternary, Even though there is no pronounced uncomformity between the Palaeogene and the beds above, no Neogene deposits have to date been positively identified. Because of the very variable nature of these sediments which appear to possess no significant seismic signature it is difficult if not impossible in places to separate them from Quaternary ablation deposits in much of the main basin. However, as the cross-section confirms an attempt has been made and two separate Upper Tertiary sequences are recognised; a basal massive section and an upper bedded section.

3. Conclusion

The geological structure of these two regions appears at first sight to be different yet both are dominated by graben structures. The one to the north extends into the Cheshire basin with its remnant Liassic whilst the southerly one is really an extension of the North Celtic Sea Basin.

Two fault patterns, one Caledonoid the other Kimmerian to Tertiary control both. The associated smaller units are similarly controlled. This basic regional framework responsible for the orientation and form of the blocks and basins, has had a profound influence on both glacial and post-glacial events.

References

AL-SHAIKH, Z. D. 1970. The geological structure of part of the central Irish Sea. *Geophys. J. R. astr. Soc.* **20**, 233–237.

AUDLEY, Charles M. G. 1970. Stratigraphical correlation of the Triassic rocks of the British Isles. *Q. J. geol. Soc. Lond.* **126**, 19–47.

BACON, M. and McQUILLIN, R. 1972. Refraction Seismic surveys in the North Irish Sea. *J. geol. Soc. Lond.* **128**, 613–621.

BLUNDELL, D. J., DAVEY, F. J. and GRAVES, L. J. 1971. Geophysical surveys over the south Irish Sea and Nymphe Bank. *J. geol. Soc. Lond.* **127**, 339–375.

BOTT, M. H. P. 1964. Gravity measurements in the north-eastern part of the Irish Sea. *Q. J. geol. Soc. Lond.* **120**, 369–394.

——1967. Geological investigations of the northern Pennine basement rocks. *Proc. Yorks. geol. Soc.* **36**, 139–168.

——and MASSON-SMITH, D. 1960. A gravity survey of the Criffell granodiorite and the New Red Sandstone deposits near Dumfries. *Proc. Yorks. geol. Soc.* **32**, 317–332.

COLTER, V. S. and BARR, K. W. 1975. Recent developments in the geology of the Irish Sea and Cheshire Basins. *In*, WOODLAND, A. W. (Editor). *Petroleum and the Continental Shelf of North West Europe*, Vol. 1, 61–73. Applied Science Publishers 501 pp.

CORNWALL, J. D. 1972. A gravity survey of the Isle of Man. *Proc. Yorks. geol. Soc.* **39**, 93–106.

CRAIG, G. Y. and NAIRN, A. E. M. 1956. The Lower Carboniferous outliers of the Colvend and Rerrick shores, Kirkudbrightshire. *Geol. Mag.* **93**, 249–256.

CURRY, D., GRAY, F., HAMILTON, D. and SMITH, A. J. 1967. Upper Chalk from the sea-bed, south of Cork, Eire. *Proc. geol. Soc. Lond.* **1640**, 134–136.

DAY, G. A. and WILLIAMS, C. A. 1970. Gravity compilation in the N.E. Atlantic and interpretation of gravity in the Celtic Sea. *Earth and Plan. Sc. Lett.* **8**, 205–213.

DOBSON, M. R., EVANS, W. E. and WHITTINGTON, R. 1973. The geology of the south Irish Sea. *Rep. Inst. geol. Sci.* No. 73/11, 35 pp.

GEORGE, T. N. 1958. Lower Carboniferous Palaeogeography of the British Isles. *Proc. York Geol. Soc.* **31**, 227–318.

GRIFFITHS, D. H., KING, R. F. and WILSON, C. D. V. 1961. Geophysical investigations in Tremadoc Bay, North Wales. *Q. J. geol. Soc. Lond.* **117**, 117–171.

JOHNSON, G. A. L. 1967. Basement control of Carboniferous sedimentation in Northern England. *Proc. Yorks. geol. Soc.* **36**, 175–194.

KELLING, G. and WELSH, W. 1970. The Loch Ryan Fault. *Scott. J. Geol.* **6**, 266–271.

KENT, P. E. 1949. A structure contour map of the surface of the buried Pre-Permian rocks of England and Wales. *Proc. Geol. Assoc.* **60**, 87–104.

MOSELEY, F. 1973. A tectonic history of northwest England. *J. geol. Soc. Lond.* **128**, 561–598.

WOODLAND, A. W. (Editor). 1971. The Llanbedr (Mochras Farm) Borehole. *Rep. Inst. geol. Sci.* No. 71/18. 115 pp.

WRIGHT, J. E., HULL, J. H., McQUILLIN, R. and ARNOLD, S. E. 1971. Irish Sea investigations 1969–70. *Rep. Inst. geol. Sci.* No. 71/19, 43 pp.

ZIEGLER, P. A. 1975. North Sea Basin history in the tectonic framework of North-Western Europe. *In*, WOODLAND, A. W. (Editor). *Petroleum and the Continental Shelf of North West Europe*, Vol. 1, pp. 131–148. Applied Science Publishers 501 pp.

M. R. Dobson, The University College of Wales, Department of Geology, Llandinam Building, Penglais, Aberystwyth, Dyfed, Wales, SY23 3DB.

Quaternary sediments of the northern Irish Sea

H. M. Pantin

This paper gives a brief account of Quaternary stratigraphy in the northern Irish Sea, based on data collected by IGS during the years 1967–73. Grab samples, cores, boreholes, and acoustic (pinger) traverses have revealed that east of the Isle of Man, the typical upward succession is rockhead—boulder clay—proglacial water-laid sediments—marine sediments; west of the Isle of Man, the proglacial water-laid sediments are commonly missing. The conditions of deposition and possible age of the sediments are discussed, together with the appearance and origin of the zones of acoustic turbidity which appear on the pinger records.

1. Introduction

The area to be described (Fig. 1) includes virtually the whole of the northern part of the Irish Sea. Apart from the various coastlines, the area is bounded to the northwest by the line Bangor (N. Ireland)—Crammag Head (Mull of Galloway); to the southwest by a line running southeast from near Kilkeel (Co. Down, N. Ireland) and then swinging south to pass about 24 miles west of Holyhead; and to the northeast by the line Balcary Point (Galloway)—Maryport (Cumbria).

During the years 1967–73, this area was investigated by the Continental Shelf Unit South* of the Institute of Geological Sciences, and was covered by a square grid of sample stations, with superficial samples (grab samples or cores) being collected every four miles. In addition, a number of shallow boreholes down to

* Formerly known as Continental Shelf Unit I.

27

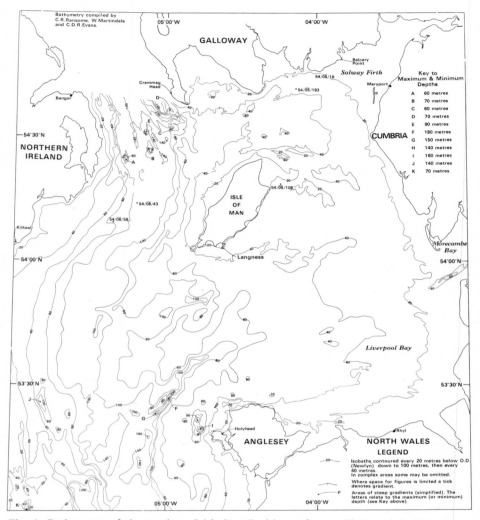

Fig. 1. Bathymetry of the northern Irish Sea. Positions of cores illustrated in Plate 6 are shown thus: 54/05/108.

rockhead was drilled. Shipek Grab samples were collected at nearly all stations, and were supplemented in most cases by vibrocores (about 9 cm diameter) ranging up to about 6 m long; in a few cases, gravity cores up to about 2 m long were taken in addition to, or instead of, the grab samples. For the purposes of description and analysis, the cores were cut by the electro-osmotic process (Chmelik 1967).

The chart of surface sediment distribution (Fig. 2) was derived from the inspection and size analysis of grab samples, together with the uppermost portions of gravity cores at the few localities where grab samples were not taken. In the region between the Isle of Man and Cumbria, this chart is essentially similar to that of Cronan (1969 fig. 3).

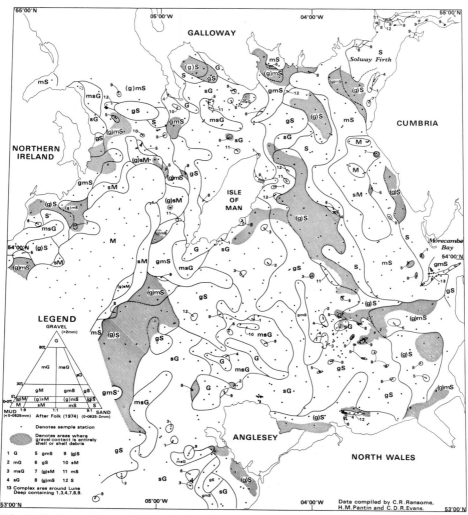

Fig. 2. Surface sediment distribution in the northern Irish Sea.

The stratigraphy of the Quaternary sediments was, to a limited extent, revealed by the vibrocore (or gravity core) samples. These findings were confirmed and extended by the shallow boreholes and by a series of acoustic profiles (pinger traverses).

2. Stratigraphy and lithological descriptions

2a. Northeastern sector (northeast and east of the Isle of Man)

This part of the area has been studied by C.S.U. South in greater detail than the other sectors, and will therefore be described first and used as a basis for comparison. The results obtained show that the rockhead, which consists mainly

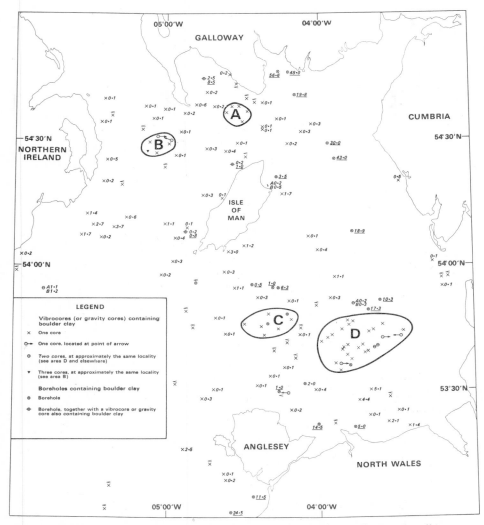

Fig. 3. Distribution of boulder clay, directly overlain by marine sediments in vibrocores, gravity cores, and boreholes from the northern Irish Sea.

Figures alongside symbols represent observed thickness of marine sediment overburden, to the nearest 0·1 m. The symbol < signifies less than 0·05 m. Plain figures refer to vibrocore/gravity core measurements, with A and B representing values for two cores at approximately the same locality. Underlined figures refer to boreholes.
(NB Thicknesses are not corrected for possible core shortening).

Overburden thicknesses in Areas A, B, C and D

Area A	Range	<0·05 to 0·6, average about	0·2
Area B	,,	<0·05 to 0·5,	,,	,,	0·2
Area C	,,	<0·05 to 0·5,	,,	,,	0·2
Area D	,,	<0·05 to 2·9,	,,	,,	0·8

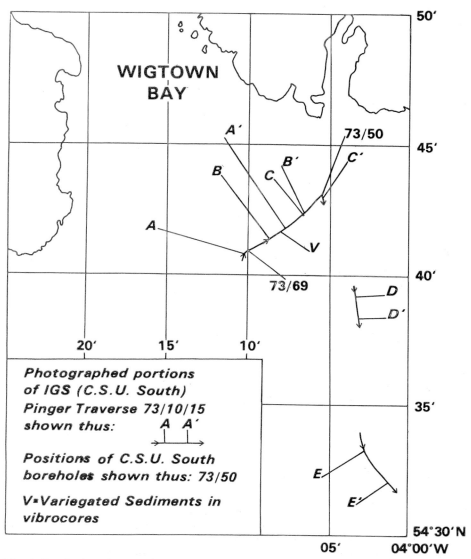

WIGTOWN
BAY

Fig. 4. Track chart for pinger traverses A–A′, B–B′, C–C′, D–D′ and E–E′. The localities
of boreholes 73/50 and 73/69 are also shown.

of Permo-Triassic red beds, is overlain over most of the area by a rather uniform
mantle of boulder clay a few metres thick. This usually shows no acoustic evidence
of internal stratification, but a faint diffuse banding occurs locally. Lithologically
it consists of stiff, typically unbedded, mainly muddy sediment, sometimes with
pebbles. Lamination occurs locally, but is intermittent and often highly distorted.
In some parts of the area the boulder clay comes within 1 m of the sea bed, being
overlain only by thin marine sediments (Fig. 3), but the pinger profiles, vibrocores

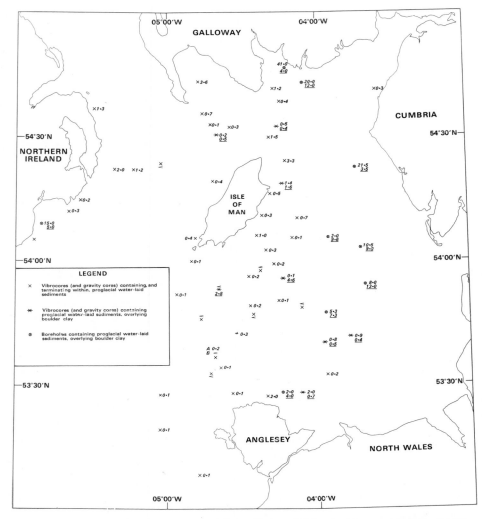

Fig. 5. Distribution of proglacial water-laid sediments (with or without underlying boulder clay) in vibrocores, gravity cores, and boreholes from the northern Irish Sea.

Plain figures alongside symbols represent observed thickness of marine sediment overburden, to the nearest 0·1 m. The symbol < signifies less than 0·05 m. A and B represent values for two cores at approximately the same locality. Underlined figures represent thickness of proglacial water-laid sediments in cores where underlying boulder clay has been penetrated. (NB Thicknesses are not corrected for possible core shortening).

and boreholes show that over much of the sector, the boulder clay is overlain by well-bedded sediments, up to several tens of metres thick (Plates 1–5, Fig. 4). The upward transition is normally abrupt. Comparison of acoustic profiles, vibro-cores and borehole evidence shows that over much of the area, the unit immediately overlying the boulder clay consists of proglacial water-laid sediments (Fig. 5 and Plate 6a); more detailed evidence (Pantin, in press) from an area lying 10–15 miles

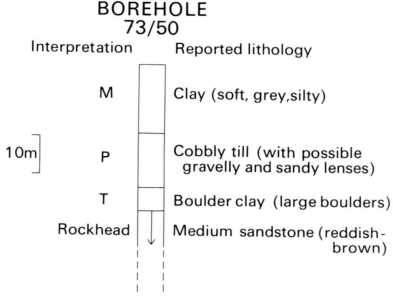

BOREHOLE
73/50

Interpretation   Reported lithology

M   Clay (soft, grey,silty)

10m   P   Cobbly till (with possible
                gravelly and sandy lenses)

T   Boulder clay (large boulders)

Rockhead   Medium sandstone (reddish-
                              brown)

Fig. 6. Diagram of IGS (C.S.U. South) borehole 73/50. M: marine sediments, P: proglacial water-laid sediments, T: till (boulder clay).

southeast of the Isle of Man indicates that here, at least, deposition of the proglacial water-laid beds took place in a lagoon, with restricted access to the sea. These particular proglacial lagoon sediments overlap into the southeastern sector (q.v.; the northern boundary of the latter may be defined as a line running east from Langness, Isle of Man).

The proglacial water-laid beds are characteristically muddy, with sand or coarse silt laminae which are frequently numerous but are usually minor in bulk. In one borehole (73/50), the zone which is here allocated to the proglacial water-laid beds (compare Fig. 6 and Plate 3) contains evidence of layers of coarser sediment (sand or gravel) with clasts ranging up to cobbles in size; such material is probably non-typical of these beds, although it must be remembered that a 9 cm vibrocorer would be effectively stopped by all but the smallest cobbles, particularly if the latter occurred in layers.

Pebbles apparently dropped by floating ice (Pantin, in press) are found within the muddy beds at many localities, but these normally form only a very small proportion of the sediment. However, discrete masses of "boulder clay", overlying water-laid beds, up to several tens of centimetres thick, and apparently also dropped by floating ice, have been found in a few vibrocores (Ovenshine 1970 pp. 891–892).

These proglacial water-laid sediments are typically well-bedded; acoustic profiles show numerous well-developed quasi-parallel reflectors (presumably bedding) on scales of 1 metre or more, while vibrocores show good banding and lamination, although the latter may be locally distorted due to small-scale slumping. These beds contain very little bioturbation and virtually no macro-organic remains, though foraminifera and organic-walled microplankton are locally present;

those from the proglacial lagoon beds lying southeast of the Isle of Man have been investigated in some detail (M. J. Hughes in press; R. Harland in press).

Overlying the proglacial water-laid beds, and extending up to the present sea floor, is a series of marine beds. These marine beds* are muddy, sandy, or gravelly, depending on locality. Over most of the sector, mainly sandy or mainly muddy sediments are comparable in bulk and form the major proportion of the sediment, with gravel occurring only in minor amounts. To the east and north of the Isle of Man, however, the proportions of sand and gravel in the sediment rise in a westerly direction, until the order of bulk dominance is gravel > sand > mud.

Acoustic profiles show that large-scale bedding occurs in these marine beds, but is usually less well-defined than in the proglacial water-laid beds. Major units separated by obvious discontinuities are sometimes visible. Vibrocores show that banding is common but lamination is comparatively rare; in contrast, bioturbation is widespread, and has presumably destroyed the lamination in many beds

---

* The term "marine beds" is used here in a broad sense, and includes possible estuarine or intertidal beds, in addition to subtidal beds deposited in water of normal marine salinity. However, the term has not been used for any of the proglacial water-laid beds; these are characterised by evidence of low salinity as well as a proglacial environment, and thus possess two features which (i) depart significantly from a 'normal' marine environment, and (ii) are not found together in any of the beds here called 'marine'.

Explanation of Plates 1 - 5 and 7 - 10 (pinger traverses).

SYMBOLS
  M = marine sediments (the deeper layers may locally include glacio-marine sediments).
  P = proglacial water-laid sediments. The upward transition to marine beds in Plates 1–3 and 8 is assumed to correspond to the level at which relatively well-bedded sediment give way to relatively poorly-bedded sediments.
  V = variegated sediments (intertidal or subtidal marine). This layer shows the "blotchy" reflection discussed in the text (p. 49).
  T = till (boulder clay), showing no more than diffuse indications of bedding, and usually none.
  H = acoustic basement (presumably rockhead).
ATZ = acoustic turbidity zone. Examples of narrow "mesoids" are seen in Plates 1–2, while more extensive mesoids with terrace-like tops are seen in Plates 2–4 and 8–9. A conspicuous "funnel" is shown in Plate 5, while a well-marked "fosoid" occurs immediately to the left of the ATZ in Plate 8. Explanations of the terms "mesoid", "funnel", and "fosoid" are given in the text (Section 7).

DEPTH SCALES.
These are based on the velocity of sound through sea water, and the zero datum line represents the level of the pinger apparatus itself, which is towed 3–4 metres below the sea surface.
  Sound velocities in the sediments are presumably higher than in the sea water, and the depth scale will therefore give only an approximate, reduced indication of sediment thickness.

TIME REFERENCE MARKS.
These are applied to the record by a manual control, and consist of thin, dark, parallel-sided lines running the whole way down (or occasionally partway down) the pinger record.

OTHER FEATURES
Sand waves, progressively increasing in amplitude, are conspicuous along the traverse S–S' (Plate 7).
  Electrical interference, represented by dark or light vertical lines running partway up the pinger record, can be seen in Plates 8 and 10.
  Crinkles in the recording paper, represented by light-coloured, diagonal tapering bands on the pinger record, can also be seen in Plates 8 and 10.
  Handwritten notes made by the operator at sea on the pinger record itself can be seen in various places. Misleading examples have been cancelled with a short black line; the remainder may be ignored.

Plate 1

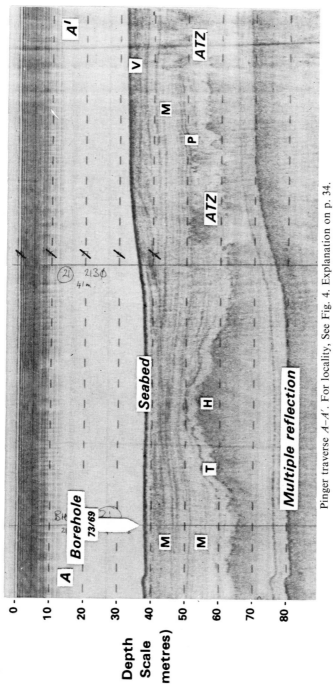

Pinger traverse *A–A'*. For locality, See Fig. 4. Explanation on p. 34.

Plate 2

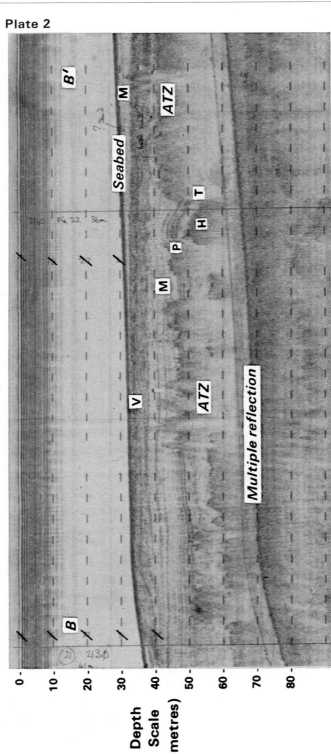

Pinger traverse *B–B'*. For locality, See Fig. 4. Explanation on p. 34.

Plate 3

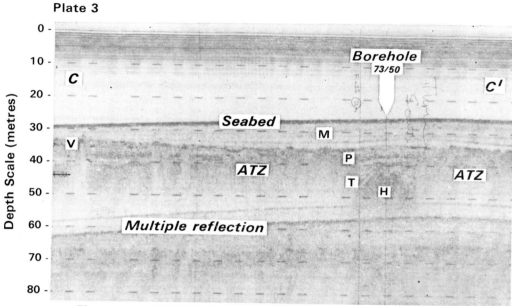

Pinger traverse *C–C'*. For locality, See Fig. 4. Explanation on p. 34.

Plate 4

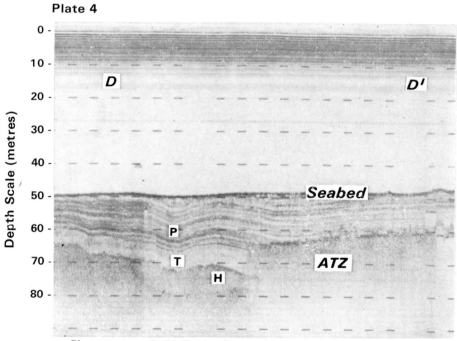

Pinger traverse *D–D'*. For locality, see Fig. 4. Explanation on p. 34.

Plate 5

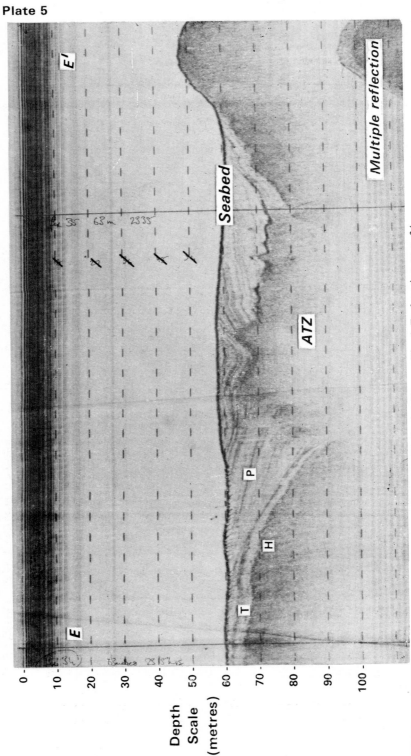

Pinger traverse *E–E'*. For locality, see Fig. 4. Explanation on p. 34

Plate 6

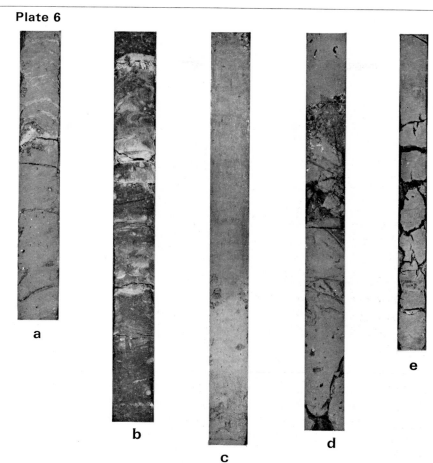

a

b

c

d

e

a Vibrocore 54/05/108, 274 to 328 cm. Proglacial water-laid beds extend from 274 to 296 cm, and consist of mud bands (colour 10R3½/2) interbedded with laminae of very fine sand (10R4/4); these beds overlie boulder clay (10R3½/2) without visible bedding. Core locality: 54°18·6′N., 04°14·1′W. (See Fig. 1).

b Vibrocore 54/05/16, 77·5 to 152·5 cm; Marine sediments, consisting of sandy mud, muddy sand, and sand (very fine). Banding and lamination are both visible, but the latter has been severely affected by bioturbation. The colour of the mud component is 9½YR4/1½, and that of the sand component 10YR5–6/1. Several types of burrow can be seen, the most distinctive being the sub-horizontal meniscus burrows at 94·5, 99·5, and 123·5 cm. The sub-horizontal orientation of many burrows gives rise to the effect of pseudolamination. Core locality: 54°45·1′N., 04°01·2′W. (See Fig. 1).

c Vibrocore 54/05/193, 79 to 156 cm. Variegated sediments (probably intertidal) consisting of muddy fine sand (dominant) and sandy mud (subordinate), with narrow diffuse banding and strong bioturbation. The colours vary as follows: 79 to about 127 cm, muddiest parts 6YR5/2, least muddy parts 8YR4½/2; about 127 cm to 143 cm, muddiest parts 10Y5½/1, least muddy parts 2Y5/1; 143 to 156 cm, 10Y5/1. The transition at about 127 cm is diffuse. Core locality: 54°41·6′N., 04°08·3′W. (See Fig. 1).

d Vibrocore 54/06/58, 308 to 384 cm. Marine mud, heavily bioturbated, containing occasional shells of *Turritella communis* Risso, extends from 308 to 321 cm; muddy sand and sandy mud with occasional pebbles and shell fragments, 321 to 344 cm; boulder clay without visible bedding, below 344 cm. Diagonal cutting-knife marks are visible in places; between 320–331 and 337–345 cm, there also occur irregular areas where the sediment was broken rather than cut.
Colours: marine mud 9YR5/2, muddy sand and sandy mud 8YR5/2, boulder clay 5YR5/2. Core locality: 54°09·2′N., 05°21·0′W. (See Fig. 1).

e Gravity core 54/06/43, 63 to 124 cm. Marine mud, heavily bioturbated, with numerous irregular cracks. A light-coloured echinoderm fragment is visible at 116·5 cm. Colours: 10YR4½/2 down to 96 cm, transition zone from 96 to 119 cm; 5Y5/1½ below 119 cm. Core locality: 54°14·7′N., 05°11·8′W. (See Fig. 1).

D

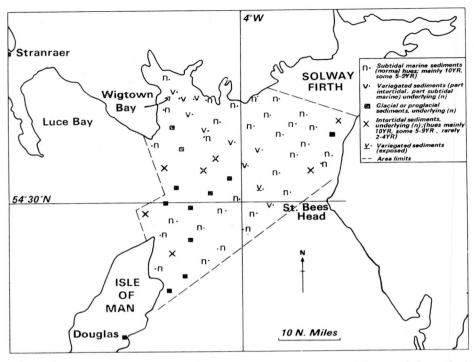

Fig. 7. Distribution of near-surface sediment types in the Solway area (vibrocore information).

(Plate 6b). Macro-organic remains are common in the marine beds, and consist almost entirely of bivalve and gastropod shells. Beds consisting dominantly of shell appear to be generally rare, very thin (10 cm or less), and limited in lateral extent; however, vibrocores have shown that the muddy sediments off Cumbria (Fig. 2), which extend to a depth of several metres (Pantin, *in press*) locally contain layers up to 0·5–1·0 m in thickness with abundant shells of *Turritella communis* Risso. Drop-stones, or discrete masses of "boulder clay", have not been identified in the marine beds.

In the SSE-trending zone east of the Isle of Man where sand, slightly gravelly sand, and gravelly sand predominate at the surface (Fig. 2), numerous sand-waves have been reported (Belderson and Stride 1969). These sand-waves, which appear to be travelling towards the ENE (*loc. cit.*, fig. 1) extend into the "southeastern sector" as defined in the present paper.

The marine beds include a local unit, situated southeast of Wigtown Bay, with a peculiar "blotchy" reflection on the acoustic trace. Vibrocores taken from an area where these beds come within 1 metre of the sea bed show that the unit is mainly muddy, and belongs to the "variegated sediments" (p. 49 and Fig. 7). They contain zones with small spots of iron-rich material, which is black when reduced but rusty-coloured when oxidised.

In localities where the marine beds are sufficiently thin to allow cores to reach underlying proglacial water-laid beds (Fig. 5), the cores always reveal an abrupt, unconformable junction between the two formations. Where the marine beds

BOREHOLE 73/69

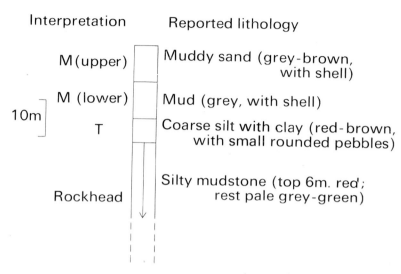

Interpretation — Reported lithology

M(upper) — Muddy sand (grey-brown, with shell)

M (lower) — Mud (grey, with shell)

T — Coarse silt with clay (red-brown, with small rounded pebbles)

10m

Rockhead — Silty mudstone (top 6m. red; rest pale grey-green)

Fig. 8. Diagram of IGS (C.S.U. South) borehole 73/69. M: marine sediments, T: till (boulder clay).

are thicker, acoustic profiles generally indicate an upward transition varying from an abrupt discontinuity to a moderately rapid change-over: the former type probably represents an unconformity, while the latter may well contain such a feature, even if intermediate lithological types occur near the junction. In various places the acoustic profiles show an apparent lateral passage from proglacial waterlaid beds into marine beds. These transitions take place within 2–3 miles, but occur at levels in the sediment which prevent their investigation by vibrocores. Along one profile, however, there occur two boreholes (73/50 and 73/69) in suitable localities (Fig. 4), which indicate that the above acoustic change is paralleled by a change in lithology from "cobbly till, with possible gravelly and sandy lenses" to "mud (grey, with shell)" (see Figs 6 and 8 for these descriptions).

The hue* of the boulder clay and the proglacial water-laid beds may be described broadly as "reddish orange"; it varies over the range 10R–9YR, although 3–8YR is more usual. That of the marine beds is typically orange (10YR), although "redder" hues (sometimes 5–9YR, more rarely 2–4 YR) or a slightly "yellower" hue (1Y) occur in some layers. Within the marine sediments in the area southeast of Wigtown Bay, there occur "variegated sediments" with more variable hues (Plate 6c); the hues range from reddish orange (5YR) through yellow (5Y) to yellow-green (5GY). In one special case, an evanescent blue colour (5B) was reported. These "variegated sediments" include the previously-mentioned unit with a "blotchy" acoustic reflection.

* Colour notations refer to the Geological Society of America Rock Color Chart 1963.

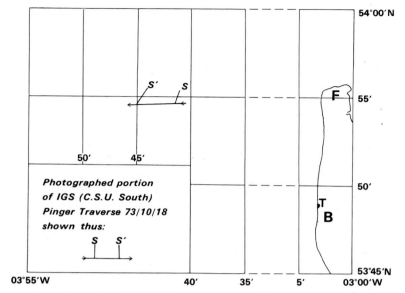

Fig. 9. Track chart for pinger traverse S–S′. F=Fleetwood, B=Blackpool, T=Tower.

The nature of the colouring materials in the various sediment facies was investigated, either by simple chemical tests or by comparison with sediments and minerals from other areas. The results are summarised below in Section 6.

2b. Southeastern sector (southeast and south of the Isle of Man)

The same general stratigraphy applies as in the northeastern sector just described. Rockhead is typically overlain by boulder clay; the latter is overlain in various places by proglacial water-laid beds, and these in turn by marine sediments. There are, however, various differences as compared to the area further north. The underlying hard rocks are again dominantly Permo-Triassic, although these are replaced in the west by large areas of Carboniferous and north of Anglesey by Pre-Cambrian or Lower Palaeozoic rocks. In at least two small areas, respectively about 18 miles, 160° T. from Langness Point (Isle of Man) and 23 miles, 342° T. from Rhyl, the bedrock comes within 0·5 m of the sea bed, boulder clay being locally absent. Elsewhere, the boulder clay appears to be fairly continuous. The proglacial water-laid beds are widespread in the western part of the sector, and include the southerly extension of the already-mentioned proglacial lagoon beds near the Isle of Man (p. 33). However, the number of cores and boreholes with proglacial water-laid beds diminishes rapidly towards Liverpool Bay; in this area, moreover, the acoustic profiles rarely show layers which can be identified with reasonable certainty as proglacial water-laid beds. This phenomenon could be due to a genuine change in abundance across the area or to a change in lithology.

Grab samples, vibrocores and boreholes, together with the appearance of pinger records, indicate that in the present sector the marine sediments possess a considerably higher sand/mud ratio than in most of the sector further north. In

Plate 7

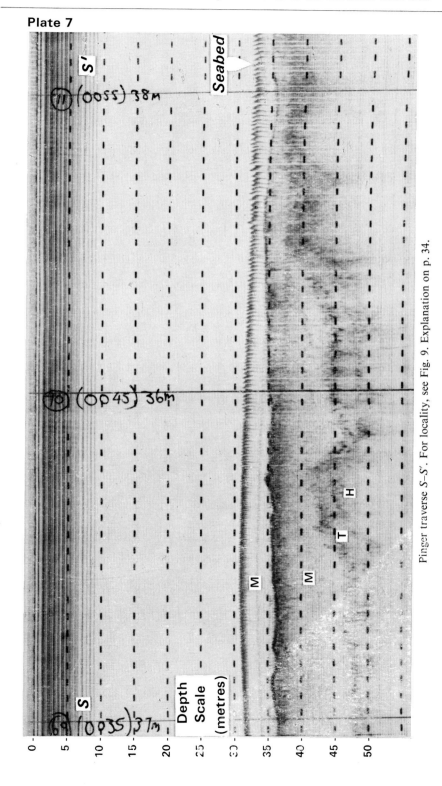

Pinger traverse *S–S'*. For locality, see Fig. 9. Explanation on p. 34.

Plate 8

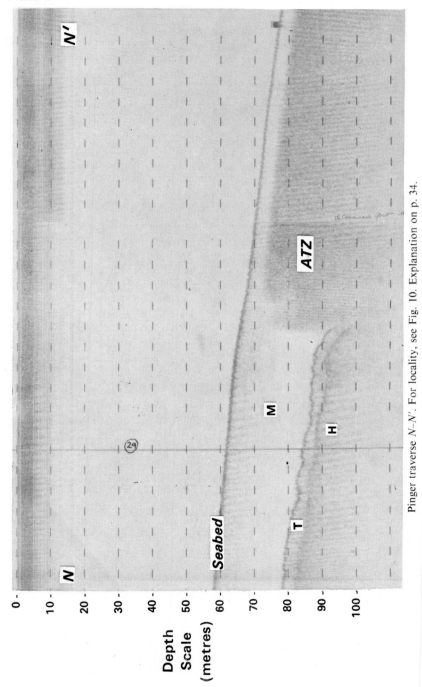

Depth
Scale
(metres)

Pinger traverse *N–N'*. For locality, see Fig. 10. Explanation on p. 34.

Plate 9

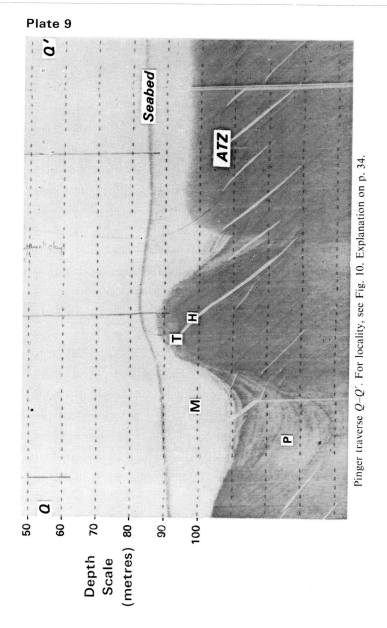

Pinger traverse *Q–Q'*. For locality, see Fig. 10. Explanation on p. 34.

Plate 10

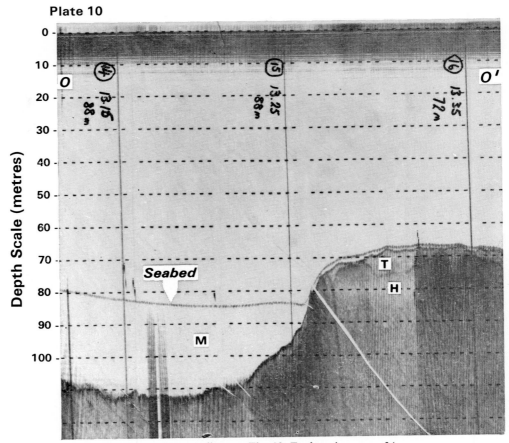

Pinger traverse *O–O'*. For ocality, see Fig. 10. Explanation on p. 34.

Liverpool Bay, dominantly sandy sediments form the bulk of the assemblage, with gravelly sediments in minor amounts; south of the Isle of Man, on the other hand, gravel predominates over sand. Pinger profiles in Liverpool Bay tend to show less penetration and less detail in depth than those in the north-eastern sector, both features which are consistent with the general dominance of sand in the sediments (Plate 7 and Fig. 9; *c.f.* Belderson 1964 p. 155).

2c. Western sector (northwest, west and southwest of the Isle of Man)
The stratigraphy over much of this area shows a significant difference from that in the two sectors previously described. There are very few IGS boreholes in the present area, but the grab samples, sediment cores, and acoustic profiles show clearly that while the rockhead is overlain by a more or less continuous cover of boulder clay, the latter is directly overlain in most parts of the area by marine sediment (Plate 8 and Fig. 10). Evidence of proglacial water-laid beds is lacking except in one or two very local depressions, and even at these localities

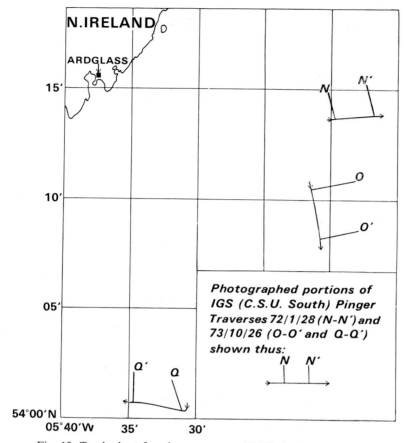

Fig. 10. Track chart for pinger traverses N–N', Q–Q', and O–O'.

there is a well-marked discontinuity (presumably an unconformity) at the junction with the overlying marine beds (Plate 9). Between the Isle of Man and Northern Ireland, there occurs a very large area in which the marine sediment is acoustically transparent throughout most or all of its thickness (Plate 10). Grab samples and vibrocores prove that this sediment consists almost entirely of mud; the latter in fact occupies a large area west of the Isle of Man (Fig. 2). Vibrocores also show that the mud (at least in some parts of the area) is separated from the boulder clay below by a layer only a few tens of centimetres thick, consisting of coarser marine sediment (Plate 6d). The upper surface of the boulder clay in many places at least obviously represents a significant unconformity.

The lithology of the marine mud in this area differs in certain respects, other than grain-size, from that of muddy marine sediments in the other sectors. The hue of the latter is typically 10YR, whereas the mud in the western sector is often somewhat "yellower", with hues lying in the range 2Y–5Y. In addition, the mud in the western sector frequently has a "fragmented" appearance (Plate 6e) due to the presence of numerous irregular cracks, a feature which has not been noted in muddy sediments in the other sectors.

In some of the local depressions where proglacial water-laid beds are apparently present, the beds presumed to be marine above the unconformity do show layering, which may be due to the presence of an alternation of sandy or gravelly layers with mud.

These results are essentially similar to those reported by Belderson (1964) from the same region. This author likewise found a large area of acoustically-transparent marine sediment (*loc. cit.* Figs 4 and 5), and commented on the apparent lack of late-glacial or post-glacial lacustrine deposits (*loc. cit.* pp. 161–162).

Outside the mud zone, the surface sediments of the western sector consist dominantly of sand and gravel in various proportions; these coarser sediments may well be laterally continuous with the layer of relatively coarse marine sediment which underlies the mud in parts of the area (see above).

3. Summary of depositional history

The ubiquitous layer of boulder clay has no doubt resulted from the melting and ablation of the Irish Sea portion of the Devensian ice-sheet. The boulder clay has not been investigated in detail, but it seems likely that the unbedded parts of the boulder clay include both ground-moraine and ablation-till components.

It is not yet possible to be certain whether the minor proportion of the boulder clay which contains distorted lamination represents intraglacial water-laid beds, or banded press-melt sediments.

The proglacial water-laid beds were presumably laid down during the late Devensian, when deglaciation was partly completed; the bodies of water in which these beds were laid down may have included arms of the sea, lagoons with access to the sea restricted by ice or topography or lakes. The almost total lack of macrofauna in these beds indicates a low salinity, as compared with the present-day Irish Sea, for all of these water bodies. The investigated foraminifera and organic-walled microplankton from the area southeast of the Isle of Man further indicate that here, at least, the proglacial water-laid beds were deposited in a body of water connected to the open sea, but colder and less saline than present-day Irish Sea water. Circulation within this water body or lagoon was probably restricted by floating ice, which may well have covered a large proportion of its surface.

This does not necessarily mean that the proglacial water-laid beds in all the other parts of the northern Irish Sea were deposited in such a lagoon. Further investigation of the microfauna and microflora in these beds will no doubt help to resolve this question.

Belderson (1964 p. 161) suggested the former existence of a "glacial lake" in the region west of the Isle of Man, the basis of his suggestion being (i) the existence of an elongated bathymetric basin in this area, between saddles at the western entrance to the North Channel and in St. George's Channel (*loc. cit.* fig. 3); and (ii) the distribution of certain animal species now found in lakes in areas bordering the Irish Sea (*loc. cit.* p. 161). No unequivocal evidence of a proglacial *lake,* as opposed to an arm of the sea or a lagoon, has yet been obtained from the northern Irish Sea. However, the general absence of proglacial water-laid beds of any kind west of the Isle of Man, together with the unconformity between the boulder clay and the marine beds, indicates a significant phase of erosion, which may have removed any proglacial lake beds in this area.

The age of the marine beds apparently ranges from the late Devensian to the

Flandrian. Most of the beds are probably subtidal, but at certain horizons there occur intertidal deposits, formed on beaches or tidal flats. Southeast of the Isle of Man, there occur intertidal beds which overlie unconformably the proglacial lagoon beds in that area (Pantin, *in press*). Those marine beds, at least, which have been sampled by vibrocore do not contain drop-stones, and were evidently deposited under temperature and salinity conditions progressively converging to those of the present day.

The local sediment unit with a "blotchy" reflection, situated southeast of Wigtown Bay, may be a salt-marsh deposit. Spots of iron-rich material, black when reduced and rusty-coloured when oxidised, occur in Flandrian salt-marsh deposits in The Wash (R. T. R. Wingfield, *pers. comm.*), as well as in the sediment unit under discussion. The "blotchy" echo might be the result of differential compaction of the mud, caused by draining of the small mud-banks which typically occur on coastal salt marshes. It is hoped that any foraminifera in these beds will provide further evidence on the conditions of deposition.

The difference in bedding definition between the proglacial water-laid beds and the marine beds may result from the much greater effects of bioturbation in the marine environment. In the first place, bioturbation would tend to blur contacts, even between major units (except in the case of beds containing a high proportion of coarse gravel). Furthermore, although pingers cannot resolve bedding units less than about 0·5–1·0 m thick, the presence or absence of lamination might well affect the overall acoustic properties of larger units; for example, the reflectivity of major beds might be reduced as a result of the bioturbative destruction of internal lamination.

The apparent lateral passages between proglacial water-laid sediments and marine sediments, indicated by some of the acoustic profiles, may well correspond to a significant real change in sedimentary facies. If a change in bedding definition can result from a variation in the degree of bioturbation, the lateral transition could be due to the local appearance of large numbers of marine animals in response to a persistent saline wedge at the bottom of the water column. The form and extent of such a wedge would be determined by factors such as bottom morphology, the distribution of glacial meltwater, the strength and direction of tidal currents, and the direction of the prevailing wind.

According to this hypothesis, the "marine" sediments formed under the saline wedge would almost certainly contain drop-stones, and these particular "marine" beds would therefore be more correctly called "glacio-marine". It has not been possible, however, to confirm the presence of drop-stones in these beds. The latter have not been penetrated by vibrocores, and the pinger profiles would not reveal the presence of individual pebbles or small cobbles; in fact, the profiles of known or presumed proglacial water-laid beds in the northern Irish Sea do not show any clear evidence of drop-stones. Evidence for drop-stones in the "glacio-marine" beds might have been expected in boreholes, but so far only one borehole (73/69, Fig. 8) has penetrated the beds in question, and drop-stones were not recorded in the log.

4. Sedimentation processes

The processes of present-day sediment transport and deposition in the northern Irish Sea are very complicated, and certain features only will be discussed here.

There appear to be two main zones of present-day sediment accumulation, one of these being the belt of mud-bearing, gravel-free sediments lying off Cumbria and Morecambe Bay, and extending northwards across the Solway Firth (Fig. 2). This accumulation zone probably also includes the narrow belt of sand (*sensu stricto*) running along the western border of the muddy sediments. Further west, however, the belt of slightly gravelly sand (the gravel fraction consisting almost entirely of shell) represents a transition into a zone where numerous sand waves have been found (Belderson and Stride 1969). This region of sand waves, some of which are isolated (*loc. cit.* p. 74), evidently represents a zone of non-accumulation, with sediments in transit moving towards the ENE, and thus towards the zone of sediment accumulation.

Both advective and diffusive processes appear to play a part in transport and deposition of the mud fraction in the accumulation zone off Cumbria (Pantin, in press), while in the sand-wave area the transport of sand is governed by the relative magnitude of the peak near-bottom ebb and flow tidal currents, with sand moving in the direction of the stronger current (Belderson and Stride, *loc. cit.* p. 74).

The accumulation zone off Cumbria probably corresponds to a zone of low relative energy; towards the ENE, wave energy increases near the Cumbrian coast, while to the WSW the energy of tidal currents increases. Although many highly variable factors are involved, it seems reasonable to suppose that zones of minimum energy will tend to correlate with areas of deposition, since it is in these particular zones that the equilibrium sediment load of the water mass is lowest, and is therefore most likely to be exceeded as a result of sediment transport. Whether deposition occurs or not will depend in a particular case on the type, magnitude and lateral variation of the various processes of sediment transport.

The second main zone of present-day sediment accumulation is the extensive belt of mud lying west and southwest of the Isle of Man. This mud belt corresponds approximately to the part of the area where the mean spring-tidal currents show a conspicuous velocity minimum (Belderson 1964 fig. 8). This part of the area is clearly a zone of relatively low tidal energy, and therefore a zone in which the equilibrium sediment load of the water mass will reach a minimum, provided that the effects of wave action or other currents do not greatly modify the situation. As previously implied, sedimentation might well be expected in such an area (*c.f.* Belderson *loc. cit.* pp. 159–160), but the sources of the mud, and the relative importance of advective and diffusive transport, have not yet been established.

5. Changes of relative sea-level

It is obvious from the available evidence that the replacement of the Devensian ice-sheet by sea in the northern Irish Sea basin was by no means a simple matter of the ice melting, accompanied by flooding as a result of rising eustatic sea-level. The widespread core and pinger evidence of a lithological discontinuity between the proglacial water-laid beds and the overlying marine beds, together with the evidence of other discontinuities within the marine sediments themselves, indicates that phases of erosion have occurred. These phases could have resulted from changes in the tidal current regime, but could also have resulted from lowerings of relative sea-level. The occurrence of intertidal sediments, interbedded with subtidal marine sediments or overlying proglacial lagoon beds, clearly indicates one or more lowerings of relative sea-level. The intertidal beds which unconformably overlie proglacial lagoon beds in the area southeast of the Isle of Man (Pantin, *in press*) appear to be lower-middle Flandrian in age (*loc. cit.*). Since a general rise in eustatic sea-level took place prior to and during this period, the indicated fall of relative sea-level further implies that the rate of isostatic uplift following deglaciation was for a time sufficient to pursue, and occasionally to overtake, the rise in eustatic sea-level.

Interaction of isostatic recovery and eustatic sea-level changes, giving rise to a relative fall in sea-level, was noted by Sissons and Brooks (1971) in the western Forth valley. One of their diagrams shows a 6 m fall in relative sea-level during the period 8500–10100 BP.

A curve showing relative sea-level changes in southern Cumbria during the Flandrian was deduced by Andrews *et al.* (1973) on the basis of local radiocarbon measurements. The curve does not show any proven lowerings of relative sea-level, but the data do not rule out the possibility that such an event occured during the late-Devensian/Flandrian period. No measurements were available for dates >9200 B.P. or 6000–3600 B.P. The measurements themselves were made on peat, wood or charcoal, which give only maximum values of sea-level, or shells, which give only minimum values. It is therefore possible that the true relative sea-level curve contains undisclosed fluctuations.

6. Sediment hues

The hues of the boulder clay and proglacial water-laid beds are probably caused by haematite and goethite adsorbed on the surface of clay-size particles or occluded within them, whereas the hue of the normal-coloured marine beds is probably due simply to goethite. In the variegated beds the yellow hues (e.g. 5Y) may be due to adsorbed or occluded jarosite, but the yellow-green hue (5GY) is probably the intrinsic colour of the mineral particles themselves, coloured adsorbates being absent. The evanescent 5B hue was probably caused by light scattering by a thin film of unstable organic material, possibly drawn to the cut surface of the sediment by the electro-osmotic process used.

The variegated sediments, which may include salt-marsh and other intertidal deposits, probably represent zones of exceptionally active biochemical diagenesis, which has exerted a decisive influence on the iron compounds controlling the sediment colour.

7. Acoustic turbidity

This feature is widespread throughout the northeastern and western sectors. The turbid zones may take the form of (a) local "pillars", sometimes accompanied by "terraces" or "platforms" along the sides or top (Plates 1–2); (b) more extensive "terraces" (Plates 3–4 and 8–9); or (c) "funnels" or local zones in which the reflections representing rockhead and bedding show a funnel-shaped cross-section (Plate 5). The acoustically turbid region in the centre of the "funnels" shows a well-defined top and sides, often with local "bright spots". In types (a) and (b) the reflections representing bedding and rockhead tend to dip towards the turbid zone, but this tendency is less pronounced and less localised than in the "funnels".

Schüler (1952) suggested and Schubel (1974) confirmed that shell beds as well as gas pockets are capable of producing acoustically turbid zones. In the present area, however, the local shelly beds found in the marine sediments do not show any correlation with the acoustically turbid zones; the latter, moreover, are frequently associated with the proglacial water-laid beds, which contain no shell beds at all. It is therefore probable that in the present area, the acoustically turbid zones are due to small pockets of gas, distributed through the sediments.

Methane and hydrogen sulphide are almost certainly important, if not dominant, constituents of the gas. The latter is probably disposed in thin, irregular fissures rather than in rounded "bubbles"; fissures of this type occur in the Chesapeake Bay sediments described by Schubel (*loc. cit.* fig. 5), while the "fragmented" nature of some of the mud west of the Isle of Man (section 2c and Plate 6e) may be due, at least in part, to the former presence of gas-filled fissures.

The terraces and platforms belonging to types (a) and (b) are probably caused by entrapment of gas by relatively impermeable beds. These features often show a very high degree of concordance with one another, and with beds visible on the acoustic trace away from the turbid zone. The observed tendency for bedding and rockhead reflections to dip towards acoustically turbid zones, conspicuous in the case of type (c), is well known from other areas (e.g. Edgerton *et al.* 1966; Hinz *et al.* 1969) and is often called the "Becken Effect" (German *Becken*=basin). This effect could well be due to preferential generation of gas by diagenetic processes in relatively thick beds. However, it must be remembered that the presence of gas pockets in sediment may significantly lower the overall velocity of sound, so that there will be an apparent fall in the level of a given reflector due to an increase in travel time. While "Becken Effect" is useful as a descriptive term, it must be realised that the "basin" appearance may not always be due to a genuine dip of the bedding or rockhead. As it is convenient to use purely descriptive terms for the features of acoustic turbidity zones, the name "mesoid" is suggested for the platform and terrace features. This does not exclude the possibility that some mesoids represent real terraces or benches cut in bedrock or consolidated sediment. For the region of apparently-dipping rockhead or bedding alongside an acoustically turbid zone, the name "fosoid" is suggested (Spanish: *foso*=ditch): again, the fosoid may in theory be either an acoustic effect or genuinely dipping rockhead or bedding.

8. Conclusions

Grab samples, cores, boreholes and pinger profiles have established that east of the Isle of Man the general upward succession in the northern Irish Sea is rockhead —boulder clay—proglacial water-laid sediments—marine sediments. West of the Isle of Man, the proglacial water-laid sediments are missing except in a few local depressions.

The sedimentary succession can be explained in terms of deglaciation, rising eustatic sea-level and isostatic recovery. The proglacial water-laid sediments were presumably laid down during the late Devensian while deglaciation was still in progress; the marine sediments range in age from late Devensian to Flandrian and up to the present day. The stratigraphic evidence indicates one or more lowerings of relative sea-level during the period considered. There is also evidence that such a phase took place during the first part of this period (late-Devensian to mid-Flandrian) when there was a general eustatic rise in sea-level. This implies that isostatic uplift was sufficiently rapid to overcome the eustatic rise for a time.

There appear to be two main areas of present-day marine deposition in the area: (i) off the Cumbrian coast, and (ii) west of the Isle of Man. Sediment transport includes both advective and diffusive processes, which depend on the local intensity of wave action or tidal currents.

Acoustic turbidity is widespread in the area frequently showing the "Becken Effect". The acoustic turbidity is probably due to the presence of small pockets of gas within the sediment; the "Becken Effect" may be due in some cases to a lowering of the overall sound velocity by the gas rather than to a genuine dip of the bedding or rockhead.

Acknowledgments. In conclusion, the writer wishes to offer his best thanks to the following colleagues, all of whom belong to the Institute of Geological Sciences (Leeds Office): Mr. C. R. Ransome, for compiling the major portions of the bathymetric chart and the surface sediment chart. Dr. B. N. Fletcher, and again Mr. C. R. Ransome, for critically reading the manuscript and making numerous helpful suggestions. Dr. W. Martindale and Dr. C. D. R. Evans for their respective contributions to the bathymetric chart and the surface sediment chart. Mr. K. E. Thornton (Senior Photographer), who supervised the preparation of the various photographs. Miss S. J. Masey prepared the lettering for Plates 1–5 and 7–10. This paper is published by permission of the Director, Institute of Geological Sciences.

References

ANDREWS, J. T., KING, C. A. M. and STUIVER, M. 1973. Holocene sea-level changes, Cumberland coast, north-west England: eustatic and glacio-static movements. *Geologie Mijnb.* **52**, 1–12.

BELDERSON, R. H. 1964. Holocene sedimentation in the western half of the Irish Sea. *Mar. Geol.* **2**, 147–163.

——and STRIDE, A. H. 1969. Tidal currents and sand wave profiles in the north-eastern Irish Sea. *Nature, Lond.* **222**, 74–75.

CHMELIK, F. B. 1967. Electro-osmotic core cutting. *Mar. Geol.* **5**, 321–325.

CRONAN, D. S. 1969. Recent sedimentation in the central north-eastern Irish Sea. *Rep. No. 69/8, Inst. Geol. Sci.* 10 pp.

EDGERTON, H. E., SEIBOLD, E., VOLLBRECHT, K. and WERNER, F. 1966. Morphologische Untersuchungen am Mittelgrund (Eckernförder Bucht, westliche Ostsee). *Meyniana* **16**, 37–50.

FOLK, R. L. 1974. *Petrology of Sedimentary Rocks.* Hemphill Publishing Co., Austin, Texas.

GEOLOGICAL SOCIETY OF AMERICA 1963. *Rock Color Chart.* Geological Society of America, New York.

HARLAND, R. *(in press).* Modern and Quaternary organic-walled microplankton from the north-eastern Irish Sea. Appendix 3 *In,* Pantin, H. M. *(in press). Bull. geol. Surv. G.B.*

HINZ, K., KÖGLER, F.-C. and SEIBOLD, E. 1969. Reflexionsseismische Untersuchungen mit einer pneumatischen Schallquelle und einem Sedimentecholot in der westlichen Ostsee. *Meyniana* **19**, 91–102.

HUGHES, M. J. *(in press).* Foraminifera from vibrocore G5 samples. Appendix 1 *In,* Pantin, H. M. *(in press). Bull. geol. Surv. G.B.*

OVENSHINE, A. T. 1970. Observations of iceberg rafting in Glacier Bay, Alaska, and the identification of ancient ice-rafted deposits. *Bull. geol. Soc. Am.* **81,** 891–894.

PANTIN, H. M. (*in press*). Quaternary sediments from the north-east Irish Sea. *Bull. geol. Surv. G.B.*

SCHUBEL, J. R. 1974. Gas bubbles and the acoustically impenetrable, or turbid, character of some estuarine sediments. *In* Kaplan, I. R. (Editor) *Natural Gases in Marine Sediments.* Plenum Press, New York and London, 275–298.

SCHÜLER, F. 1952. Untersuchungen über der Mächtigkeit von Schlickschichten mit Hilfe des Echographen. *Dt. Hydrograph. Z.* **5,** 220–231.

SISSONS, J. B. and BROOKS, C. L. 1971. Dating of early Postglacial land- and sea-level changes in the western Forth valley. *Nature Phys. Sci.* **234,** 124–127.

H. M. Pantin, Institute of Geological Sciences, Ring Road, Halton, Leeds LS15 8TQ.

A late-glacial drainage pattern in the Kish Bank area and post-glacial sediments in the Central Irish Sea

R. J. Whittington

Four distinct units can be recognised on continuous seismic profiles in the Kish Bank area. Above the pre-Pleistocene bedrock is an extensive till sheet. Into this is cut a drainage pattern which is infilled by thick well-bedded sediments. Very recent sand features complete the sequence.

From correlations with offshore borehole information the till sheet is considered to be Devensian; the drainage pattern to be of a sub-aerial origin in Late Devensian times; the well bedded sediments which occur over much of the Central Irish Sea to be Holocene deposits, predominantly of sand.

1. Introduction

As part of the geological investigations of the Central Irish Sea by the Department of Geology, University College of Wales, Aberystwyth, extensive shallow penetration continuous seismic profiling has been carried out in the Western Central Irish Sea. The bedrock geology mapped out by this method has been summarised by M. R. Dobson in this volume and will be discussed in detail elsewhere. A contribution on the nature and distribution of Quaternary deposits as a whole is also being prepared. Some aspects of Quaternary history form the basis of this contribution which arises from the interpretation of the seismic records linked with the minimal offshore borehole information. No attempt has been made to correlate in detail with the complex history of glacial and post-glacial events in the Irish Sea and the surrounding land as described elsewhere in this book. The type and quality of the original information warrants only the most general comparisons.

E

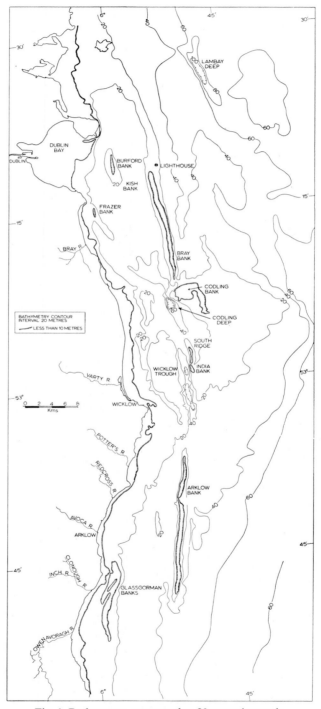

Fig. 1. Bathymetry, contoured at 20 metre intervals.

Published subsurface information in the Kish Bank area consists of the series of boreholes put down for the Dublin Port and Docks Board in Inner Dublin Bay and described by Naylor (1965) and a short borehole at the site of the Kish Bank Lighthouse tower. Contours on the Lower Carboniferous bedrock given by Naylor show a southwestward aligned channel which cuts across the mouth of the present River Liffey and reaches a maximum depth of about 35 m below present sea-level. This channel is a continuation of a bedrock channel mapped by Farrington (1929) who showed that a pre-glacial drainage system in eastern Central Ireland had cut down from a westward sloping peneplain considered to be certainly post-Cretaceous and probably Miocene in age. Above the bedrock Naylor (1965) describes nine episodes, the lowest two of which are a boulder clay succeeded by fluvial periglacial gravels. The boulder clay was presumed to be ground moraine of Midland General (Devensian) affiliations. The overlying periglacial gravels were considered to be washed and re-worked boulder clay and were not found in the outer borings. The third episode was a period of marked erosion by water using the pre-glacial exit route of the Liffey which channelled down almost to the floor of the pre-glacial valley. The succeeding six episodes consist of four units of post-glacial sediments followed by a period of uplift and erosion prior to the accumulation of the recent sediments of the present River Liffey. The four post-glacial sediment units are of muds and sands with gravel and shelly sand ranging from Boreal age and with near-shore environments of deposition.

The borehole at the site of the Kish Bank Lighthouse tower penetrated only 27 m below the sea-bed. Wash samples together with the range of penetration was used to give a sequence of fine sand with silt to 15 m followed by 1 m of stiff blue clay and 11 m of dark grey cohesive silt with some sand.

Outside the Kish Bank area there are two boreholes in Caernarvon Bay, to be referred to later, and the series of boreholes in Cardigan Bay the results of which are dealt with in section 3 of this contribution.

2. Bathymetry

Figure 1 has been prepared from standard Admiralty charts. Generally the contours show an offshore grade from the coast to 60 m. In the Wicklow Head region the 20, 40 and 60 m isobaths show a pronounced Caledonoid trend indicating the continuation of the subcrop of Ordovician rocks which form the coastal region between Wicklow and Cahore Point. Imposed on this grade are a number of positive and negative features.

The positive features consist of a series of banks, which, with the exception of the Codling Bank, are linear and sub-parallel to the coast. Continuous seismic profiles, side scan sonar and direct sampling techniques have shown these to be essentially sand bodies resting on a flat floor close to the level of the surrounding sea-floor. (Dobson *et al.* 1971).

The negative features consist of a number of troughs and pits. The most northerly of these is the Lambay Deep. As defined by the 80 m contour this is a 12 km long trough aligned NNW–SSE reaching to a maximum depth of 132 m from a surrounding sea-floor of 60 m. The Deep shows evidence of bifurcation at its northern end. Immediately west of the Codling Bank is a linear depression again with a NNW–SSE alignment here called the Codling Deep. The maximum depth of 97 m occurs in the form of a pit. The feature closes simply at its southern end

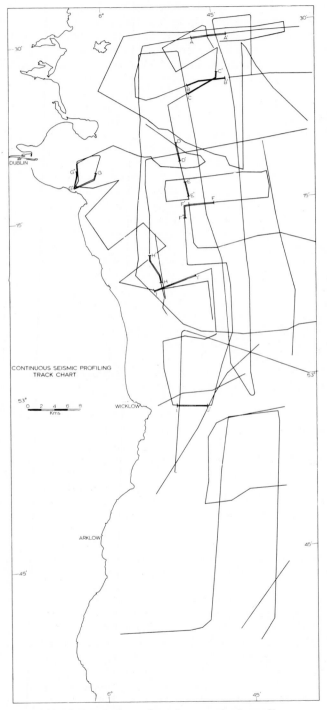

Fig. 2. Track chart of continuous seismic profiles.

whilst trifurcating at its northern end. Further south there occurs the Wicklow Trough with a maximum depth of 70 m from a surrounding sea-floor depth of 20 m. As defined by the 40 m contour the Wicklow Trough is some 12 km long with a north–south trend. Though an obvious connection between the Wicklow Trough and the Codling Deep can be suggested there is no hint in the bathymetry of a connection between these and the Lambay Deep.

3. Investigations

Continuous seismic profiling (C.S.P.) studies were conducted using 1000 Joule sparker discharges twice per second as the seismic source. Standard E. G. & G. components were used to give geological information down to $\frac{1}{2}$ second two way travel time with a resolution of 2–4 m. The majority of the results were obtained on the N.E.R.C. research vessels R.R.S. *John Murray* in July 1974 and R.V. *Edward Forbes* in December 1975. The total C.S.P. coverage used in this study is shown in Figure 2 and forms part of a wider study of the geology of the Central Irish Sea.

4. Interpretations

4a. Main element shown on the profiler records

Interpretation of the C.S.P. records allows a division into four distinct units. The lowest unit is that of the pre-Pleistocene bedrock of varying age and hence seismic character. This seismic character together with gravity anomaly, magnetic anomaly and direct sampling evidence has been used to produce the summary pre-Pleistocene geology map discussed by M. R. Dobson in this volume.

The second unit is that generally identified on profiler records in the Central and South Irish Sea as glacial till. The interface between it and the bedrock is obviously erosional and there is a strong correlation between the depth of erosion and the age of the bedrock, the younger softer rocks having been extensively over-deepened while the older harder rocks remain as upstanding blocks. Southward direction of the ice movement is also indicated to some extent by the contours on the bedrock surface. Reflecting horizons from within the unit are few whilst those that do occur are weak, discontinuous and suggestive of lensing. The background signal return from within the unit is high and is most probably caused by back scatter of small boulders in the clay matrix. Larger boulders give rise to point source diffraction hyperbolae within and at the surface of the unit. Morainic deposits are inferred east of Wicklow and the Codling Bank, whilst in the deeper water area east of the Kish Bank the hummocky surface of the unit may represent a drumlin tract. From the network of intersecting C.S.P. traverses which cover the Central and South Irish Sea this unit may be correlated from the Kish Bank area into Cardigan Bay where the Institute of Geological Sciences' boreholes have shown the unit to consist of Devensian Irish Sea till. No sub-division of the unit which might be indicative of pre-Devensian deposits can be attempted from the profiler records in the Kish Bank area.

The third unit is characterised by internal strongly reflecting horizons which are laterally persistent over a large area. Several reflecting horizons are so well marked and identifiable from record to record that contouring of the individual reflecting horizon could be carried out. The beds are horizontal in general, but follow the

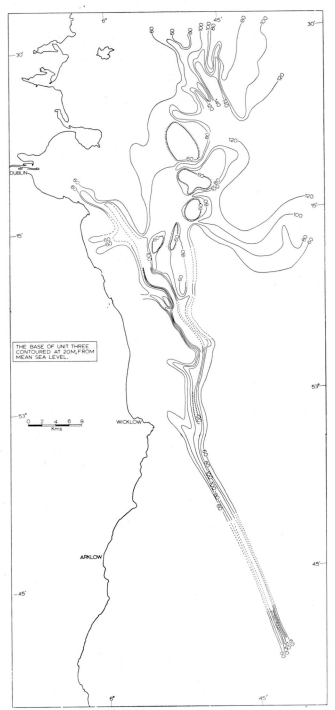

Fig. 3. The base of unit three contoured at 20 metre intervals below mean sea-level.

topography of the underlying unit surface where that topography is rugged. Slumping into negative features and drape thinning over positive features in the underlying unit is shown on some records. Point source diffraction hyperbolae occur within the unit though in a restricted locality and immediately above a particular reflecting horizon. The unit appears entirely depositional in origin, no erosional phases can be recognised within it.

The topmost unit consists of the banks and other sand wave bodies, and recent muds and silts which infill some hollows. The banks have a pronounced flat transgressive base in the inshore regions though in the deeper water areas (< 40 m) large sand wave complexes are found sitting with little or no apparent base on unit three.

This fourfold succession is not present everywhere. Isolated patches of Lias outcrop at the sea-bed and have been sampled, whilst in some areas the well-bedded deposits of unit three rest directly on the pre-Pleistocene bedrock.

4b. The base of unit three

Figure 3 is a map of the base of unit three contoured at 20 m intervals. Velocities of 1·5 m/millisecond for seawater and 1·7 m/millisecond for unit three and the overlying unit four, where it is present, were used to convert the time sections of the profiles into depth sections. Errors arise from non-vertical incidence due to the step out of the seismic source and the receiving hydrophone and from the fact that the velocity of 1·7 m/millisecond found for the uppermost layer in a refraction survey at the site of the Kish Bank Lighthouse tower is a composite velocity for unit two, glacial till, and unit three. These errors to some extent cancel each other out while tidal variation introduces a further error of up to 4 m. An individual depth estimate might therefore be 5% in error, probably on the high side.

Figure 3 shows that the sediments of unit three infill a complex valley system. The Lambay Deep is fed by five channels, only the two at the northwestern end having present day bathymetric expressions. The Deep opens southward and eastwards into a large area where the thickness of unit three exceeds 60 m in 60 m of water. This lake-like area continues eastwards out of the area considered. A north-south ridge forms the western margin to this lake-like area. The ridge is cut by three narrow channels which then coalesce to form a wider shallower channel which trends southwesterly and joins a major southeastwards trending channel, the bathymetric expression of which is now the Codling Deep. This major channel extends down from Dublin Bay and has minor tributaries from the onshore rivers. A connection occurs to the Wicklow Trough though the connection is now infilled and in part overlain by the South Ridge Bank. A further tributary, from the onshore River Varty, joins the system in the Wickow Head region which then turns southeasterly by the top of the Arklow Bank and continues on this trend out of the area surveyed. It is not known if any of the onland rivers between Wicklow and Arklow have tributaries to the channel. The total length of this linked channel system is approximatley 100 km.

In order to appreciate better the significance of the channelling Figures 4 and 5 show cross sections from the locations shown in Figure 2. The sections are line drawings traced from the original records and thus show considerable vertical exaggeration. From Figures 4 and 5 it can be seen that the present Lambay Deep, Codling Deep and Wicklow Trough are relatively minor remnants of an erosional period which post-dates the deposits of glacial till. The bedrock profile beneath these features indicates that the later channelling was probably influenced by pre-glacial or en-glacial channelling of the bedrock. The channels cut through the north-south ridge profiles, DD', EE', FF' F', are eroded into the till and sometimes through into the bedrock which is otherwise flat, so that there is no pre-glacial or en-glacial control of the later channelling.

5. Discussion

5a. Correlation with the Inner Dublin Bay boreholes

The main channel has been shown to extend down from Dublin Bay. The shallowness of the water prevented a direct survey connection to be made to the Inner Dublin Bay boreholes. Offshore from Dun Laoghaire (section G G' G'' on Fig. 5) a channel has been eroded into bedrock to more than 80 m below sea-level. This channel has a noticeable asymmetric profile, the southern margin being a scarp. The morphology of the channel in this area may reflect the northern limit to the offshore continuation of the Leinster Granite into Dublin Bay. From these profiles it is about 7 km to the area investigated by the Inner Dublin Bay boreholes. There is thus a correlation suggested over this 7 km between unit two, the glacial till, in the Kish Bank area and episode one boulder clay of assigned Devensian age in the boreholes. The erosional surface which forms the base of unit three and is contoured in Figure 3, correlates with episode three, the period of marked fluviatile erosion, whilst the sediments of unit three correspond in part to the post-glacial sediments encountered in the boreholes.

5b. Pre-glacial and en-glacial channelling prior to that at the base of unit three

The bedrock channel proved by the Inner Dublin Bay boreholes confirmed the earlier work by Farrington (1929) on the pre-glacial topography of the Liffey drainage basin. The later channelling proposed to correspond to the erosional surface at the base of unit three used the same exit route as the pre-glacial, probable Miocene river, cutting down almost to the pre-glacial valley floor. Some 7 km to the southeast and along the same trend, section G G' G'' on Figure 5 shows that the later channelling has removed any evidence for a pre-existing channel in the bedrock. This is also true for other profiles further southeast where the bedrock is thought to be Triassic in age. The channel here exploits the margin of the sedimentary basin where the Triassic rocks are faulted against the Ordovician. Section H H' shows that the later channelling exploits a pre-existing bedrock channel as does section I I' in the Wicklow Head area. Further southeastwards the later channelling has again obliterated any evidence for a pre-glacial channel in well-bedded presumed Mesozoic rocks. Thus, it seems probable that the later phase of channelling extending down from Dublin Bay re-excavated an older channel system which could be a glacially modified pre-glacial valley from the Liffey Basin.

The bedrock valley which underlies the Lambay Deep is closed at its northwestern end. Its southeastward extent is unknown as the seismic profiler does not record bedrock in this area where glacial over-deepening into mesozoic rocks was particularly effective. From the northernmost profiles across the Deep, unit two can be seen to occur in the bottom of the bedrock valley (sections AA' and BB') however on the other more southerly profiles a thick sequence of unit three masks the underlying reflectors. Though the subsidiary channels which lead into the Lambay Deep are cut down though the till unit and thus belong to the later phase of channelling it appears that the main Lambay Deep bedrock valley was formed earlier and at least partially infilled with till before the drainage system shown in Figure 3 was established. An explanation for its origin is afforded by its position and orientation which is at and along a fault margin between proven Lias and presumed later Mesozoic rocks. The Lias rock samples obtained are of a hard grey mudstone whilst the profiler records show that the rock as a whole has been sufficiently resistant to erosion so as to remain an upstanding block. Deflection of ice moving from the north or northwest around this block with consequent preferential erosion along the margin of the Lias may account for the formation of the Lambay Deep bedrock valley.

5c. The possible north and southward extensions of the channel system

From Admiralty charts the Lambay Deep can be seen to have a bathymetric expression for another 4 km in a northerly direction beyond the limits of the area considered in this paper. However, there are no further troughs or hollows which might indicate remnants of this drainage system. Sea floor topography north and east of Lambay Island is much more subdued than it is to south. This fact was noted by Dobson *et al.* (1971) who suggested that this might be caused by a mantle of Holocene deposits blanketing irregularities in the Pleistocene surface. If this is so a completely infilled channel system may connect with the Lambay Deep. The southeastwards continuation of the channel system is also unknown. Sparker profiling traverses are as shown in Figure 3 of Dobson *et al.* (1973) from which it can be seen that a large gap in coverage exists along the line of a possible extension. Those traverses which do exist further to the south are not of sufficient quality to determine whether the channel-like depressions in the bedrock are cut through the glacial material. Further profiling will be undertaken to resolve these problems.

5d. The mode of origin of the channel system

Three modes of origin might be proposed for the Kish Bank channel system. These are:

(a) Tidal scour during the Holocene transgression.

(b) Sub-glacial stream erosion.

(c) Sub-aerial stream erosion at a time of low sea-level.

Examples of troughs thought to be formed by tidal scour have been given by Donovan (1973). Strong tidal currents associated with restricted areas are required which may then erode linear closed basins. The bathymetry of the Central and South Irish Sea has been considered by Dobson *et al.* (1971). As defined by the 40 fathom (75 m) contour the region consists of two coastal shelves with a central channel. During periods of low sea-level the two coastal shelves would have formed a considerable restriction to the width of St George's Channel. High tidal velocities and the formation of scour basins might then be expected to form in the central channel and indeed there are a number of closed elongated basins northwest of Anglesey. The Kish Bank area would be part of the Irish coastal shelf even when the contour map Figure 3 is taken into consideration. It therefore seems unlikely that the high tidal velocities required to give scour would be present in this area, moreover simple elongated closed basins are characteristic of tidal scour formation whereas the sinuous branching character of the Kish Bank system is not at all in keeping with a tidal scour mode of origin.

Sub-glacial stream valleys have been described on land by Woodland (1970) and Howell (1973) and below the sea-floor by Flinn (1967), Dingle (1971) and Destombes *et al.* (1975). These "Tunnel Valleys" are characterised by elongated closed basins which occur in a closely spaced network and may be interconnected. The depth of erosion is very variable with irregular depressions separated by sills and threshholds. Though the southwestwards trending channel network which occurs in the Lambay Deep area might well be tunnel valleys it seems a much less likely mode of origin for the main inshore system which extends down from Dublin Bay and has tributaries from the established onshore rivers.

A sub-aerial fluviatile origin is supported for the main inshore channel by its general morphology, the tributaries and the overall downstream grade. A correlation has been proposed between this erosional period and episode three from the results of the Inner Dublin Bay boreholes. Episode three which was considered to be a period of fluviatile erosion at a time of low sea-level cut down through a thin deposit of stony gravel which is "probably washed and reworked boulder clay of fluvial periglacial origin formed during ice retreat" (Naylor 1965, p. 186).

A sub-aerial fluviatile origin during a period of low sea-level is therefore proposed for the main channel system though the channelling in the Lambay Deep area may have been instigated sub-glacially.

5e. Age of the channel system

There is no direct evidence for the age of the channel system. However, the system manifestly post-dates the till sheet which extends over the Central and South Irish Sea. The till sheet can be correlated with the I.G.S. boreholes in Cardigan Bay, whilst correlation with the Inner Dublin Bay boreholes is not direct but appears extremely likely. The till sheet is thus considered to be Devensian (Naylor 1965, and section 3 above) and by correlation with Bowen's (1973) discussion of the Pleistocene of the Irish Sea region is more specifically Middle to Late Devensian. A sub-aerial fluviatile origin for the channel system implies that ice withdrawal had at least reached the latitude of Lambay Island. Ice withdrawal from northwest Wales had been completed by at least 14468 B.P. (Coope *et al.* 1971) and in the Firth of Clyde by 13020 years B.P. (Bishop and Dickson 1970). Thus this drainage pattern may have established itself by 14000 B.P. The rate of erosion was vigorous cutting down through the till sheet into the underlying bedrock to a base level which is approximately −130 m below present sea-level. Some erosion took place below this level where there was both a confluence of the drainage channels and underlying soft rocks of Mesozoic age.

5f. The distribution of the post-glacial deposits

It follows from the foregoing discussion on the age of the channel system that the sediments of unit three are post-glacial in age. It is not known if they continue north of the area though the subdued bathymetry commented upon by Dobson *et al.* (1971) suggest that they do so. In the South Irish Sea they do not occur. Eastwards of the Kish Bank area the post-glacial sediments thicken with increasing water depth reaching a maximum of 100 m in 105 m of water. The eastern margin of this area is the northeastward trending basement ridge, the Mid Irish Sea Uplift (see M. R. Dobson, this volume). Deposits of both unit two, the glacial till, and unit three, the post-glacial sediments, thin on the flanks of this feature and indeed bedrock outcrop occurs in some small areas. The sequence then continues eastwards across the Central Irish Sea Basin into Caernarvon Bay. In his study of the eastern Central Irish Sea, Al-Shaikh (1970) recognised a division on his C.S.P. records which he proposed as the junction between "old" pre-Devensian drift and "new" Devensian drift. The southern margin of the "new" drift was approximately east-west against the basement ridge of the Irish Sea Geanticline (see M. R. Dobson, this volume). Further and better quality profiling over the Central Irish Sea Basin and connecting into the Kish Bank Basin has established that the "new" drift is identical with the post-glacial deposits of the Kish Bank, though the reflecting horizons are less well developed in the east. This sequence is further confirmed by two I.G.S. boreholes in Caernarvon Bay.

Borehole No. 71/53 Latitude 53° 1'N Longitude 4° 35'W

Water depth 27·5 m
Sand and gravels 24·5 m thick
Grey sandy boulder clay 2 m thick
Bedrock.

Borehole No. 71/54 Latitude 53° 4·73'N Longtitude 4° 24·40'W

Water depth 18 m
Sand and gravel 11·5 m thick
Stiff grey boulder clay 4 m thick
Sandy gravels 2·5 m thick
Reddish brown sandy boulder clay 8 m thick
Bedrock.

The post-glacial deposits thus form a blanket over much of the Central Irish Sea, the southern margin being formed by the basement ridges of the Wicklow Head Shelf in the west, and the Irish Sea Geanticline in the centre and east.

5g. The nature and origin of the post-glacial sediments

The C.S.P. records indicate that the nature of the post-glacial sediments is one of individual members a few metres thick of varying composition. Each member is flat-lying and laterally persistent over many kilometres. Such a character implies at least a lacustrine or more likely, in view of the widespread occurrence over the Central Irish Sea of these deposits, a marine environment of deposition. The Holocene transgression provides the marine environment, but direct evidence of the nature of the sediments is restricted to the few boreholes already mentioned. The post-glacial deposits in the Inner Dublin Bay boreholes consist of muds and sands with some shells and pebbles and are mainly nearshore in origin (Naylor 1965). They have a gentle seaward slope which Naylor suggests is the original sedimentation slope. Sands and silts dominate the sequence at the Kish Bank Lighthouse tower whilst sands with gravel form the supra-boulder clay unit in the Caernarvon Bay boreholes. It therefore seems likely that the bulk of the post-glacial sediments consists of marine sands, muds, silts, shells and gravels in admixtures forming the bedded sequence. This suggestion is supported by the presence of large sand wave complexes in the deeper waters These sand waves arise out of the post-glacial deposits with little or no indication of a base and thus seemed to have been formed *in situ*. There is thus a ready provenance of material for the sand banks which fringe the coastline.

The bulk of this material must be outwash derived during the main phase of ice retreat in the northwest Irish Sea and central and northeast Ireland. Some ice rafting of material, which included large boulders, occurred as witnessed by the diffraction hyperbolae within the unit.

Such a thick accumulation of deposits (up to 100 m) in so short a time covered by the post-glacial, reflects the bedrock geology of this area whereby glacial over-deepening into the softer rocks of the sedimentary basins bounded by the resistant rocks of the basement ridges left a natural sediment trap for the influx of material from the northwest Irish Sea and off the Central Ireland landmass during deglaciation.

5h. Present day conditions: infill or scour?

Parrish (1973) considered the recent sediments of the western Irish Sea. From analyses of bed forms and sediment and tidal information he suggested that the Wicklow Trough and Codling Deep are at present undergoing active erosion whilst the Lambay Deep may be infilling slowly. Some profiles across the Lambay Deep (see Section BB' on Fig. 4) do show evidence of recent infilling giving rise to a smooth bottom profile while profiles across the Wickow Trough and Codling Deep show a rough profile (see Section G–G' on Fig. 5) in keeping with present day erosion.

6. Conclusion

Though much of the foregoing discussion is based upon indirect evidence and may be open to interpretations other than those presented, the following salient points arise out of this study:

(1) Contours on the pre-Pleistocene bedrock show a strong correlation between the depth of erosion and the disposition of basement ridges and the Mesozoic-Tertiary sedimentary basins. The contours further indicate southward direction of ice movement.

(2) A till sheet which is seismically a single unit blankets the pre-Pleistocene surface.

(3) A period of intense erosion which cut a linked channel system post-dates the emplacement of the till sheet.

(4) Further thick sedimentation of well-bedded material occurred to the north of a line between Wicklow Head and the Lleyn Peninsula.

Acknowledgments. This contribution is based on the results obtained mainly on N.E.R.C. research vessels. The award of ship time is gratefully acknowledged as is the assistance of the Officers and crew and colleagues who took part in the cruises. Dr. D. Q. Bowen kindly commented upon the manuscript.

References

AL-SHAIKH, Z. D. 1970. *Geophysical Investigation in the Northern Part of Cardigan Bay and part of the Central Irish Sea.* Unpublished Ph.D. Thesis, University College of Wales, Aberystwyth.

BISHOP, W. W. and DICKSON, J. H. 1970. Radiocarbon Dates related to the Scottish Late-Glacial Sea in the Firth of Clyde. *Nature Lond.* **227**, 480–482.

BOWEN, D. Q. 1973. The Pleistocene Succession of the Irish Sea. *Proc. Geol. Ass.* **84**, 249–272.

COOPE, G. R., MORGAN, ANNE and OSBORNE, P. J. 1971. Fossil Coleoptera as Indicators of Climatic Fluctuations during the Last Glaciation in Britain. *Palaeogeog. Palaeoclim. Palaeoecol.* **10**, 87–101.

DESTOMBES, J. P., SHEPHARD-THORN, E. R. and REDDING, J. H. 1975. A buried valley system in the Strait of Dover. *Phil. Trans. R. Soc. Lond. A.* **279**, 243–256.

DINGLE, R. V. 1971. Buried Tunnel Valleys off the Northumberland Coast, Western North Sea. *Geologie Mijnb.* **50**, 679–686.

DOBSON, M. R., EVANS, W. E. and JAMES, K. H. 1971. The sediment of the floor of the Southern Irish Sea. *Mar. Geol.* **11**, 27–69.

——, —— and WHITTINGTON, R. J. 1973. The Geology of the South Irish Sea. *Rep. Inst. Geol. Sci.* No. 73/11, 35 pp.

DONOVAN, D. T. 1973, The geology and origin of the Silver Pit and other closed basins in the North Sea. *Proc. Yorks. geol. Soc.* **39**, 267–293.

FARRINGTON, A. 1929. The pre-Glacial topography of the Liffey Basin, *Proc. R. Ir. Acad.* **38B**, 148–170.

FLINN, D. 1967. Ice front in the North Sea. *Nature Lond.* **215**, 1151–1154.

HOWELL, F. T. 1973. The sub-drift surfaces of the Mersey and Weaver catchment and adjacent areas. *Geol. J.* **8**, 285–296.

NAYLOR, D. 1965. Pleistocene and Post-Pleistocene sediments in Dublin Bay, *Sci. Proc. R. Dub. Soc.* Series, A. **2**, 175–188.

PARRISH, J. G. 1973. *Recent sediments and shelly fauna of the Western Irish Sea.* Unpublished
    Ph.D. Thesis, University College of Wales, Aberystwyth.
WOODLAND, A. W. 1970. The buried tunnel valleys of East Anglia. *Proc. Yorks. geol. Soc.* **37,**
    521–578.

R. J. Whittington, Department of Geology, The University College of Wales, Llandinam
Building, Penglais, Aberystwyth, Dyfed, Wales, SY23 3DB.

The sediments of the South Irish Sea and Nymphe Bank area of the Celtic Sea.

R. A. Garrard

Offshore drilling and coring together with high resolution seismic surveys have established that the central portion of the South Irish Sea contains one of the largest successions of Quaternary sediments outside the North Sea. In the east, the offshore glacial sediments are restricted to those of Devensian age, which in turn are covered by well-developed estuarine sediments of late glacial to Flandrian age. Selective erosion of soft Tertiary strata has allowed the accumulation of at least two units of Irish Sea till, separated by thick sequences of temperate marine interglacial sediments of possible Ipswichian age. South of Ireland only small and isolated outliers of glacial drift have been found.

1. Introduction

The South Irish Sea is known to be thickly mantled by post-Tertiary sediments (Fig. 1). The subdued nature of the topography in this area since late Tertiary times appears to have been one of the most important criteria in the deposition and preservation of these sediments. This region probably remained a flat low-lying depression throughout the Quaternary, contrasting sharply with the positive crustal blocks, or mountainous highlands, of the Welsh Massif and southeast Ireland. These offshore deposits appear to have sufferered very little solifluction or sub-aerial weathering which has complicated the picture so much on land, a factor which makes them ideal for detailed study. Therefore an investigation of the Quaternary sediments in the South Irish Sea was conducted between 1970 and 1974 by the Continental Shelf Unit 1 of the Institute of Geological Sciences, and the Department of Geology, University College of Wales, Aberystwyth. Over

230 separate localities were sampled by either rotary drilling, or shallow gravity and vibrocoring. The cores were later sectioned and analysed using standard sedimentological techniques. Horizons indentified in cored samples were correlated between sample stations using high frequency seismic equipment (O.R.E. and Edo-Western Pingers working at 2·4 and 3·5 kHz, and multielectrode and trielement sparkers working at $\frac{1}{4}$ and $\frac{1}{2}$ second sweep rates). As most of the sampling was confined to British territorial waters, seismic profiling only was conducted in the west of the area. Where possible the successions recorded offshore were always related to important coastal sections close-by.

2. Pre-Devensian Quaternary sediments

2a. Glacial

Underlying the marine sediments of last interglacial age (see 2b) is a thick (>9 m) unit of Irish Sea till, the base of which has only been penetrated by one borehole (Fig. 2 Borehole 71/55). This pre-Devensian till covers an extensive area of St George's Channel coincident with the distribution of Tertiary strata (see M. R. Dobson, this volume). In the east thinning takes place towards the more resistant subcrops of Mesozoic age, while over most of Cardigan Bay this formation is absent (Fig. 3).

Although stratigraphically separated from the Irish Sea till of Devensian age, the two tills are surprisingly similar. The older till is a fresh mature glacial sediment showing no signs of having been extensively weathered, decalcified or oxidised. The erratics are of Irish Sea derivation and are all small. The abundant clay matrix contains marine sands and shell fragments, and has a dark yellowish brown (10 YR 4/2) to olive grey (5 YR 4/1) colour. The carbonate content ranges from 7% to 13%.

Relict glacial sediments, some showing evidence of weathering and recycling, are present only within certain restricted localities in other areas of the South Irish Sea, and parts of the Celtic Sea and Bristol Channel (Fig. 9). At Tonfannau, till of Irish Sea type is present at the base of the coastal cliff section where it is overlain by glacial drifts of Welsh origin. Although the till retains its dark yellow brown colour, in contrast to the dark greys of the Welsh sediments, it has been largely decalcified (<3%). The upper surface has deformed into southwesterly orientated recumbent folds. At borehole 71/57 (Figs 2 and 3) the Welsh glacial drift does not rest directly on bedrock but 18·5 m of Irish Sea type till. Further south isolated exposures of Irish Sea till have been identified in several coastal exposures between Llanrhystud and Aberayron (Watson 1970). These are no longer in a primary position but interbedded with glacial drift of Welsh origin. The till, which is red brown in colour as a result of weathering and oxidation, has been decalcified (<2%).

2b. Interglacial

Sequences of marine clastic sediments underlying the Devensian Irish Sea till sheet were identified in at least six boreholes in St George's Channel (Figs 2 and 4). Thicknesses vary between 3 m and at least 56 m at depths below the sea-floor of between 9 m and 41 m. Continuous seismic profiles suggest each marine interval to be laterally continuous; for example extending for a north-south distance, between boreholes 73/43 and 71/56, of over 37 km (Fig. 4). The base of these

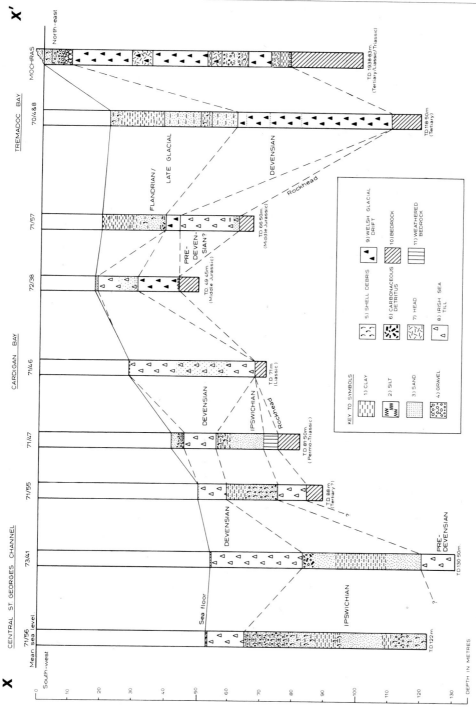

Fig. 2. East/west correlation of I.G.S. boreholes off the west coast of Wales. Section x–x¹ is shown on Figure 4.

F

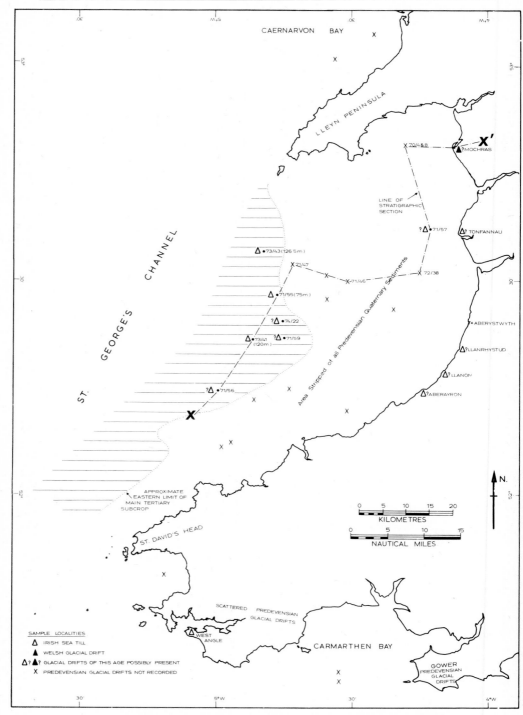

Fig. 3. Distribution of pre-Devensian Glacial sediments in the South Irish Sea.

marine sediments, where penetrated, consists of either bedrock or the lower Irish Sea till.

The sediments, which are predominantly arenaceous, consist of a bimodal mixture of sands and gravels and contain thin layers of silt and clay. The gravel fraction includes an assortment of different rock types including the diagnostic Middle Jurassic limestones. Certain pebbles exhibit "flat iron" surfaces and rare striations. The sand fraction consists of moderately well sorted, subrounded, highly polished, quartz grains.

Many of the sediments contain a rich marine assemblage of both macro- and microfauna. Much of the macrofauna is, however, worn and fragmentary. Certain of the larger pebbles exhibit scars from the attachment of epifauna especially barnacles and bryozoans. Within the sand fraction are small bivalve fragments including *Astarte* sp. and *Nucula* sp. besides barnacle plates and echinoid spines. Preliminary results from microfaunal analysis have shown a rich, dominantly marine boreal assemblage of both Ostracods and Foraminifera (Haynes *et al. pers. comm.*) In addition, there is a small, very cold water element, the most noticeable being *Robertsonites tuberculata*. Brackish water species are always rare and make up less than 2% of the total population. In general the microfauna is fresh, with the Ostracods especially showing no signs of having undergone extensive reworking.

Two samples from borehole 71/56 were subjected to micropalaeontological analysis by Dr. R. Harland of the I.G.S. at Leeds. The deeper sample differed markedly from samples of Flandrian age recovered further east in borehole 71/57. Dr. Harland noted three main differences. 1. Lack of *Lingulodinium machaerophorum* (Deflandre and Cookson) Wall in the present assemblage. 2. Presence of cysts of the genus *Achomosphaera* Evitt. 3. Higher proportion of cysts of the genus *Spiniferites* Mantell.

Continuous seismic profiles across the Central area show a conformable sequence of horizontally stratified deposits indicative of marine sedimentation, which in turn rest on a lower non-stratified unit, the base of which is not always visible. The stratified section, which varies in both depth and thickness is laterally continuous with the pre-Devensian marine sediments seen in the outer boreholes. Overlying the marine sediments is a non-stratified unit containing only a few inpersistent reflectors and identified as Devensian Irish Sea till. This succession is well displayed by a $\frac{1}{4}$ second sparker profile 8 km to the southwest of borehole 74/43:

Sparker Profile EF. 05·73
Fix 27/33 Lat. 52° 33·16′ N.
    Long. 04° 56·6′ W.

Borehole 73/43
Lat. 52° 33·68′ N.
    Long. 04° 51·07′ W.

		Velocity		
Water depth	65 m	1·5 km/s	Water depth	60·0 m
Non-stratified	41 m	1·8 km/s	Upper till	40·0 m
Stratified	24 m	1·8 km/s	Marine formation	26·5 m
Non-stratified	31 m	1·8 km/s	Lower till	10·5 m +
	161 m			137·0 m

2c. Discussion of the Pre-Devensian Quaternary sediments

The lower group of glacial drifts, which underlie the marine interglacial formation in the central St George's Channel, exhibit all the characteristics of a typical

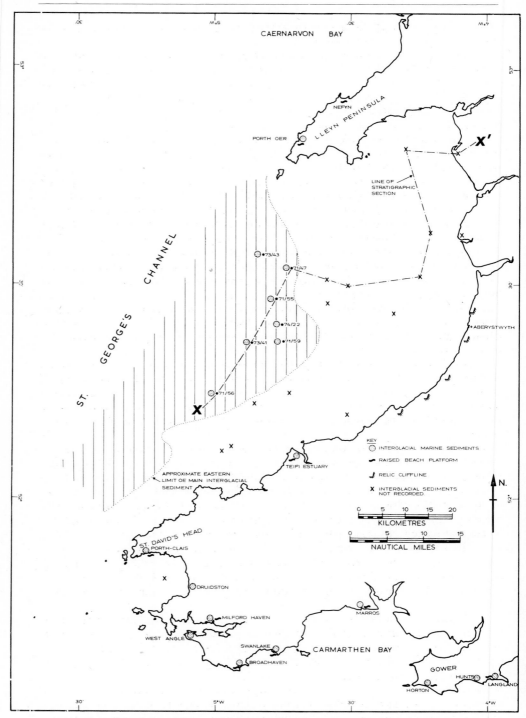

Fig. 4. Distribution of Ipswichian Interglacial sediments in the South Irish Sea.

Devensian Irish Sea till. The age of the deposits immediately below the lowermost till unit in this area is not known.

Although the stratigraphic position of the relict glacial drifts further to the east is not always clear they are considered for a number of reasons to be of pre-Devensian age. Also associated with deposits of this type are accumulations of Irish Sea erratics, probably representing the eroded remains of pockets of pre-Devensian glacial drift. Derivation by marine action from the Devensian Irish Sea till sheet to the west is considered unlikely.

At Shortalstown, in southeast Ireland, a lower till unit of Irish Sea origin underlies marine sediments of Ipswichian age (Colhoun and Mitchell 1971). Further west, along the coast of Ballycotton Bay, at Ballycroneen, Shanagarry and Garryvoe (Fig. 9) Irish Sea till overlies a raised beach formation and has been assigned to both the Wolstonian (Mitchell 1972) and the Devensian (Bowen 1973a) glaciations. Detailed marine investigations off the south coast of Ireland have shown the rockhead to be exposed at the sea-floor covered by rare patches of Irish Sea type till or overlain by a thin veneer of recent marine sediments.

Further south in the main Celtic Sea area, till, tentatively recorded as Irish Sea type, has been recovered in cores during recent I.G.S. surveys (Evans *pers. comm.*). The till, which is present only in isolated patches occurs as far south as 50° 25′ N in the vicinity of the Haig Fras granite (Fig. 9). In the Bristol Channel, Carmarthen Bay and Barnstaple Bay no glacial sediments have been found in the I.G.S. boreholes or on the seismic sections examined. The lag gravels present in the Bristol Channel and in parts of the Celtic Sea may well represent the eroded remains of a pre-Devensian till.

Detailed particle size analysis of the pre-Devensian marine sediments reveals similarities to the recent marine sediments on the present day sea-floor, and also to those of known marine interglacial age exposed onshore. The gravel fraction, like that of the recent, is glacial in character and has almost certainly originated in a similar fashion to the present day lag gravels, by the marine erosion of underlying glacial deposits.

Microfaunal analysis of the marine sediments suggest certain similarities to Ipswichian assemblages recorded from sites in both Ireland and the United Kingdom (Haynes *pers. comm.*). Comparison with present day assemblages living in the Irish Sea suggest temperatures were slightly colder, probably similar to those of the Minches or the Malin Sea off the west coast of Scotland. The presence of very cold water forms is not fully understood although similar assemblages have been recorded from sub-recent sediments in Tremadoc Bay (Wall and Whatley 1971). The macrofauna which is worn and fragmented as a result of deposition under high energy conditions is less useful as an indicator of age and environment, although the presence of *Turritella communis* Risso (Plate 3b) suggests muddy marine conditions, similar to those of present day Tremadoc Bay.

No detailed palynological investigation of the pre-Devensian marine sediments has yet been undertaken. The brief analysis of samples from borehole 71/56 indicates either a particular climatic phase within the Flandrian or an interglacial (Ipswichian?). As this marine formation underlies sediments of glacial origin, a Flandrian age is not stratigraphically possible.

The deposition of the pre-Devensian marine sediments is thought to have been in response to a major temperate episode of considerable duration. The sediment characteristics are not those of fluvioglacial origin while an interstadial origin seems unlikely in view of their thickness and temperate fauna. This and other evidence would suggest a last (Ipswichian) interglacial age.

3. Devensian Glacial sediment

3a. Glacial sediments of Irish Sea origin

Glacial drifts of Irish Sea origin form by far the most abundant Quaternary sediments in the South Irish Sea covering, with the exception of eastern Cardigan Bay, almost the entire area. Irish Sea till, a mature uniform terrestrial boulder clay with a highly calcareous matrix and distinctive erratic content of both Mesozoic and Upper Palaeozoic rocks, is the commonest and most widely distributed of these sediments. Large quantities of marine sands and shell fragments are also diagnostic of this till. Fluvioglacial sediments are locally developed but never abundant, while periglacial deposits are apparently absent. Offshore there is no evidence to suggest that the deposition of glacial sediments was followed by extensive episodes of solifluction.

Irish Sea glacial sediments were recovered from a total of 22 separate borehole localities, and 108 shallow core stations. Thicknesses vary between 5 m and 50 m and average 30 m to 40 m, while seismic profiles suggest this may locally reach 70 m. Southwards they become progressively thinner, averaging only 10 m off the West Pembrokeshire coast, where it forms a narrow lobe of glacial sediment protruding into the northern part of the Celtic Sea (Plate 1a). The main Irish Sea till sheet eventually terminates due west of St. Govan's Head, Latitude 51° 34′ N (Fig. 1). South of this limit only small and isolated outliers of glacial drift have been found.

In the east, the Irish Sea glacial sediments extend up to the seaward termination of the Sarns, pebble ridges associated with glacial sediments of Welsh origin. Here the rate of thinning is more rapid with Irish Sea sediment finally overlying glacial drifts of Welsh origin (Plate 1c). Across the western tip of the Lleyn Peninsula and in northwest Pembrokeshire glacial sediments of this type and age have been deposited on land (Fig. 1).

Plate I

Pinger profiles from the South Irish Sea. The horizontal time lines are 12 milliseconds apart and represent approximately 9 metres. The vertical fix lines are ten minute lines and the horizontal distance approximately 1·5 km.

a. Pinger profile (O.R.E., 2·4. kHz)
The main Irish Sea till sheet west of St David's Head. Till up to 10 m thick rests on a glacially smoothed rock platform. The sea floor is covered by lag gravel giving rise to acoustic defraction patterns in the till.

b. Pinger Profile (O.R.E., 2·4. kHz)
A north-south section west of St David's Head showing the southward termination of the main Irish Sea till sheet.

c. Pinger Profile (Edo-Western 3·5 kHz)
An east-west section west of Aberyron showing the eastward termination of the main Irish Sea till sheet. Part of the till sheet overlies glacial drifts of Welsh origin while along the contact laminated sediments of late Glacial/early Flandrian age have been deposited. Location of section A–A′ shown on Figure 5.

d. Pinger Profile (Edo-Western 3·5 kHz)
Laminated sediments of late Glacial/early Flandrian age postdate an eroded ridge of glacial debris (Sarn-Y-Bwch) associated with the maximum extent of the Mawddach piedmont glacier. Location of section B–B′ shown on Figure 5.

Plate 1

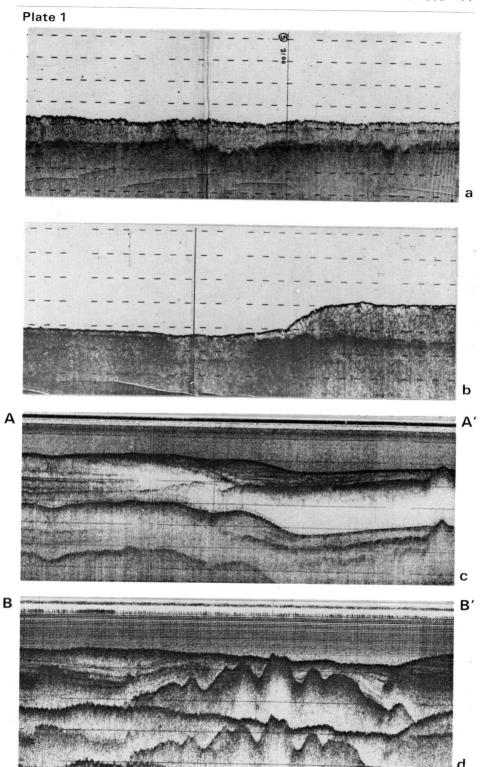

a

b

A   A′

c

B   B′

d

Over much of Cardigan Bay, Irish Sea Devensian deposits directly overlie Palaeozoic and Mesozoic bedrock. Further west, in St George's Channel, they are underlain by marine formations of Quaternary age (Fig. 2). At the surface, glacial sediments are normally present within one metre of the sea-floor. In Cardigan Bay the cover of recent marine sediments is often in the order of 0·10 m or less while in some localities till is exposed at the sea-floor.

Lodgement till is the commonest and most widely distributed of the glacial sediments. Locally silts, sands and gravels of fluvioglacial origin, and laminated ablation till may be present. On continuous seismic profiles the till is represented as a single acoustically homogeneous unit, showing no evidence of internal stratification except for occasional faint diffuse banding caused probably by impersistent layers of silt, sand or gravel.

In comparison with the majority of tills of Welsh origin the Irish Sea till exhibits a high degree of textural maturity. The texture of the till collected from different areas of the South Irish Sea, and at different depths, remains constant (Plate 3a). A fine grained clay matrix forms between 70% and 90% by weight of the total sediments and contains up to 20% sands and shell fragments. Erratics, which normally represent less than 10% by weight, are small and form an open framework within the clay matrix. Distinctive erratics include: Carboniferous limestones, Permo-Triassic sandstones, Middle Jurassic limestones, Cretaceous flints and Tertiary lignite. The Middle Jurassic limestones, which are the most important and distinctive group of erratics, are easily recognisable on account of their oolitic and bioclastic texture, and differ in both colour and hardness from limestones of Carboniferous age.

One of the most diagnostic features of the Irish Sea till is a high carbonate content of between 12% and 27% of the total weight. This is present mainly as a fine grained rock flour in the clay fraction. Colour is another constant characteristic with the majority of samples being either dark yellowish brown (10 YR 4/2) or olive grey (5 Y 5/1). The olive grey coloured till is concentrated more in the east. Close to the surface recent oxidation has increased the hues of red and yellow.

The upper part of the till sheet in some areas is laminated and contorted by folds and slump structures (Plate 3c). On a much larger scale distinct sand and gravel horizons are present at certain depths but these were not found to be laterally continuous. Diapiric structures of silt and sand occasionally penetrate the base of overlying lodgement till.

West of the Lleyn Peninsula Irish Sea till recorded on seismic profiles is characterised by a series of offlapping till sheets. The largest can be traced from Bardsey Island westwards and extends as far south as Latitude 52° 40·6′ N. The base of this upper till sheet is flat, and truncates the deeper main unit of Irish Sea till.

3b. Glacial sediments of Welsh origin

Eastern Cardigan Bay is covered by glacial drifts which are unlike the Irish Sea till. They are characterised by light to medium grey (N5), non-calcareous units of immature glacial and periglacial sediments, and are markedly similar to those present in north and central Wales. Because of this and their rich content of locally derived Welsh erratics, they are considered to be of Welsh origin.

Sediments of this type are exposed extensively around the coast of Cardigan Bay where many of the sections are complicated by solifluction and localised sedimentary conditions. Offshore the examination of Welsh glacial drifts has been hampered by a thick cover of Flandrian sediments. The sedimentary evidence

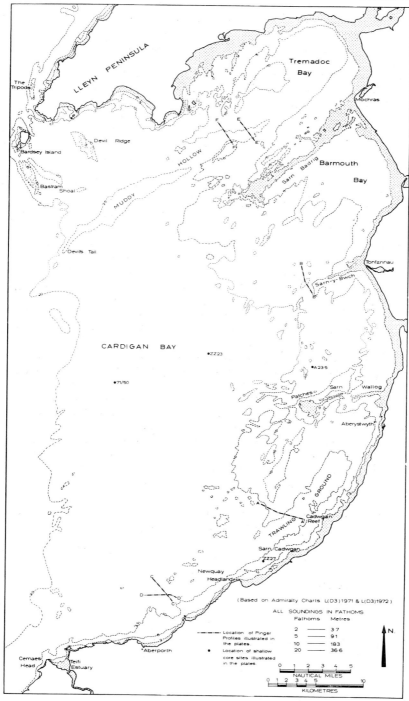

Fig. 5. Bathymetry of Cardigan Bay, showing location of some of the seismic sections and sample stations illustrated in the plates.

is confined to six successful core sites and 11 coastal and nearshore boreholes. Geophysical information is also limited because of lack of penetration caused by the rubbly nature of the drift, and because of multiple and noise problems as a result of shallow water.

The Welsh glacial drifts, like those of Irish Sea origin, may be locally very thick, especially where selective erosion of soft Tertiary formations has taken place. In Tremadoc and Barmouth Bays much of the Quaternary is underlain by Tertiary strata. (Wood and Woodland 1968, and the offshore boreholes 70/8 and 71/49.) Four other boreholes failed to penetrate the base of the Quaternary in this area although the deepest had a terminal depth of 149 m below mean sea-level. Sparker profiles in Tremadoc Bay suggest the Quaternary may locally reach a thickness of approximately 200 m. The absence of a strong impedence contrast between the Quaternary and the Tertiary in this area has made the base of the Quaternary very difficult to recognise. In the south and to the southwest of Barmouth Bay, Welsh glacial drift is underlain by Mesozoic formations of Middle and possibly Upper Jurassic age. Bathonian limestones were penetrated in boreholes 71/57 and 72/38 at depths of 62 and 44 m respectively. Elsewhere Lower Palaeozoic rocks subcrop beneath the Quaternary. Across the entrances to the Glaslyn/Dwyrd, Mawddach, Dysynni and Dovey estuaries, geophysical investigations have indicated overdeepening (Blundell *et al.* 1969).

Unlike the Irish Sea till the Welsh glacial drifts do not form a homogeneous group of sediments. There are at least four separate areas of sedimentation separated by the Cardigan Bay Sarns, shallow ridges extending seaward from the interfluves between the major valley systems onshore (Fig. 5). The Sarns, which are composed of glacial drift of Welsh origin, are covered by layers of reworked boulders (Plate 1d). To the west they are truncated and overlain by till of Irish Sea origin. Due to fortunate differences of acoustic impedence existing between the Welsh and Irish Sea glacial drifts, it is possible to record by seismic profiling the nature of their contact. In regions where both these drift types are encountered those of Welsh derivation are consistently recorded as underlying those of Irish Sea origin (Plate 1c). At borehole site 72/38 situated to the west of the Dovey Estuary this stratigraphic succession was proved by the recovery of 13 m of Irish Sea till resting on 13·5 m of Welsh glacial drift (Fig. 2). Each area is characterised by its own group of sediments which can include a combination of tills, heads, varved clays and fluvioglacial outwash. To the south of Newquay headland only periglacial sediments of Welsh origin are present. The texture and maturity of the till varies not only from one area to another but also at different depths. Outside Tremadoc Bay, Welsh till is normally composed of a gravel fraction in excess of 50%. Unlike the Irish Sea till the sand portion consists of mainly lithic fragments. The carbonate content is low, usually less than 5%, but may be locally high in the vicinity of limestone subcrops. In Tremadoc Bay the till differs by having a rich clay matrix.

3c. Discussion of the Devensian glacial sediments

It has not been possible to determine an exact age for the main unit of glacial drift although the extent, faunal content and sedimentary characteristics all suggest it to be the product of the last (Devensian/Weichselian) glaciation. In the west, the Irish Sea till sheet overlies temperate marine sediments considered to be of last (Ipswichian) interglacial age (Fig. 2). In the east the same glacial formation can be correlated with an onshore succession dated by several authors

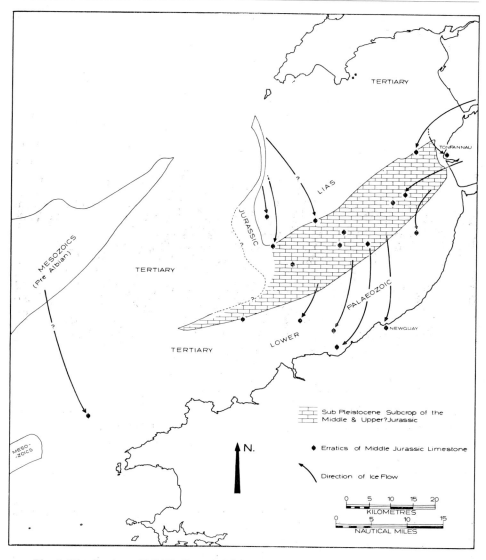

Fig. 6. Distribution of Middle Jurassic Limestone erratics off the west coast of Wales.

as Devensian (John 1970 and Bowen 1973a and b). Much more uncertainty surrounds the glacial drifts of Welsh origin now preserved in the east of Cardigan Bay (Mitchell 1972; Watson 1970; Bowen 1973b; Garrard and Dobson 1974). In this paper it is considered that extensive deposition, together with some recycling of older glacial sediment, took place during the Devensian.

The distribution and maturity of the Devensian Irish Sea till is indicative of extensive continental glaciation from ice of northern origin. The abundance of Upper Palaeozoic, Mesozoic and Tertiary erratics are now known to have been scoured from the floor of the Irish Sea (Wright *et al.* 1971; Dobson *et al.* 1973). Those of Middle Jurassic limestone form the most important indicators of ice flow because of derivation from restricted localities in Cardigan Bay (Fig. 6). The high calcareous content of the clay matrix probably represents the eroded remains of carbonate rocks especially Carboniferous and Middle Jurassic limestones, and Permo-Triassic and Tertiary marls. The sand fraction, which is not glacial but marine in character, consists of polycyclic quartz grains derived partly by the erosion of Upper Palaeozoic and Mesozoic sandstones and partly by the incorporation of preceding Quaternary marine sediments. Marine sands, rich in shell detritus, were stranded in the Irish Sea at the onset of the last glaciation (Plate 3b).

The presence of laminated till may suggest supra-glacial deposition during times of ice ablation. Sediment of this type can be classified as "flow" or "melt out" till. Much larger diapiric density structures, present in lodgement till, may have been produced post-depositionally as a result of ice loading.

In the south, both the sedimentary character of the till and the rate of thinning suggested on the seismic profiles indicates that Irish Sea sediments of Devensian age were unlikely to have been deposited much further south than their present limit (Plate 1b). In the north the presence of an upper till sheet is thought to represent minor readvance episodes associated with the retreat of Devensian ice. Other evidence supporting a similar readvance has been reported onshore in the Lleyn Peninsula (Saunders 1968). Offshore most of the superficial glacial structures connected with the Irish Sea glacial sediments have been either removed or smoothed over. This appears to have taken place during a brief episode of marine erosion associated with the passage of the Flandrian surf zone.

Glacial erosion by Welsh ice was generally less powerful, and transportation more limited, than that of Irish Sea type. This is reflected by the immature texture of the glacial drifts which were deposited close to their site of generation. The

---

Plate 2

Pinger profiles from the South Irish Sea. The horizontal time lines are 12 milliseconds apart and represent approximately 9 metres. The vertical fix lines are ten minute lines and the horizontal distance approximately 1·5 km.

a. Pinger Profile (Edo-Western 7 kHz)
Cross section of the Trawling Ground meltwater channel due north of Aberporth. The channel is cut into the main Irish Sea till sheet to just above the Lower Palaeozoic rockhead and is filled by late Glacial/early Flandrian sediments. Location of section C′–C shown on Figure 5.

b. Pinger Profile (Edo-Western 3·5 kHz)
Longitudinal section of the same meltwater channel showing two separate episodes of channel fill. Location of section D–D′ shown on Figure 5.

c. Pinger Profile (Edo-Western 3·5 kHz.)
Cross section of the Muddy Hollow meltwater channel in Tremadoc Bay filled by late Glacial/early Flandrian and recent marine argillaceous sediments. The central section is acoustically impenetrable due to the presence of trapped interstitial gases. Location of section E–E′ shown on Figure 5.

d. Pinger Profile (Edo-Western 3·5 kHz)
Cross section of the Muddy Hollow meltwater channel at the entrance to Tremadoc Bay which has been kept almost clear of sedimentary fill by tidal scour. The channel is cut into glacial sediments of Welsh origin consisting of two till sheets to the north but only one to the south. Location of section F′–F shown on Figure 5.

Plate 2

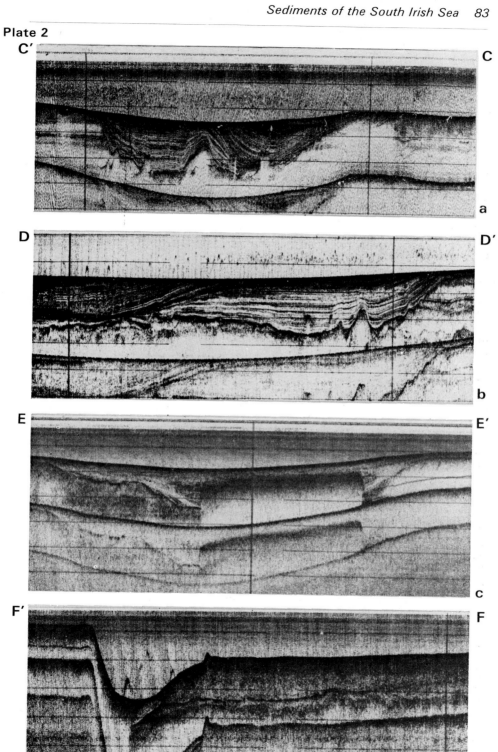

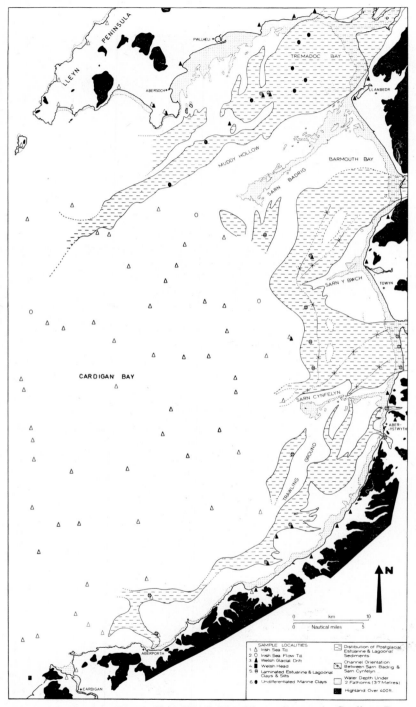

Fig. 7. Distribution of Late Glacial/Early Flandrian sediments in Cardigan Bay.

is the result of the erosion of Lower Palaeozoic predominantly grey coloured rocks in north and mid-Wales. In Tremadoc Bay the till differs by being more clay rich in response to the selective erosion of argillaceous Tertiary formations. The overdeepening at the entrances to many of the valleys entering Cardigan Bay was probably produced by the lateral confinement of local Welsh ice.

In Tremadoc Bay an upper unit of Welsh till (Plate 2d) may be stratigraphically equivalent to that recorded west of the Lleyn Peninsula and therefore represents the same episode of readvance. Similar evidence has been reported from around the coast of Tremadoc Bay (Bowen 1973b). No upper till sheet was identified further south and it is assumed that these areas underwent only erosion and periglacial activity at this time.

4. Post-Devensian sediments

4a. Late Glacial/Early Flandrian sediments

The post-Devensian channels are among the most prominent geomorphological features associated with this period. The shape of the channels varies according to the nature of the underlying Quaternary sediments. Where till of Irish Sea type is present, the channels are narrow and steep sided, while those cut in glacial drifts of Welsh origin are broad but shallow. In some cases the channels lead to and from gaps in the Sarns, while others wind their way sub-parallel to the coast but obliquely to the gradient of the sea-bed. Although most of the channels are now obscured by recent sediment till, their effect on the present day bathymetry may still be detected in certain localities today. Muddy Hollow to the south of the Lleyn Peninsula, and the Trawling Ground to the southwest of Aberystwyth are the partially filled remains of two such channels (Fig. 5).

The Trawling Ground system is the largest of the Cardigan Bay channels, covering a distance of 68 km between Barmouth Bay and Cemaes Head. Off Aberporth the channel is now totally obscured by more recent sediment fill (Plates 2a and b). The second largest channel is the Muddy Hollow system, which commences in Tremadoc Bay and terminates south of Bardsey Island, 43 km to the southwest. South of St Tudwal's Peninsula recent tidal scour, has kept this channel partially clear of sediment (Plate 2d). In Barmouth Bay the glacial drift has been eroded in one area to form a narrow, linear, steepsided depression which closes both to the southwest and northeast. Borehole (74/25), situated in the centre of this depression penetrated over 120 m of recent sediment fill without reaching bedrock.

Both the channels and the shallower areas between the Sarns contain laminated argillaceous sediments of marginal marine type (Fig. 7, Plates 2a and b). Samples taken from such regions consist predominantly of clays, faintly laminated with thin layers of silt, fine sand and carbonaceous detritus (Plates 3c and d). In certain deeper cores isolated pebbles are present in the clay matrix. In freshly sectioned cores the laminated clays and silts exhibit alternating bands, up to 5 cm in thickness, of light oxidised and dark reduced sediment. These are considered to be varves. Detailed faunal analysis indicates the sediment now closer inshore to have been deposited under estuarine and lagoonal conditions. The most diagnostic species include the molluscs *Hydrobia ulvae* (Pennant), *Cardium edule* Linné and *Littorina littorea* (Linné). In some areas thin peats have been developed. A peat (Plate 3c), recovered in the Trawling Grounds to the northeast of

Newquay Headland 18·5 m below present-day mean sea-level gave a radiocarbon age of 8740±100 years B.P. (Birm. 400). As this peat is located near the top of the channel fill most of the sediments must precede this date.

Further seaward microfaunal analysis (Kitely 1975) has shown that sediments become progressively more marine while their texture remains constant. Cross-sections obtained from "Pinger" profiles in the deeper areas indicate several episodes of sedimentation separated by further periods of channelling (Plate 2c). In many cases, sequences of sediment are seen draped into channels, themselves superimposed on older channels. In others, especially in Tremadoc Bay, the presence of trapped interstitial gases within the sediment has rendered them acoustically impenetrable at high frequencies (Plate 2c). Not only do the sediments become more marine laterally in a seaward direction but also vertically towards the surface of the channel fill. In some cores the change is gradational as more and more layers of marine sand replace layers of estuarine sediment. Sand eventually becomes so dominant, that the upper part of the sequence passes undisturbed into the marine arenaceous sediments of the present day (Plate 3d). In other cores the transition between true marine and estuarine sediment is abrupt.

4b. Recent Marine sediments

Over much of the South Irish Sea the upper surface of the glacial and post-glacial drifts is smooth and is now covered by a thin layer of recent marine sediments which seldom exceed thicknesses of more than 1 m. Wherever glacial drifts approach the sea-bed, reworked pebbles and cobbles, of glacial type litter the sea-floor. These extensive covers of lag gravel, which are little more than a few centimetres in thickness, are heavily encrusted by recent epifauna. Beneath the lag gravels the upper part of the till has been moderately bioturbated by recent marine organisms (Plate 3a).

In places the lag gravels are partially covered by thin semi-mobile sand bodies, and in a few localities where conditions are favourable by large sand wave complexes. Some of the larger tidal sand ridges and supporting waves, including the Tripods, Bastram Shoal, Devil Ridge (Fig. 5) and Bias Bank reach thicknesses of over 30 m.

In Tremadoc Bay and the Trawling Grounds argillaceous deposits predominate. Here the surface sediments consist of non-stratified intensely bioturbated, semi-

Plate 3

a. Sectioned Vibrocore 71/50
Recent sands and gravels rich in shell debris rest on typical Irish Sea till. The top few centimetres of till have been bioturbated and the burrows filled with recent sediment.

b. Part of Vibrocore ZZ23
A complete gastropod, *Turritella communis* Risso, contained within Irish Sea till. Diameter of core 8 cm.

c. Part of sectioned Gravity Core ZU23.
Laminated (flow or meltout) till of Irish Sea origin.

d. Part of sectioned Hydrocore ZZ27.
Laminated clays and silts of estuarine origin containing a layer of peat (C_{14} age 8740 ± 100 years B.P.) rest on local Welsh till. Scale interval 10 cm.

e. Part of sectioned Vibrocore A23·5
Late Glacial/early Flandrian laminated clays and silts of estuarine origin showing an increasing number of layers of coarser grained marine sediment towards the surface. Scale interval 10 cm.

Plate 3

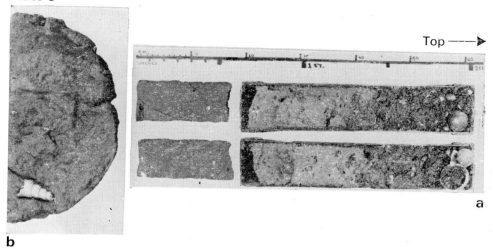

Top ———▶

a

b

Top ———▶

c

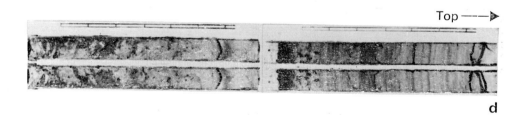

Top ———▶

d

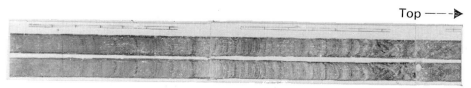

Top ———▶

e

STAGE	EVENTS	1.) SOUTHERN ST. GEORGES CHANNEL	2.) CENTRAL ST. GEORGES CHANNEL	3.) NORTHERN ST. GEORGES CHANNEL	4.) CENTRAL CARDIGAN BAY	5.) EASTERN CARDIGAN BAY	6.) CENTRAL WALES COAST	7.) TREMADOC BAY
FLANDRIAN	RECENT MARINE		Marine sands locally developed.	Marine sands locally developed. Lag gravel.	Marine sands resting on lag gravel.	Marine sands		Marine clays & silts.
FLANDRIAN	LATE GLACIAL / EARLY FLANDRIAN	Lag gravel.	Lag gravel.			Laminated clays & silts of marginal marine origin with locally developed peats	Sub-aerial erosion	Laminated clays & silts of marginal marine origin
DEVENSIAN	LATE DEVENSIAN GLACIATION	Erosion of meltwater channels.		Fluvioglacial outwash. Upper Irish Sea till sheet	Erosion of meltwater channels	Erosion of meltwater channels.	Head	Fluvioglacial outwash Upper Welsh till sheet.
DEVENSIAN	MAIN DEVENSIAN GLACIATION	Irish Sea till sheet terminating to the south	Main Irish Sea till sheet	Main Irish Sea till sheet.	Main Irish Sea till sheet.	Irish Sea till terminating to the east. Welsh glacial drift terminating to the west. Head	Welsh glacial drift. Head	Main Welsh till sheet Fluvioglacial outwash.
IPSWICHIAN	PRE-DEVENSIAN INTERGLACIAL	Bedrock. Raised beaches of S.W. Wales.	Marine sediments containing a rich fauna.	Bedrock. Marine sediments now preserved in Irish Sea till.	Bedrock. Marine sediments now preserved in Irish Sea till.	Bedrock		Bedrock
PRE-DEVENSIAN	PRE-DEVENSIAN GLACIATION		Lower Irish Sea till sheet (base not seen).				Recycled till & erratics of Irish Sea origin. Bedrock	

Fig. 8. Summarised Quaternary stratigraphy of the South Irish Sea.

fluid, marine muds characterised by the gastropod *Turritella communis* Risso. In other nearshore areas well-sorted marine sands form continuous sheets of sediment.

4c. Discussion of the Post-Devensian Quaternary sediments

The period following the Devensian glaciation was characterised at first by rapid but localised sub-aerial erosion and redeposition of the preceding glacial sediments. The formation of the channels is thought to have taken place during the closing stages of the Devensian glaciation by meltwater from retreating ice sheets. The different shape of these meltwater channels, in the east as opposed to those in the west of Cardigan Bay, reflects the textural differences between the two main groups of glacial sediment. The location of each channel is associated with differing rates of glacial retreat and minor readvance. The Trawling Ground meltwater system may have been initiated in response to a more rapid retreat of Welsh rather than Irish Sea ice in the south of Cardigan Bay. On the other hand, the Muddy Hollow system was eroded by meltwater associated with the readvance of Welsh ice in Tremadoc Bay. The Barmouth Bay channel, because it forms a closed depression, cannot have been eroded by simple fluviatile action. Channels of this type are probably associated with some form of sub-glacial erosion.

During the early stages of the Flandrian Sea transgression, the channels, especially, acted as traps for the deposition of estuarine sediments. The presence of varved clays and dropstones suggest that conditions were initially very cold. The contact between these formations and the marine sediments above can be either gradational or abrupt and erosional.

The surface of the glacial and post-glacial drift in most areas was reworked by marine action primarily during the passage of the Flandrian surf zone. The lag

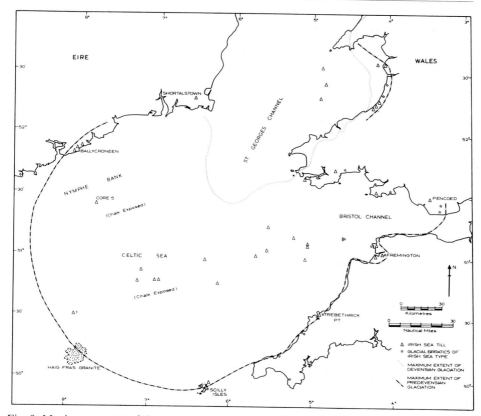

Fig. 9. Maximum extent of the pre-Devensian Irish Sea Glaciation.

gravels produced during this brief interval rest directly on a relatively uniform plane of marine erosion. In some localities, especially Tremadoc Bay, the argillaceous sedimentation has continued unabated since early Flandrian times.

5. Sequence of events and general discussion

The basic stratigraphy can be summarised into six main sedimentary units covering a total of four separate Quaternary stages (Fig. 8).

A major glaciation, larger and more erosive than that of the Devensian, predates an interglacial marine episode considered to be of Ipswichian age. This glaciation which covered all of the South Irish Sea, and much of the Celtic Sea and Bristol Channel, extended further to the south, the east and the west than the Devensian ice sheet (Fig. 9). In St George's Channel it also eroded deeper into the underlying sediments. During the last cold stage ice failed to remove all of the pre-Devensian glacial sediments especially where they had been deposited in overdeepened areas left by the selective erosion of Tertiary strata.

It is not possible at present to date the older glacial event more accurately than the pre-Devensian. Stratigraphically it underlies the marine sediments of

Ipswichian age in the South Irish Sea and is probably equivalent to the glaciation antedating the raised beaches of South Wales (Bowen 1973b). This older glacial episode gave rise to sediments which are similar to those of the Devensian, suggesting both ice sheets to have flowed along much the same lines and to have eroded similar underlying strata. Because of the absence of more conclusive data the possibility that the pre-Devensian cold stage, in fact, represents more than one older glaciation, cannot be ruled out.

The pre-Devensian glaciation was followed by a temperate episode of considerable duration, which resulted in the deposition of thick formations of marine sediment over much of the South Irish Sea. Both the litho- and bio-stratigraphical evidence supports a last (Ipswichian) interglacial age rather than that of an interstadial. In the centre of the area the thickness of Ipswichian sediments protected the older glacial drifts from erosion by Devensian ice, while those present in exposed localities, especially on land, underwent extensive weathering, recycling and erosion during the last glaciation.

In the Devensian the South Irish Sea became glaciated by a large continental ice sheet flowing southwards from Scotland towards the southern entrance to St George's Channel. Local ice was restricted to the peripheral areas only (Fig. 10). The structural depression which forms the Irish Sea was probably the most important factor influencing the movement of ice in the west of the British Isles. The positive physical features of the Lleyn Peninsula, West Pembrokeshire and the Cornubian Peninsula all formed natural barriers at certain times, restricting ice movement southwards.

In the east of the area Welsh ice, in the form of local piedmont glaciers, fanned out from the mountains of north and central Wales onto the old sea-floor of Cardigan Bay before the arrival of Irish Sea ice, a view supported by the stratigraphical position of their associated sediments. This was to be expected due to the proximity of each of these small glaciers to their feeder ice-caps. The Sarns are the most prominent morphological features to remain from this episode of glaciation, and represent each piedmont glacier's maximum terminal and lateral extent. Subsequently western extensions of the Sarns became truncated by the Irish Sea ice, which incorporated and removed Welsh morainic material along its contact (Fig. 10). The Welsh ice being unable to escape, and derived from considerable smaller feeder ice-caps than those of the Irish Sea ice, stagnated on the floor of Cardigan Bay. The only reason that Welsh ice could compete at all for space within Cardigan Bay was by virtue of its location in the lee of the Lleyn Peninsula which contained sufficient highland to form a natural barrier to the southward movement of Irish Sea ice.

In Tremadoc Bay the Welsh ice may well have re-advanced a second time as indicated by a second succession of Welsh drifts, which in places overlie those from the Irish Sea. Likewise to the west of the Lleyn Peninsula a lobe of Irish Sea ice at the same time probably re-advanced to the south of Bardsey Island. It is tempting to correlate both these events with a similar episode marked by the Ellesmere/Whitchurch moraine of the English Midlands.

During deglaciation near the end of the Devensian stage, the recently deposited drifts underwent a short period of erosion. Drainage channels fed by meltwater from retreating ice sheets, quickly excavated localised areas of drift. Barriers of ice also played a large part in the formation of certain channels causing them to run obliquely to the general gradient, by blocking the most obvious route seaward. The Trawling Ground system was formed in such a fashion by meltwater flowing

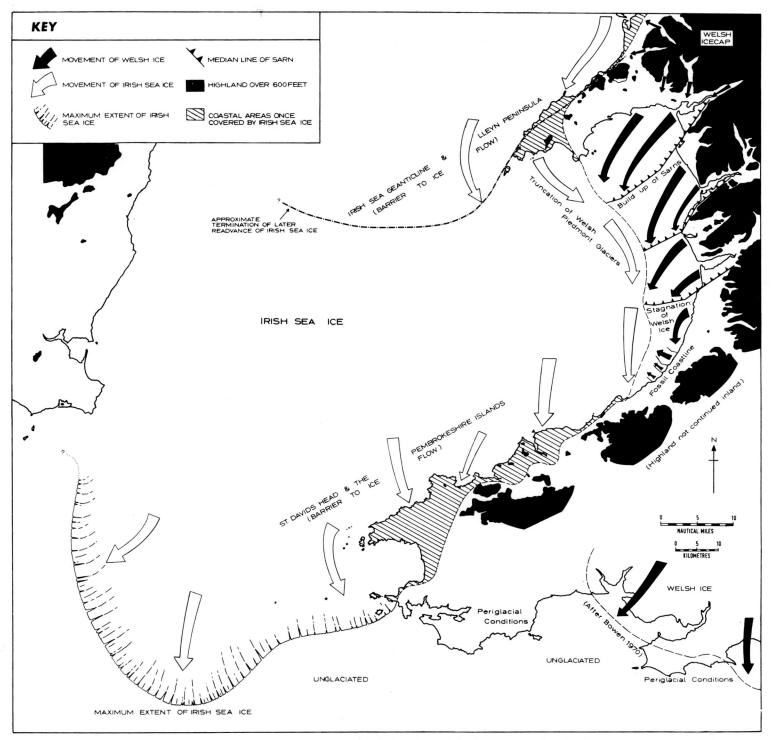

KEY

◣ MOVEMENT OF WELSH ICE	◄ MEDIAN LINE OF SARN
⇦ MOVEMENT OF IRISH SEA ICE	■ HIGHLAND OVER 600 FEET
MAXIMUM EXTENT OF IRISH SEA ICE	COASTAL AREAS ONCE COVERED BY IRISH SEA ICE

WELSH ICECAP

IRISH SEA GEANTICLINE & (BARRIER TO ICE)

LLEYN PENINSULA FLOW)

Truncation of Welsh Piedmont Glaciers

Build up of Sarns

APPROXIMATE TERMINATION OF LATER READVANCE OF IRISH SEA ICE

IRISH SEA ICE

Stagnation of Welsh Ice

Fossil Coastline

(Highland not continued inland)

PEMBROKESHIRE ISLANDS FLOW)

ST DAVIDS HEAD & THE (BARRIER TO ICE)

N

0   5   10
NAUTICAL MILES

0   5   10
KILOMETRES

WELSH ICE

Periglacial Conditions

(After Bowen 1970)

UNGLACIATED

UNGLACIATED

Periglacial Conditions

MAXIMUM EXTENT OF IRISH SEA ICE

Fig. 10. West Wales and the South Irish Sea during times of maximum Devensian Glaciaton.

between a lobe of Irish Sea ice to the north and the old coastline to the south. As this episode of channel erosion was confined to the relatively short period of time between the retreat of the Devensian ice sheets and the rise of the Flandrian Sea, it is not surprising to find their maximum development today in regions of shallower water, where sub-aerial conditions prevailed longest.

Both the meltwater channels and the shallower areas between the Sarns later formed areas of paralic sedimentation as they became submerged by the transgression of the Flandrian Sea. The rise of sea-level in relationship to these sediments may not have always been continuous, so that further reactivation of the channels resulted in subsequent sequences of fill being superimposed in older channels. The radiocarbon date of almost 9000 years B.P. would suggest sea-level rose relatively rapidly following deglaciation, with the meltwater channels becoming quickly filled. At a later date because of either a more rapid rise in sea-level, or a decrease in the supply of sediment from the land, the estuarine and other related sediments were unable to compete against increasing influxes of marine sediment. Eventually true marine conditions took over completely and have continued up until the present day.

During the advance of the Flandrian Sea across the exposed drift the high-energy environment of the surf zone briefly reworked the surface of the glacial sediments, creating a thin but extensive cover of lag gravel, resting on a distinct plane of marine erosion. As this lag gravel is now heavily encrusted with recent epifauna, the erosion of the glacial sediment offshore in the South Irish Sea now appears to be over.

Acknowledgments. The assistance received during this research project from members of the Continental Shelf Unit 1 of the Institute of Geological Sciences, and the Department of Geology, Aberystwyth, is gratefully acknowledged.

References

BLUNDELL, D. J., GRIFFITHS, D. H. and KING, R. F. 1969. Geophysical investigations of Buried River Valleys around Cardigan Bay. *Geol. J.* **6**, 161–180.

BOWEN, D. Q. 1973a. The Pleistocene Succession of the Irish Sea. *Proc. Geol. Ass.* **84**, 249–274.

—— 1973b. The Pleistocene History of Wales and the Borderlands. *Geol. J.* **8**, 207–224.

COLHOUN, E. A. and MITCHELL, G. F. 1971. Interglacial Marine Formation and Lateglacial Freshwater Formation in Shortalstown townland, Co. Wexford. *Proc. R. Ir. Acad.* **71B**, 211–245.

DOBSON, M. R., EVANS, W. E. and WHITTINGTON, R. J. 1973. The Geology of the south Irish Sea. *Rep. No.* 73/11. *Inst. geol. Sci.* 35 pp.

GARRARD, R. A. and DOBSON, M. R. 1974. The Nature and Maximum Extent of Glacial Sediments off the West Coast of Wales. *Marine Geol.* **16**, 31–44.

JOHN, B. S. 1970. Pembrokeshire. *In,* LEWIS, C. A. (Editor). *The Glaciations of Wales and adjoining regions.* Longmans, London. 229 pp.

KITELY, R. J. 1975. *Recent and Postglacial Foraminifera in cores from Cardigan Bay.* Unpublished M.Sc. Thesis. University College of Wales.

MITCHELL, G. F. 1972. The Pleistocene History of the Irish Sea: Second Approximation. *Sc. Proc. R. Dublin Soc.* **4**, 181–199.

SAUNDERS, G. E. 1968. Glaciation of possible Scottish Re-advance age in North-West Wales. *Nature. Lond.* **218,** 76–78.

WALL, D. R. and WHATLEY, R. C. 1971. The Ostracoda of the Subrecent Deposits of Tremadoc Bay, Southern Irish Sea. *Bull. Centre Rech. Pau-SNPA* **5,** 295–309.

WATSON, E. 1970. The Cardigan Bay area. *In*, LEWIS, C. A. (Editor) *The Glaciations of Wales and adjoining Region,* Longmans, London 229 pp.

WOOD, A. and WOODLAND, A. W. 1968. Borehole at Mochras, west of Llanbedr, Merioneth. *Nature, Lond.* **219,** 1352–1354.

WRIGHT, J. E., HULL, J. H., McQUILLIN, R. and ARNOLD, S. E. 1971. Irish Sea investigations 1969–70. *Rep. No.* 71/19 *Inst. geol. Sci.* 55 pp.

R. A. Garrard, Exploration Department, Burmah Oil Development Limited, 38, Hans Crescent, London, SW1X 0ND.

The history of the Irish Sea Basins

M. R. Dobson

The Irish Sea basins which first developed in the Permo-Triassic are thought to preserve, if only in part, evidence of passive margin tectonics and sedimentation. The major phases recognised include:

Permo-Triassic sedimentation and erosion; Liassic regional marine transgression; Late Kimmerian movements uplift and erosion; Late Cretaceous Early Tertiary transgressions; Mid-Tertiary uplift and erosion; Quaternary sedimentation. Where possible attempts have been made to relate these phases to the broad regional evolution.

1. Introduction

The basins around the British Isles, which became established during the final stages of the Hercynian Orogeny, probably owed their siting to regional down-warping and their subsequent persistence to block faulting and graben develop-ment. This appears to be the nature of the control not only for the eastern part of the North Irish Sea and the associated Cheshire Basin, but also for much of the Central and South Irish Sea. Yet it should not be forgotten that the history of the area is not one of continued sinking and basin fill; important periods of erosion affected the Irish Sea basins in, for instance, the Lower Cretaceous and the Palaeocene. Erosion stripped away the Upper Cretaceous rocks east of the St. Tudwal's Arch; in the Mochras Borehole Upper Oligocene freshwater beds overlie directly the Upper Liassic. Further to the south Middle Jurassic beds subcrop and Upper Cretaceous chalk is absent throughout Cardigan Bay. A chalk absence pertains in the Bristol Channel also. This fact points to a similar tectonic history simply defined as early active subsidence which was later replaced by quiescence.

The following account attempts to trace the sequence of events since the early Permian of this region. Necessarily approximations and generalizations have to be made from time to time as on the one hand the record is not everywhere complete whilst on the other much of the available data is indirect being limited to deep geophysical evidence.

2. The Permo-Triassic

The Permo-Triassic, deposited on a deeply dissected topography, is dominated throughout by successions of continental type sediments that include dune sands, alluvial fan and playa lake material and fiuvial sequences. Marine incursions are prominent in the Upper Permian and above the Hardegsen unconformity and laid down marls, shales and evaporites particularly thick salt.

Deep seismic data principally for the southern half of the region reveal that much of this post-orogenic sequence occurs as a uniformly thick blanket between narrow positive ridges. Axial thickening along these downwarps occasioned no doubt by basement fault movements appears as a rather late feature. During the early part of the long Permo-Triassic period it was the Caledonoid trending ridges like the Southern Uplands — Longford Down massif and the Irish Sea "Geanticline" that were prominent. By Keuper times the topography was very subdued, such that virtually all of what is now Wales and Ireland including the region to the Scottish borders lay beneath a Triassic blanket. It should, however, be noted that overstep of Keuper over Bunter from the English Midlands towards central Wales implies that the blanket over the eastern margin of the Irish Sea area consisted almost entirely of Keuper.

3. The Jurassic

The subdued landscape was rapidly inundated by the Jurassic transgression, the first major Mesozoic marine invasion; it is thought that the whole region was covered. In addition continuous regional subsidence throughout the Liassic allowed for thick sediment accumulation. It is apparent from mapping the facies types in the Liassic for northwest Europe and across to the Scotian Shelf in Canada using a pre-drift fit that the Irish Sea occupied a transitional zone between essentially terrigenous facies to the east and carbonate evaporite facies to the west. The hybrid sediment that is typical of the Irish Sea region is a mixture of silts and clays and carbonate detritus, the latter derived from shoals located on the intervening ridges first established in the Permo-Triassic and now assuming a clear positive role. Regional subsidence, nevertheless, was the feature of Lower Jurassic tectonic activity, with virtually no land areas from Scotland to the Bristol Channel, excepting the South Wales archipelago, a part of the Pembroke Arch. But whilst the deposition was widespread present day persistence is limited to the Solway Basin, Northern Ireland, the Cheshire Basin, the Kish Bank Basin (unpublished core data) and the St George's Channel Basin complex extending northeast to Mochras. Small but yet to be confirmed outliers may occur in the east Irish Sea basin.

As in the North sea, early Kimmerian movements caused local warping and erosion of the Liassic. It is probable that the Cheshire graben and its northern extension developed at this time, certainly these movements appear to have initiated the central graben in St George's Channel and ensured lensing of the Middle Jurassic sequence.

These extension stress patterns may be linked to early transform fault movements; the North Atlantic south of the Azores had begun to open at the beginning of the Jurassic (180 m.y. ago) (Pitman and Talwain 1972) and by 130 million years the mid-Atlantic ridge extended as far north as the Western Approaches. A NE/SW trending transcurrent fault has been proposed for this period to accommodate the speading to the north (Williams 1975). These events may have contributed to the vertical movements and consequent tensional phase apparent in the British Isles and the North Sea, and most particularly the associated rifting and local mid-Jurassic uplift of the St Tudwal's Arch.

More significant in terms of vertical movements and their effects on sedimentation was the Upper Jurassic–Lower Cretaceous tectonic activity. These later Kimmerian events may be viewed, perhaps profitably, in terms of the Vogt-Ostenso model (Vogt and Ostenso 1967). The suggestion is that linear doming allows substantial erosion to take place before significant lateral spreading on the mid-oceanic ridge occurs. The material eroded would be carried in part eastwards giving thick essentially non-marine clastics. Although faulting and rifting on domes is both radial and tangential reactivation of old lines of weakness would be more likely. Consequently the major Caledonoid/Hercynian fractures would determine the size and extent of the subsequent blocks and their alignment.

The faulting restricted the area of recordable Upper Jurassic sedimentation to the axial part of the now clearly defined St George's Channel graben. There is no evidence of other sites of accumulation although it is probable that a thin veneer of sediments of this age existed in the Cardigan Bay Basin. It is also likely, although by no means certain, that the Lancashire coast faults were active at this time. The defining of the grabens by faulting extended into the Lower Cretaceous notably during the latter part when major events were occurring in the Bay of Biscay and the Rockall Trough (Williams 1975). These movements are considered to have affected the Mochras fault complex so that for the first time, at least since the Permian, North Wales became with the Lleyn and the Mid-Irish Sea "Geanticline" a positive zone. It follows from this that Lower Cretaceous erosion over North Wales was severe such that only a remnant of Mesozoic remained. It may be further supposed that much of the Mesozoic rock recorded for the North Irish Sea area was also lost at this time.

4. The Cretaceous

The record of the regional tilting and attendant erosion incurred by much of the British Isles during the Lower Cretaceous is now only preserved where deposits introduced by the Cenomanian transgression remain. Marginal unconformities in the rifts (see Dobson *et al.* 1973 fig. 10) best demonstrate this, as elsewhere Tertiary regressions in particular have blurred the legacy. That the positive blocks as we know them today had been established by Mid-Cretaceous now seems increasingly likely especially when the Upper Cretaceous deposits are considered. To the south a post-Wealden period of folding is apparent in Dorset where Albian Upper Greensand rests on rocks that range in age from Oxfordian to Wealden. The deposits of Upper Cretaceous age on land in these Western areas are strongly transgressive, often of shallow water marginal type (greensands).

In the South Irish Sea and the North Celtic Sea, sedimentation was continuous along the axis of the basin, but not on the positive ridge to the south (the Pembroke Arch) or to the north (Irish Sea "Geanticline") and northwest. Thus in the intervening basins pure chalk of considerable thickness was deposited. The contrast between positive ridges and subsiding basins was at this time more marked than ever before. The concidence of thick Upper Cretaceous sedimentation adjacent to a spreading ridge as occurs in the Celtic Sea provides an additional complication although collapsed margins of this type are not unusual.

The North Irish Sea probably acquired a chalk cover at the time of the Cenomanian transgression in southeast Antrim. The regression and erosion of the Irish deposits during the Lower and Middle Turonian (Hancock 1961) supports the views of Reid (1973) that much of Ireland was emergent until the transgression that occurred during the *mucronata* zone. The Upper Campanian transgression was widespread and incorporated southern Ireland (Barr 1966) probably submerging the whole of the island block with the possible exception of Donegal. The Campanian chalk of Kerry resting on Carboniferous together with the non-sequences recorded for Northern Ireland reflect at least in part the Atlantic spreading convulsions that were occurring immediately west of Ireland. The "Hiberian Greensands" of the Irish Block may be correlated with the marginal facies on the flanks of the "Geanticline" which were deposited during most of Upper Cretaceous and may only be a remnant of a once widespread sediment.

5. The Tertiary

Salt activity was important in the St George's Channel basin, and as a deeper feature in Cardigan Bay was probably markedly active at this time. Apart from local warping and erosion, sedimentation was continuous and rapid up to the top of the Oligocene, particularly in the south. Lower Tertiary sedimentation was not important in the Kish Bank Basin, whilst north of this there is only an erosion surface covered by the thick flows of the Ulster lavas. Uplift towards the northwest probably restricted early Tertiary sedimentation. By contrast growing faults in the main southern rifts ensured sediment preservation and a thickness in excess of 1000 m has been recorded for the Lower Tertiary.

The top of the Palaeogene in the southern graben and in the Celtic Sea is characterised by a marked erosion surface of probably Oligocene age and much was removed where not faulted down, for example regional uplift especially of the rift margins initiated the bevel that is a characteristic of the offshore region and eroded both Mesozoic and Tertiary deposits from the uplifted blocks. In addition the active basement faults caused gentle folding in the Cretaceous and Tertiary beds. In Atlantic terms the Palaeogene was a period of great activity and change with the development first of a triple point junction to be followed by spreading on the Reykjanes ridge. This would almost certainly have produced marked linear doming which especially affected the northwest, with its associated igneous activity, and later the more southerly regions.

Throughout virtually the whole of western Britain there was, particularly in the late Palaeogene early Neogene, a period of intense tectonic activity producing uplift and eastward tilting of basement blocks, a regime that particularly distorted the Continental margin. Widespread regressions which produced a sudden end to marine deposition on the shelf initiated regional erosion and ultimately caused an increase in continental margin sedimentation. In Upper Oligocene–Lower Miocene times continental deposits, the products of deep weathering, formed a widespread blanket of clays and silts derived at least in the South Irish Sea area mainly from eroded Palaeozoic rocks.

Known Oligocene deposits on the west side of the British Isles lie on Palaeozoic and early Mesozoic rocks in the south and warped basalts in the region off northwest Scotland. These are restricted to non-marine sequences. They occur at Bovey Tracey, Petrockstow, east of Lundy Island (the Stanley Bank Basin), Flimston, Mochras, Lough Neagh, in the Colonsay basin and south of the Isle of Skye. Lignitic clays have also been cored in the English Channel south of Plymouth.

The thick offshore Tertiary sequences in the South Irish Sea and the Celtic Sea do contain marine Oligocene although evidence from the well-sampled English Channel contrasts with this. The distribution of the non-marine sequences is noteworthy for four are aligned along one fault—the Sticklepath. The sequences equally divide in number by present elevation. Half are on the offshore platform and half at the "200 ft" level. Perhaps of equal significance is the thickness of the successions at Bovey (> 750 m), Petrockstow (> 661 m), Lundy (340 m), Mochras (> 525 m), Lough Neagh (> 1000 m), South of Skye (1000 m) and the Colonsay basin (500 m) compared with those Oligocene sediments preserved in the Hampshire Basin (< 100 m) and the Southern North Sea (100 m) (Rhys 1974).

Preservation of these thicknesses together with the nature of the material combine to give the impression that the warping and accompanying faulting was incremental, but resulting in cumulatively massive throws.

The material preserved, notably well rounded Palaeozoic pebbles and no flint, lends support to earlier erosion probably of Lower Cretaceous and Palaeocene age. In addition the clay mineralogy reflects the development of a very deep weathering profile in moist conditions and well established drainage. Lateral reworking by rivers would have produced a very subdued landscape with a thick mantle of sediment that extended westwards over much of the Continental Shelf. This landscape persisted into the Miocene and, although worldwide there was a gradual transgression recorded for much of the early Upper Tertiary, thermal linear doming northwest of the British Isles ensured that marine penetration occurred only in the SW Approaches providing a mainly pelagic shelf deposit. By late Miocene–Pliocene times the transgression became extensive and must have removed landscape irregularities caused by the faulting that accommodated the lignitic rich sequences, and trimmed the weathered landscape at least as far across as the existing "200 ft" peneplain and the present day land-bound Oligocene deposits.

The existence of trimmed Oligocene sediments at both the "200 ft" level and on the submerged marine platform implies that the transgressive event was extensive. Major vertical movements on the coastal faults had largely ceased by the early Pliocene as evidenced by the Neogene overstep across the south of Ireland fracture and the structure west of St Bride's Bay. It must, therefore, be assumed that uplift in North Wales southeast of the Dinorwic fault probably occurred by block rotation during the Upper Miocene.

The transgression that scoured and trimmed up to the limits of the present "200 ft" platform was brief as a minor regression occurred during the Pliocene which allowed the development of a fresh cliff line. The volume of rock that must be removed to establish a cliff line on a sloping (5°) suface is relatively small and could have been accomplished during the 2 million year period of the late Pliocene. Pliocene sedimentation in the region is largely restricted to St George's Channel and the Celtic Sea. Tidal scour under a regime little different from that of today ensured the continued exposure of much of the offshore platform. The Pliocene now present may be only a remnant of its former thickness, the irregularities of the upper surface testifying to the effects of the Quaternary.

6. The Quaternary

Although the Pliocene–Pleistocene appears to be a tectonically quiet period there is growing evidence to suggest that the major basin faults not only in St George's Channel, but also in the Kish Bank area have been active. This activity has encouraged the preservation of thick Quaternary deposits the base of which approaches 200 m in the Central areas. These thick deposits which are the subject of the preceding three contributions occur throughout the Irish Sea and represent a major phase of sedimentation. In St George's Channel pre-Devensian tills are recognised beneath a thick cover of interglacial marine sands and muds which in turn are covered by tills of Devensian age. The upper glacial deposits terminate at the latitude of St David's Head. Towards the north the glacial sequences appear more complex (see H. M. Pantin, this volume).

During the late glacial sea-level rise the shallower north Irish Sea together with parts of the central region supported depositional environments of the estuarine-deltaic type. In the south these environments were restricted to the coastal region of Cardigan Bay. The Flandrian transgression gradually overwhelmed these, imposing a marine sediment veneer.

Along the east Irish coast a rare combination of high tidal velocities and small tidal ranges has ensured that through lack of estuaries, the mobile sand persists principally as large sand banks. By contrast on the east side large estuaries have abstracted the marine sand leaving only a thin offshore veneer. Thick mud deposits characterise the tidal slack regions.

References

BARR, P. T. 1966. Upper Cretaceous foraminifera from the Ballydeanlea Chalk, County Kerry. *Palaeontology* **9**, 492–510.
DOBSON, M. R., EVANS, W. E. and WHITTINGTON, R. 1973. The Geology of the South Irish Sea. *Rep. Inst. Geol. Sci.* No. 73/11, 35 pp.
HANCOCK, J. M. 1961. The Cretaceous system in Northern Ireland *Q. J. geol. Soc. Lond.* **117**, 11–36.
PITMAN, N. C. and TALWAIN, M. 1972. Seafloor spreading in the North Atlantic. *Bul. geol. Soc. Am.* **V. 83**, 619–645.
REID, R. E. H. 1973. The Chalk Sea, *The Irish naturalist's Journal.* **17**, 357–375.
RHYS, G. H. (Compiler) 1974. A proposed standard lithostratigraphic nomenclature for the southern North Sea and an outline structure nomenclature for the whole of the (U.K.) North Sea. *Rep. Inst. Geol. Sci.* No. 74/8, 14 pp.
VOGT, P. R. and OSTENSO, N. A. 1967. Steady state crustal spreading. *Nature Lond.* **215**, 810–817.
WILLIAMS, C. A. 1975. Sea floor spreading in the Bay of Biscay and its relationship to the North Atlantic. *Earth and Planet. Sci. Lett.* **24**, 440–456.

M. R. Dobson, University College of Wales, Department of Geology, Llandinam Building, Penglais, Aberystwyth, Dyfed, Wales, SY23 3DB.

The Quaternary marine record in southwest Scotland and the Scottish Hebrides

W. G. Jardine

The earliest record of Quaternary marine action in southwest Scotland and the Scottish Hebrides is preserved as remnants of high-level rock platforms and associated coastal cliffs in parts of the Inner Hebrides. Originally regarded as pre-Quaternary, the high-level shore-lines are now considered to be interglacial (Hoxnian and/or Ipswichian) in age and composite in nature. During Quaternary low stands of sea-level the sea coast lay near the edge of the continental shelf: possibly around 40000 years B.P. much of the present shelf and land area was subaerially exposed; probably around 18000 years B.P. the western part of the shelf was subaerially exposed but much of the eastern part of the shelf and the present land area of western Scotland was ice-covered. Initial penetration by the late-Devensian sea into the Solway Firth, Firth of Clyde, Malin Sea and Sea of the Hebrides areas was marked by the accumulation of marine muds that overlie glacial and fluvio-glacial deposits. Later, coastal deposits fringing the Firth of Clyde, Clyde Estuary and many of the islands of the Inner Hebrides were mainly coarse-grained gravels. The low-level marine platform of the Oban area, formerly regarded as interglacial in age, may have been cut at the time of the Loch Lomond Readvance, but other low-level platforms, e.g. on Lewis in the Outer Hebrides, are not necessarily contemporaneous with the platform in the Oban area. Along most parts of the coast of southwest Scotland and on several of the Hebridean islands, the shore-line at the maximum of the main Holocene marine transgression lay a short distance inland from the present coast; in a few places much larger embayments occurred. Culmination of the main Holocene marine transgression was diachronous from place to place due to differential adjustments between eustatic water movements on the one hand and regional (mainly isostatic) and local land movements on the other.

1.   Introduction

The Quaternary sub-era was a time during which alternations of cold and temperate climate were accompanied by periodic growth and decay of ice sheets in Antarctica, North America, Fennoscandia, Greenland and, to a lesser extent, in other parts of the Northern Hemisphere including Britain. On theoretical grounds it has been contended that during cold periods the surface level of the oceans was lower than now and that during temperate periods the ocean surfaces were at approximately the same level as now or, occasionally, somewhat higher. Observation around the coasts of Britain, as in other glaciated regions, substantiates the theoretical concept especially as regards temporary higher stands of sea-level; direct evidence of low stands is less readily obtainable, in many places being concealed by sediments, terrestrial or marine in origin, or by the transgressive sea. The record considered here is the imprint that the sea has left of its fluctuations in position and its temporary high and low stands in the area comprising the present land surface and sea floor of the Scottish mainland southwest of a line running approximately from Oban through the Glasgow area to the eastern end of the Solway Firth, together with the vast area incorporating and surrounding the archipelago of the Scottish Hebrides (Fig. 1).

Evidence of former high levels of the sea in relation to the land takes various forms. Remnants of old sea-cliffs now some distance inland or high above present sea-level, raised shore platforms cut in solid rock or in glacial till, river terraces graded to heights above present sea-level, raised estuarine, beach and other coastal sediments of saline or brackish-water environments all bear witness to former sea-levels relatively higher than the present. Former low sea-levels are suggested by the occurrence of buried valleys, etched in solid rock, underlying the present courses of several rivers draining the area and by the "drowned" digitate form of parts of the coast. In addition, the occasional occurrence along the coasts of the mainland and islands of southwest and western Scotland of former vegetation-covered land surfaces and of tree stumps in position of growth at levels below present high water mark is convincing evidence of former relatively lower sea-level.

Because the movements of sea-level were superimposed on a surface of variable relief, former stands of the sea, whether higher or lower than the present level, frequently produced shore-line configurations markedly different from those at present. In this essay the changes that occurred in the position and form of the Quaternary shore-lines and coastal deposits of southwest Scotland and the Scottish Hebrides are discussed insofar as these changes are known. The completeness of the Quaternary marine record varies greatly from one locality to another within the area discussed, as does the published account of the record.

In an area as extensive as that covered here, present tidal range varies greatly from place to place (as, presumably, it did in the past), locally being as small as < 1 m and as great as 9 m. Because of this, and for other reasons discussed elsewhere (Jardine 1975 pp. 187–192), it is inadvisable to use the raw data given in this paper to attempt construction of graphs showing movements of former sea-level on a regional scale. Below, heights recorded in relation to Ordnance Datum Newlyn (O.D.) are shown (for example) as $+16$ m or -8 m. Heights recorded in relation to a local High Water Mark of Ordinary Spring Tides are shown (for example) as $+12$ m HWM.

Fig. 1. Map of southwest Scotland and the Scottish Hebrides, showing the main locations mentioned in the text, the area in which the high-level marine platforms originally were identified in the Inner Hebrides by W. B. Wright (1911), and the additional area within which they were identified by S. B. McCann (1968). Inset: map of Colonsay and Oronsay showing the locations on these islands mentioned in the text.

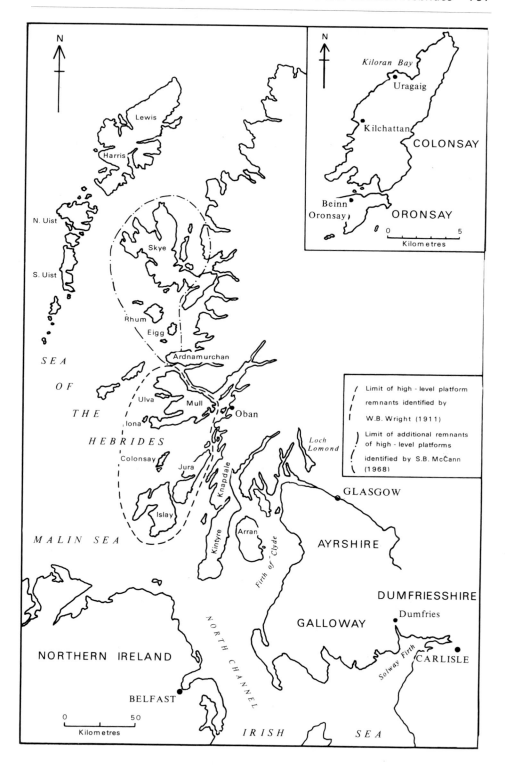

2. The high-level rock platform of the Inner Hebrides

An early high-level stand of the sea in western Scotland during the Quaternary sub-era appears to be represented by remnants of a rock platform and associated cliffs that Wright (1911) identified as the traces of a "pre-glacial" shore-line at levels of c. +27 to +41 m HWM in parts of the Inner Hebrides (Fig. 1). It is widely believed now that the shore-line features described by Wright and his contemporaries are "pre-glacial" only in the sense that their formation preceded at least one glaciation—that is, the features are pre-Devensian in age but not necessarily pre-Quaternary, and they may have been produced in the course of one or more interglacial intervals (McCann 1964 pp. 1–3; *cf.* Sissons 1974b p. 326).

Remnants of the high-level rock platform have been identified and described on Islay, Oronsay, Colonsay, Mull, the Treshnish Islands and (less satisfactorily) on Iona (Wright 1911; Cunningham Craig *et al.* 1911 pp. 62–63; Bailey *et al.* 1924 pp. 386–391), and on Jura, the Ardnamurchan peninsula (included here for completeness although strictly not part of the Hebrides), Rhum, Skye, and (less satisfactorily) on Eigg, Raasay and the Applecross peninsula (McCann 1964 fig. 4; McCann 1968 pp. 23–27; McCann and Richards 1969 pp. 17–20; Richards 1969 p. 124) (Fig. 1).

Remarkably, although the authors cited made extensive reference to the platform and the cliffs which commonly back it as marine in origin none provided conclusive evidence of such an origin. Descriptions of the platform were concerned mainly with the effects of glacial erosion and deposition upon it, and its marine origin was inferred from the proximity of its remnants to the present coasts and the combination of its approximately planar (though dissected) surface and adjacent near-vertical cliffs. Convincing evidence of the influence of the sea in moulding the platform does exist, however, and is presented here.

At Uragaig on Colonsay (Fig. 1, inset), at location (approximately) NR 395978, sea-eroded pot-holes similar to those being excavated at present in the inter-tidal rock surface of nearby Kiloran Bay were exposed temporarily between 1964 and 1974 in a small road-side cutting transecting the rock platform. The pot-holes were infilled with medium- to coarse-grained angular gravel, the origin of which, marine or fluvio-glacial, is uncertain. The existence of the pot-holes themselves, however, is proof of the marine nature of the platform in which the vertical-sided holes are cut. Additional evidence of the marine origin of the platform also is to be found at Uragaig in the form of remnants of several degraded sea-stacks which project five to ten metres above the general level of the rock platform at distances of twenty to fifty metres in front of the old sea cliff. The stacks are surrounded now by a thin cover of glacial till which rests upon the rock platform, but there can be no doubt that the upstanding stacks and the platform from which they rise owe their origin to the action of the sea.

The character of the high-level platform of the Inner Hebrides and of the cliffs that back the platform varies from place to place, depending partly on the type of rock in which the features are cut, partly on the direction in which the features face in relation to the sea that cut them (*cf.* McCann and Richards 1969 p. 18). In general, the surface of the platform, although smoothed overall by the passage of ice, is rather irregular, being traversed by fissures some of which may pre-date the cutting of the platform (Wright 1911 p. 100), and the platform varies in altitude by a few metres over a small area as the results both of its inter-tidal or shallow-water origin and of its post-formational erosion by ice and other abrasive agents. The cliffs that back the platform vary in height and gradient from place to place. Commonly the slope is very steep and the crags high and imposing; frequently the cliffs exceed 25 m, occasionally reaching 50 m or more in height and, in places, as in western Colonsay and western Rhum, they are higher than the adjacent cliffs of the present shore-line.

Not only does the height of the surface of the platform vary locally within any one remnant and within the individual islands where its remnants are preserved, but the height of the old shore-line which the inner margin of the platform is believed to represent also varies throughout the area where fragments of the rock bench have been identified (Fig. 1). The heights quoted for the level of the inner margin of the platform over the relevant area of the Inner Hebrides range between +21 m and +43 m HWM, and on this evidence it was suggested by McCann (1968 pp. 26–27) that the platform is warped, declining in height both to the north and to the south away from a centre in Ardnamurchan, heights of +49 m HWM at the head of Loch Scridain on Mull, and at +41 m HWM at Uragaig on Colonsay being considered anomalous.

The range in height of the surface of the high-level marine "platform" may indicate more than warping of the feature after its formation. It may indicate that what frequently has been regarded in the past, and what so far has been treated in this discussion, as a single feature in fact may be composite. Wright and his colleagues hinted at this in noting the anomalous height of the platform remnant at the head of Loch Scridain (+49 m HWM) and of the rock platform in the Ulva Cave (+45 m HWM) compared with the height of +35 m HWM for the inner margin of the rock platform over most of western Mull (Bailey *et al.* 1924 pp. 386–388), and McCann and Richards (1969 p. 19) identified two platforms, with inner margins at +27 m HWM and at +23 m HWM, in eastern Rhum.

The belief that the so-called high-level "platform" may be composite is confirmed by the writer's own observations on the island of Oronsay. Wright (1911 p. 102 and fig. 4) quoted the height of the platform on the southern side of Beinn Oronsay as +37 m HWM, and levelled heights of +36·5 m to +38·5 m have been obtained recently. The dissected bench at c. +38 m, however, is but the highest of several rock benches which occur on the southern side of Beinn Oronsay and cut across the intricate fold-structures in the Torridonian greywackes which are exposed there. In particular, a pronounced though narrow ledge occurring at c. +32 m to +35 m appears to be part of the composite rock platform mapped around Beinn Oronsay by Wright and his colleagues (Cunningham Craig *et al.* 1911 fig. 19) and seen to have a distinct eastward dip on the north side of the Beinn. The 1911 map of what is shown as a single rock platform also appears, on field evidence, to include a prominent bench-like feature, at c. +18 to +21 m, which occurs as a narrow ledge on the southern side of Beinn Oronsay but extends much more widely on the northern side of the Beinn (Fig. 1, inset). In the field the two major platforms, one at c. +38 m, the other at +18 m to +21 m, are separated by a cliff which in places is steep, but for the most part is relatively subdued. The place of the lower of these two benches in the Quaternary coastal history of Oronsay is still under investigation. In the present context, the significance of its recognition in recent years is that its presence demonstrates that the "high-level platform", originally identified in Oronsay by Wright, is composite. Together with the evidence from Rhum presented by McCann and Richards (1969 pp. 17–20) and the more general points raised by McCann (1968 pp. 23–27), the evidence from Oronsay suggests a much more complex history of shore-line development associated with

H

the "high-level platform" of the Inner Hebrides than hitherto has been recognised.

The ages of the high-level platforms of western Scotland are uncertain as yet. McCann (1968 p. 24) implied that collectively the platforms date either from the Ipswichian (Eemian) or from the Hoxnian interglacial but, because they are composite, the platforms may comprise elements formed in Hoxnian and/or in Ipswichian times. It is interesting to note that if similar glacio-isostatic effects operated in the British area in late-Anglian to early Hoxnian times and in late-Wolstonian to early-Ipswichian times as operated in the same area in late-Devensian to early-Holocene times (*cf.* Godwin 1963), an interglacial marine platform with a present maximum elevation of c. +35 m in the western isles of Scotland might be contemporaneous with shore-line features close to present sea-level in northern England and Northern Ireland (*cf.* Wright 1937 fig. 127). The maximum extent of the British ice-sheet during both the Anglian and Wolstonian Ages, however, is believed to have been greater than that of the Devensian ice-sheet. If the greater extent of the earlier ice-sheets is an indication of greater total mass of these ice-sheets than that of the Devensian ice-sheet, isostatic recovery in northern Britain on melting of the Anglian and Wolstonian ice-sheets probably exceeded recovery on melting of the Devensian ice-sheet. It follows that marine-cut platforms now reaching maximum elevations of c. +45 m in western Scotland, and currently of uncertain age but perhaps in the future proved to be Hoxnian or Ipswichian, may be the correlatives of remnants of Hoxnian coastal deposits at c. +25 m in East Anglia and southern England or of Ipswichian coastal deposits at c. +5 m on the coasts of Yorkshire and the southern Irish Sea area (*cf.* Mitchell 1972 fig. 2 and fig. 4).

3. Effects of low stands of sea-level

Evidence of low stands of the sea during the Quaternary sub-era in the area considered here takes a variety of forms, some of them only now beginning to be revealed in the course of underwater exploration of the continental shelf. In the Glasgow area buried channels below the present beds of the Rivers Clyde, Kelvin and Cart are cut in solid rock to levels well below Ordnance Datum. For example, at Drumchapel in western Glasgow, the floor of the valley of the former River Kelvin occurs at −75 m, whilst farther west, at Dumbarton, the buried valley of the River Leven descends to at least −68 m (Macgregor 1941 pp. 121 and 130), and in the Firth of Clyde a number of winding channels in rockhead at −70 m to −80 m occur in an area that includes occasional glacially-eroded rock basins descending to c. −160 m and in extreme cases −320 m (Deegan *et al.* 1973 fig. 9). The relevant valleys are sinuous, narrow and deep, and extend outwards from the centre of the Midland Valley of Scotland. These facts suggest that they were cut by running water rather than by glacier ice, which is known to have been moving eastwards between Dumbarton and Glasgow, certainly during the Devensian glaciation and possibly during earlier glaciations.

Additional possible evidence of a Quaternary low stand of sea-level occurs in western Galloway. There, the isthmus between Luce Bay, facing the Solway Firth, and Loch Ryan, opening into the Firth of Clyde (Fig. 2), consists of an expanse of superficial deposits (Jardine 1966 p. 8). The Quaternary tills and fluvio-glacial sands and gravels vary in thickness and extend to different depths from place to place, but near the south-central extremity of the isthmus two

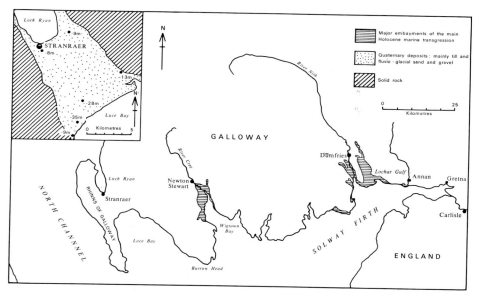

Fig. 2. Map of the northern shore of the Solway Firth, showing the major embayments occupied by the sea at the maximum of the main Holocene transgression. Contemporaneously, along much of the remainder of the northern coast of the Solway Firth the shore-line was located a few tens of metres inland from the present shore-line. Inset: map of the isthmus between Loch Ryan and Luce Bay showing the altitude of rockhead as established in boreholes at the sites indicated.

boreholes penetrated drift to depths of −28 m and at least −35 m, and elsewhere boreholes indicate rockhead at −8 m or deeper close to both the western and the eastern edges of the isthmus (Fig. 2, inset). Although borehole coverage of the area is poor, it seems reasonable to infer that were the Quaternary sediments not present between Luce Bay and Loch Ryan, the Rhinns of Galloway now would be separated from the remainder of Galloway by a deep salt-water channel, and the Rhinns may have been an island at times during the Pleistocene epoch.

The presence of the low-level buried valleys of the Glasgow area, the Firth of Clyde and western Galloway poses other, greater, problems in interpretation. Theoretically the main cutting of the channels may have been performed by water at the time of the maximum of a glaciation when sea-level was at or near to its lowest stand, but the glacial maximum also was the time when much of northern Britain may have been covered by a large ice-sheet which would have prevented contemporaneous river action. Such an apparent paradox ceases to exist when (in reference to events of the Last Glaciation as examples) it is realised that the height of sea-level in the British area as elsewhere is controlled by the growth and decay of ice masses on a global scale, whereas the inhibition of river action in northern Britain by the presence of an ice-sheet is merely a local phenomenon.

In North America several glacial stades are known to have occurred in the course of the Last Glaciation (Dreimanis and Karrow 1972 fig. 2). During an early- or mid-Wisconsinan stade at (?) c. 40000 years B.P. sea-level stood at c. −145 m (Curray 1965 p. 724 and fig. 2). Contemporaneously in Britain the Upton Warren interstadial complex was in process of formation (*cf.* Coope 1975 pp. 160–162) and possibly the major part of Britain, including much of Scotland, was ice free from then until at least 27000 years B.P. (*cf.* Sissons 1974b p. 312). Between c. 35000 and 22000 years B.P. global sea-level was fairly high, possibly c. −10 m around 30000 years B.P., being followed by a low-level stand at approximately −130 m c. 18000 years B.P. (Curray 1965 p. 724; Milliman and Emery 1968 p. 1121) during a late-Wisconsinan stade that was penecontemporaneous with the maximum of the Weichselian glaciation in northern Germany and Scandinavia and with the maximum of the Devensian glaciation in the British Isles.

The now-buried river channels of central and southwest Scotland may have been cut c. 40000 years B.P. when the greater part of the continental shelf surrounding the area was exposed and the present land area was largely, if not entirely, ice-free. During the later period of low sea-level c. 18000 years B.P., although much of the continental shelf again was free of marine waters, the adjacent land areas and parts of the shelf were covered by glacial ice. As a result river excavation was minimal or impossible, but it would be at this time that the previously-excavated low-level river valleys would be infilled with the glacial debris that they now contain. Contemporaneously extensive tracts of the shelf areas now constituting the floors of the Solway Firth, Firth of Clyde, Malin Sea area and Sea of the Hebrides were covered with variable thicknesses of glacial and fluvio-glacial deposits (Pantin 1975; Deegan *et al.* 1973 pp. 30–34; Dobson and Evans 1975; Binns *et al.* 1974). Similar sequences of events may have occurred on the continental shelf in the course of pre-Devensian glaciations.

It is important to note that although *global* sea-level may have stood at c. −130 m around 18000 years B.P., in the area considered here *relative* sea-level would be markedly higher than this because of ice loading of the surface of the present land area and continental shelf. Also, ice loading would vary considerably from place to place, being greater in the east than in the west of the area concerned. Accordingly, at the time of the glacial maximum *relative* sea-level would be considerably lower on the western part of the continental shelf and in the vicinity of the Outer Hebrides than in the Firth of Clyde, in the Glasgow area and in the area around Oban (Fig. 3). These points should be borne in mind when the late-Devensian history of penetration of the sea into southwest Scotland and the Hebridean area is considered.

4.   The Devensian marine record

At the maximum of the last (Devensian) glaciation of the British Isles, the major part of the Scottish area was covered by a thick ice-sheet while contemporaneously the marine shore-line was located on the present continental shelf several tens of kilometres to the west of the Outer Hebrides. Marine influence in southwest and western Scotland, therefore, was at a minimum while glaciation was at its maximum c. 18000 years B.P. In contrast, the effects of the sea in the transitional phases from the inception of formation of the Devensian ice-sheet to its maximum, and from the glacial maximum to the commencement of the Holocene epoch, must have been considerable because during these phases a vast area was being progressively uncovered and covered by the sea. The details of the earlier phase are very obscure, but some of the effects discussed in the preceding section on low stands of sea-level may have been produced then.

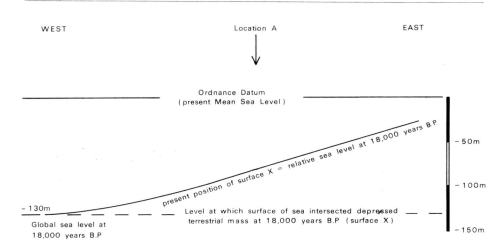

Fig. 3. Hypothetical section from near the western edge of the Continental Shelf (left) to the vicinity of the locus of maximum Devensian ice loading and unloading in Scotland (right). Vertical displacements since 18000 years B.P. are chosen arbitrarily. The vertical scale is greatly exaggerated.

Note: relative sea level at a given location at a given time is the altitude (positive or negative) of the present position of the relevant shore, i.e. the line of intersection of the contemporaneous mean sea surface and the adjacent terrestrial mass. At any given location, at any given time: relative sea level = contemporaneous global sea level + the vertical component of terrestrial movement at the given location since the given time. For example, at location A, where vertical terrestrial movement since 18000 years B.P. is + 50 m, relative sea level at 18000 years B.P. is − 80 m compared with the contemporaneous global sea level of − 130 m.

4a.   The evidence of late-Devensian marine deposits

(*i*) *Stratigraphy and Sediment Distribution.* Penetration of the sea into the area of southwest Scotland and the Hebridean archipelago was dependent largely on the balance between the rate of rise of sea-level on a global scale as "Devensian" ice sheets diminished in various parts of the world and the rate of recovery, from loading of ice, of the continental mass on a regional scale, minor modifications being imposed locally by barriers produced by remnants of the ice sheet or its depositional products.

Evidence of early phases of marine influence is provided by the recent under-water work by the Institute of Geological Sciences in four main areas: the Solway Firth (Pantin 1975), the Firth of Clyde (Deegan *et al.* 1973 pp. 30–34), the Malin Sea area (Dobson and Evans 1975) and the Sea of the Hebrides (Binns *et al.* 1973 pp. 22–23; Binns *et al.* 1974). The submarine Quaternary sequence varies in detail from area to area, but in broad terms it comprises three units, in descending order: true marine sediments, glacio-fluvial and semi-marine sediments, glacial till (presumed to be the product of the Devensian glaciation). On the floors of Wigtown Bay and the Solway Firth east of Burrow Head (Fig. 2), the till is succeeded by lagoonal deposits, typically well-stratified, showing few signs of bioturbation and almost bereft of organisms other than a few foraminifers. In several places the beds contain pebbles, apparently dropped from floating ice. The deposits are ascribed to late-Devensian sedimentation in a water body connected with the open sea, but colder and less saline than the present waters of the Irish Sea. The overlying marine beds, in which bioturbation is common and molluscan valves frequently occur, are thought to range from late-Devensian to Holocene in age. In places there may be lateral passage from lagoonal beds to marine beds. The sequence *in toto* suggests the proximity of marine waters to the Solway Firth area at a level below present sea-level while the late-Devensian ice-front lay in the vicinity or a short distance to the north of the present northern shore of the Solway Firth, followed by penetration into the Firth slightly later.

On the floor of the Firth of Clyde the late-Devensian tripartite sedimentary sequence is imperfectly developed, but there are indications nevertheless that the period of glaciation was succeeded by a period during which calcareous silty-clays, locally laminated and in places containing occasional molluscan shell fragments, foraminiferid tests and marine-ostracod valves, accumulated in the waters of the transgressive sea (Deegan *et al.* 1973 p. 30). In the Malin Sea area and the Sea of the Hebrides the glacial till is succeeded by poorly-sorted sandy muds with variable amounts of coarse-grade rock fragments ("Formation 2") and a dino-flagellate cyst population which indicates that the sediments were deposited in a marine environment closely associated with ice. The overlying deposits ("Formation 3") also are poorly-sorted sandy muds, but they are softer than the sediments of "Formation 2" and contain only rare coarse-grade rock fragments; they are thought to have been deposited a considerable distance from the ice-front probably sufficiently late in the Devensian age for the ice mass (if, in fact, in existence at that time) to have been restricted almost entirely to the present main-land area of western Scotland.

Evidence of later phases of marine influence, when the sea impinged on the present land surface to give a now-recognisable beach or shore-line, occurs at a large number of places. Some examples are discussed below. Along the northern seaboard of the Solway Firth scattered remnants of late-Devensian "raised beach deposits" were recorded in Geological Survey maps published originally between 1871 and 1885 but virtually unrevised since then. It may be shown (e.g. Jardine 1971 fig. 4), however, that most of the deposits concerned are fluvio-glacial in origin (*cf.* Wright 1928 p. 102) and it is doubtful if late-Devensian shore-lines ever existed in the Solway Firth area at heights above present sea-level (*cf.* Jardine 1971 p. 103).

In the Firth of Clyde late-Devensian marine deposits occur at numerous localities at levels between a few metres below O.D. and c. +37 m. The bulk of the sediments are of fine sand or of finer grade and they constitute the "Clyde Beds", intensively studied in the late 19th Century, which characteristically are finely-laminated or non-laminated and contain macro- and microfaunas of arctic aspect together with occasional ice-rafted striated stones (Brady *et al.* 1874 pp. 22–71). The Clyde Beds were deposited in full-marine conditions shortly after the ice of the main Devensian glaciation melted in the Firth of Clyde but in sufficient depth of water or sufficiently far from the shore for disturbance to be minimal. At some localities the Clyde Beds are overlain by horizontally- or cross-laminated sands and gravels that accumulated as beach deposits along the late-Devensian shores.

In the area between Dumbarton and Glasgow (Fig. 4) a similar sequence of late-Devensian deposits occurs from levels low in the buried channel of the River Leven (at c. −50 m) up to c. +25 m at some localities. From top to base the sequence broadly is: stratified sands and gravels up to 3 m thick; clay and silt with locally-abundant remains of mollusc valves and occasional remains of tests of foraminifers and valves of ostracods up to 47 m thick; laminated clays and silty clays (varves accumulated in local hollows at various heights above the con-temporaneous sea-level as the late-Devensian ice gradually melted) up to 2 m thick.

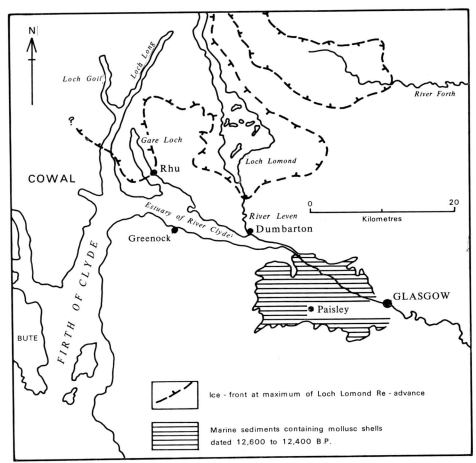

Fig. 4. Map of the northern part of the Firth of Clyde, the Clyde estuary and the environs of Glasgow, showing the main locations in that area mentioned in the text. The approximate position of the ice-front at the maximum of the Loch Lomond Readvance and the approximate extent of marine sediments that were deposited in the Paisley-Glasgow area between c. 12600 and 12400 years B.P. are also shown.

Some of the most magnificent developments of beach deposits of the late-Devensian sea are to be found on Islay, Jura and Colonsay. In northeast Islay, marine-gravel beds up to 5 m thick rest in places on the high-level coastal rock platforms (p. 102), elsewhere on glacial till, at levels up to c. +29 m HWM (McCann 1964 pp. 3–7). On the west coast of Jura immediately north of Loch Tarbert (Fig. 5), late-Devensian beach ridges composed of cobbles of the local highly-resistant quartzite extend almost continuously for 8 km, in places resting on remnants of the high-level rock platform(s). The ridges occur between c. +37 m HWM and c. +12 m HWM and are interpreted by McCann (1964 pp. 7–13) as having formed on the shore of a late-Devensian sea whose (relative) high water level dropped progressively from c. +37 m to c. +17 m HWM as land uplift outpaced sea-level rise. The evidence on Jura is paralleled to some extent by that on Colonsay, but there are some interesting contrasts also. The deposits of the late-Devensian sea that have been preserved on Colonsay are almost entirely cobble-sized beach sediments of quartzite. At Kilchattan, on the west coast (Fig. 1, inset), ridges occur up to heights of c. +33 m HWM, but the highest deposits are regarded as storm debris (Cunningham Craig *et al.* 1911 pp. 65–68) and the maximum height of high water of the late-Devensian sea on Colonsay appears to have been c. +29 m HWM. As a result, and in contrast with Islay and Jura, nowhere on Colonsay is there evidence of late-Devensian transgression across the highest of the pre-existing high-level rock platforms. Indeed, at Uragaig (Fig. 1, inset) the lowest part of the surface of the highest pre-Devensian rock platform stands more than ten metres above the surface level (c. +23 m HWM) of the highest late-Devensian marine gravels. Below these gravels, the solid rock, exposed only occasionally in section, not at all in plan, may be a remnant of a rock platform corresponding with the +18 m to +21 m pre-Devensian platform in Oronsay (see page 103). In Oronsay no late-Devensian marine deposits have been found resting on this pre-existing platform, circumstances that may be explained by the more exposed situation of Oronsay compared with Colonsay in late-Devensian times. The extent of both Oronsay and Colonsay was considerably less then than now, each area almost certainly comprising several islands where now there is one, as was the case on a larger scale in Islay and Jura.

On the east coast of northeastern Skye, there is evidence that the late-Devensian sea extended beyond the shores of the present sea. There, coarse-grade beach deposits occur up to c. +18 m (Richards 1969 p. 126) and in places they may rest on remnants of ancient rock platforms, but few details are available as yet. On Rhum, late-Devensian cobble- and gravel-sized rounded beach material occurs as abandoned storm ridges at heights between c. +12 m and +30 m HWM on the western side of the island where in many places it rests directly on remnants of earlier high-level coastal platforms. In addition, on the northwestern coast of the same island, where high-level platforms are missing, a gently-sloping terrace of late-Devensian beach gravels occurs at c. +15 m HWM. In Rhum, as elsewhere in western Scotland, marine erosion during late-Devensian times was minimal, being restricted to the removal of older drift deposits the constituents of which shortly after were deposited as beach material (McCann and Richards 1969 p. 21), except perhaps about the time of the Loch Lomond Readvance when exceptionally cold conditions may have occurred (see below, and Sissons 1974a p. 46).

(*ii*) *Chronology*. The detailed chronology of late-Devensian marine penetration into western Scotland is not well known. According to Penny *et al.* (1969), the receding Devensian ice-front may have cleared the Solway Firth area by c. 13000 years B.P. (*cf.* Mitchell 1972 fig. 4), and the southern half of the island of Arran in the Firth of Clyde was free of Highland ice by that time, whilst by 12500 years B.P. the receding ice-front was clear of the north of Arran (Gemmell 1973 p. 31). More recently it was suggested, on the basis of biological evidence from the Grampian Mountains, that by 13000 years B.P. most of the ice had disappeared from Scotland, and total deglaciation may have occurred by 12500 years B.P. (Sissons 1974b p. 315), ideas that are consistent with the suggestion by Coope (1975 p. 164) that in Britain there was a relatively sudden change in climatic conditions from cold to very warm c. 13000 years B.P., a change which may have led to exceptionally rapid melting of the late-Devensian ice mass.

Radiocarbon dates for various types of biogenic material from a number of widely-distributed sites in western Scotland give a slightly more detailed indication of the chronology of marine penetration, but there still are many gaps and uncertainties. Evidence from a borehole (Number 71/9; Binns *et al.* 1974 p. 753 and fig. 1) sunk by the Institute of Geological Sciences a few kilometres east of Colonsay suggests that *relative* sea level at that location was at c. −60 m shortly after 16470 ±300 years B.P. (SRR–118, Harkness and Wilson 1974 p. 242; see also Binns *et al.* 1973 pp. 22–23), but there is a dearth of published information concerning the chronology of later Devensian marine transgression in the Hebridean area. The radiocarbon age of a marine shell bed, approximately at O.D., at a site near Lochgilphead, 12360 ±85 years B.P. (SRR–63, Harkness and Wilson 1974 p. 240), supports Gemmell's suggestions regarding marine incursion farther south in the Firth of Clyde, and indicates that the sea had penetrated into southern Loch Fyne by this time. Also, penecontemporaneously marine waters appear to have been present in the Paisley-Glasgow area, radiocarbon assay of shells from the Clyde Beds there giving dates of c. 12600 to 12400 years B.P. (Peacock 1971; SRR–62, Harkness and Wilson 1974 p. 239), but the mode of penetration of the sea into that area and the position of the ice-front at the time of marine penetration is not agreed (*cf.* Sissons 1974b p. 330 with Peacock 1971).

Whatever the detailed history of marine penetration in relation to ice-front retreat in the Glasgow area (and much remains to be investigated), it is certain that by c. 11800 to 11300 years B.P. the sea occupied the Dumbarton area (Peacock 1971), had penetrated the present Vale of Leven area and extended into the southern part of the Loch Lomond basin (Jardine 1973 pp. 166–167; see also Fig. 4) and, in the vicinity of Oban (Fig. 5), was present at South Shian on Loch Creran (Peacock 1971) and in Loch Spelve on Mull (Gray and Brooks 1972 p. 98).

4b. The low-level marine platforms of Western Scotland

Rock platforms, backed by steep cliffs, at levels ranging from a few metres below O.D. to c. +11 m to +12 m occur at numerous localities around the coasts of the mainland and islands of southwest and western Scotland. Some of the very small remnants of platforms may be the remains of Holocene marine-cut benches, but the widest and best-preserved remnants, occurring in the vicinity of Oban and on the Kintyre peninsula, commonly have been regarded as pre-Holocene in age since McCallien (1937 p. 197) suggested that marine erosion was minimal during the comparatively short high stands of relative sea-level of the Holocene epoch. The marine platforms at low elevations on the islands of Islay and Jura (McCann 1968 p. 28), Oronsay and Colonsay (Cunningham Craig *et al.* 1911 p. 64; Wright 1911 p. 102), Mull (Gray 1974), Rhum (McCann 1968 fig. II.1) and Skye (McCann 1968 fig. II.1; Richards 1969 p. 126), to mention the best known Hebridean examples, also are regarded now as being pre-Holocene in age although the authors who originally described them did not necessarily take this view. Until recently the platforms at all these locations were regarded as being pre-Devensian in age, the evidence for this being that in places the platforms show indications of having been scratched by glacier ice whilst elsewhere till is reputed to rest on their remnants. Gray (1974 pp. 86–87) suggested that the low-level platform of the Oban area was formed in an interglacial interval probably fairly late in the Pleistocene epoch. Previously McCann (1968 p. 29) had noted that the same feature need not necessarily be younger than the (presumed interglacial) high-level platforms of the Scottish Hebrides (discussed above).

Recently, Sissons (1974a pp. 44–45) took a more positive view by equating the time of formation of the rock platform of the Oban area with that of an extensive buried and/or submerged erosion feature, previously detected in southeast Scotland, which he claimed dates from immediately prior to c. 10500 to 10300 years B.P. i.e. about the time of the cold episode during which the Loch Lomond Readvance occurred on the western Scottish mainland. It was suggested that the critical factor leading to extensive platform-cutting at that time was the severe periglacial climate that is believed to have characterised the stadial represented by the deposits of the Loch Lomond Readvance. The rather scanty evidence suggesting a pre-Devensian age for the rock platform was dismissed on the grounds of the excellent preservation of the feature as a wide bench on the islands of Lismore, Kerrara and Mull and on the mainland near Oban (cf. Gray 1974) in an area which otherwise shows intense glacial erosion including a rock basin descending to −220 m.

Sissons's suggestion is attractive because it provides a solution to many problems which hitherto have remained unsolved but, if the suggestion is valid, it does not follow that *all* remnants of low-level platforms that occur in western Scotland are contemporaneous with the platform whose remnants are to be found in the Oban area. For example, in the northern part of the Outer Hebridean Island of Lewis, a marine-cut rock platform up to 150 m in width with associated cliffs occurs at heights between O.D. and +8 m. McCann (1968 pp. 30–33) hesitated to given an age to the platform, but von Weymarn (1974 pp. 61–111) claimed that it is pre-Devensian in age because in places the rock bench is covered by beach deposits (attributed to the Ipswichian interglacial interval) which occasionally are underlain by glacial drift deposits resting on the platform. In places the beach sediments are covered by or interstratified with soliflucted till, and elsewhere have been subjected to periglacial disturbance. In extension of the reasoning used by Sissons for the Oban area, it is arguable that the beach deposits in Lewis date from the time of the Loch Lomond Readvance rather than from the Ipswichian interval. The soliflucted till and other periglacial phenomena provide evidence of the effects of a late-Devensian cold episode, but the presence of undoubted glacial drift above the platform but below the beach deposits precludes contemporaneity of the rock platform and the gravels, implying that the platform dates from the Ipswichian or an earlier interglacial interval.

During the Loch Lomond Readvance the ice-front reached the sea at only a few places in western Scotland (cf. Sissons 1974b fig. 1) notably, in the vicinity of Oban, at the seaward ends of Lochs Etive and Creran, at the head of Loch Feochan (Fig. 5), and on the island of Mull where a local ice mass extended to the coast on the Sound of Mull and almost to the coast on the Firth of Lorne. Farther south ice occupied the northern parts of Loch Fyne and Loch Long (including Loch Goil), and occupied the greater part of Gare Loch north of Rhu (Fig. 4), so that broadly the extent of the sea at that time may have been slightly less than now. The relative height of the sea shortly after the maximum of the Loch Lomond Readvance, however, was about ten metres higher than now (slightly below +12 m to +14 m at Lochs Etive and Creran, slightly less than c. +11 m at Loch Feochan and in Mull; Gray 1975 p. 237), and there is evidence that at Girvan on the southern Ayrshire coast the sea may have occupied a small embayment about this time also (Jardine 1975 pp. 176–177), so that on aggregate the extent of the sea may not have differed greatly from the present despite the marked differences in coastal configuration in some places. In the Oban area, the maximal position of the ice-front at a short distance to the east of the remnants of the low-level rock platform, together with the absence of remnants of the platform within the mapped limit of the Loch Lomond Readvance in that area

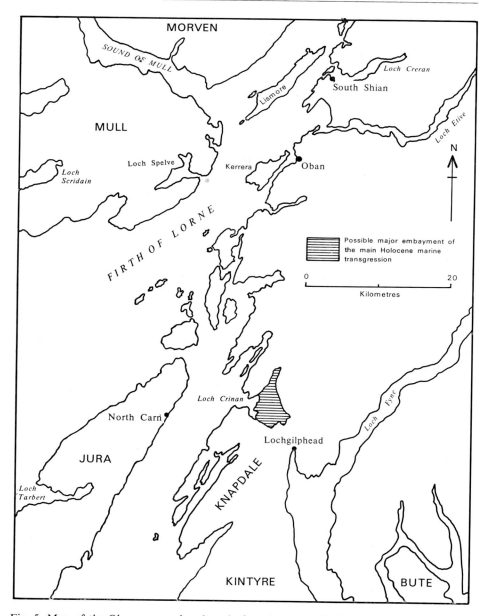

Fig. 5. Map of the Oban area and seaboard of northern Argyll, showing the main locations in that area mentioned in the text and the possible extent of the embayment east of Loch Crinan occupied by the sea at the maximum of the main Holocene transgression.

(*cf.* Gray 1974 fig. 1 with Gray 1975 fig. 1), lends support to Sissons' suggestion that the platform dates from the time of the ice readvance.

Evidence mainly from outside western Scotland shows that relative sea-level in Scotland dropped shortly after the end of the Loch Lomond Readvance, to be followed by a major marine transgression.

5.   The Holocene marine record

Numerous occurrences of beds of organic matter, some of which have been radiocarbon-dated or subjected to pollen analysis, and occasional tree stumps in position of growth at levels below present high water mark at localities widely distributed throughout the area (Sissons 1967 fig. 85; von Weymarn 1974 fig. 8–1) indicate that about the beginning of the Holocene epoch (c. 10000 years B.P.) the sea encroached less on the land that at present. Thereafter marine waters began to cross the present position of much of the coastline of the mainland of south-west Scotland and parts of the coastlines of the Hebridean islands in the course of the main Holocene marine transgression. At any particular location, the time at which the transgression commenced and culminated, and the rate of progress and lateral extent of marine penetration at any one time, depended mainly on the balance between eustatic water movement and contemporaneous (mainly isostatic-rebound) movements of the land area and to a lesser extent on the local occurrence of pre-existing natural barriers some of which were modified or destroyed in the course of transgression. As a result, the record of Holocene marine transgression and regression in southwest Scotland and the Scottish Hebrides exhibits diachronism from place to place, and the chronology of Holocene coastal events in the eastern Solway Firth area (for example) is not precisely the same as that in the Glasgow area, just as it is not identical in southwest Scotland, the Forth Valley and Cumbria (Jardine 1975 pp. 187–194).

Broadly, there is evidence that the shore-line of the sea at the local maximum of the main Holocene transgression was located a few tens of metres inland from its present position along much of the mainland coast from near Gretna at the eastern end of the Solway Firth, *via* the Firth of Clyde and the Clyde estuary, the eastern and western shores of the Kintyre peninsula and Knapdale to the vicinity of Oban. In some places, however, the sea penetrated considerably farther beyond the present coast. For example, embayments extending up to 10 km inland from the present shore-line were occupied to the southeast of Dumfries (the Lochar Gulf), at the heads of Wigtown Bay and Luce Bay in Galloway (Jardine 1975), in southern and central Ayrshire to the north of Girvan and both to the north and south of Troon, in the Glasgow area in the vicinity of Renfrew (Jardine 1971 pp. 109–112), probably to the east of Loch Crinan in Knapdale (Fig. 5), and possibly there was also brief penetration of the sea into the southern end of the Loch Lomond basin (J. Rose, personal communication).

Details of the chronology of marine transgression and regression are scarce except along the northern shore of the Solway Firth (Jardine 1975) and to a lesser extent along the Ayrshire coast and in the Clyde estuary (Bishop and Dickson 1970; Jardine 1971 pp. 110–112). In summary, along the Solway Firth shore, after local penetration beyond the present coastline at a number of places between c. 9400 and 7800 years B.P., more extensive marine transgression occurred c. 7500 to 7200 years B.P. In southern Ayrshire the main transgression took place shortly after 8400 years B.P., whereas in central Ayrshire it probably was somewhat later (at Troon organic detritus dated 8015±120 years B.P. is overlain by beach sands and gravels; Welin *et al.* 1975, p. 159), and in the Renfrew area initiation of marine penetration was c. 8000 years B.P. The time of commencement of regression of the Holocene sea after its maximum extension and height had been reached has not been determined as yet for most parts of the mainland coast of southwest Scotland. In the Solway Firth area, however, it has been shown that the sea was excluded from the Lochar Gulf c. 6600 years B.P., probably because of local growth of gravel and sand bars (Jardine 1975 p. 183), but it also is demonstrable that marine transgression continued in the eastern Solway Firth until c. 5600 years B.P., and at the head of Wigtown Bay the transgression may have continued until c. 5000 years B.P.

At many places on the mainland of southwest Scotland the location of the shore-line at the culmination of the main Holocene marine transgression was determined by the position of the junction of the pre-existing low-level rock platform and associated cliffs (*cf.* McCallien 1937 p. 197). Nevertheless, in some places along parts of the Solway Firth and Ayrshire coasts, the Holocene sea lapped against (Devensian) glacial till and apparently cut cliffs in the till (Jardine 1971 p. 107).

If, however, it is the case that the low-level coastal rock platform of much of southwest Scotland was formed at the time of the Loch Lomond Readvance, rather than in pre-Devensian times, as Sissons (1974a, p. 46) implied, it is possible that the Holocene sea lacked the power to erode even the unconsolidated till deposits which in places formed its shore, and the cliffs in till like the rock platform and rock cliffs may then be attributed to late-Devensian marine action. This suggestion requires testing, but perhaps there is some support for it in the former Lochar Gulf area and in the vicinity of Annan (at Newbie Cottages) on the northern seaboard of the Solway Firth (Fig. 2). At these locations medium- to fine-grained sediments of the main Holocene marine transgression rest directly on or above late-Devensian fluvio-glacial sand and gravel ridges and mounds (Jardine 1971, p. 108 and fig. 4), and there appears to have been virtually no erosion of the fluvio-glacial deposits by the Holocene sea which quietly penetrated these areas c. 7500 to 7200 years B.P. This is in marked contrast with the activity of the sea at present which is vigorously attacking the same esker ridges near Annan that were inviolate towards the culmination of the main Holocene marine transgression.

Less is known of the detailed chronology of Holocene coastal events in the Hebrides than of those on the mainland of southwest Scotland, although there are written accounts of the presence of "fossil" Holocene coastal deposits on many of the islands including, for example, Colonsay (Cunningham Craig *et al.* 1911, pp. 67–70), Mull (Bailey *et al.* 1924, pp. 410–411), Rhum (McCann and Richards 1969, p. 22) and Lewis (von Weymarn 1974, pp. 127–143). In the Inner Hebrides, at the maximum of the main transgression the extent of the Holocene sea commonly was greater than at present, but recent detailed observations by von Weymarn (1974) suggest that in Harris and Lewis there is no unambiguous evidence that the relative level of the sea around that island rose above its present level during the Holocene epoch, ridges of gravel on the northwest coast of Lewis located a few metres above, and extending a short distance inland from, the present shore being attributed to the action in the comparatively recent past, of violent storms similar to those which occasionally still occur in western Scotland.

There is evidence, however, in the form of now-submerged organic layers (e.g. in South Uist, Ritchie 1966; in Lewis, von Weymarn 1974) that in the Outer Hebrides during the Holocene epoch the rate of rise of sea-level at times slightly outpaced the rate of rise of land. Also, it has been suggested (Ritchie 1972, p. 33) that the marked eustatic rise of sea-level which took place broadly between 8400 and 5500 years B.P. was the major event responsible for sweeping vast quantities of (mainly shell-) sand and coarser-grained (mainly inorganic) fragmental debris landwards from the adjacent shallow continental shelf area to give the extensive beach deposits which now fringe many of the Hebridean islands.

In the Hebrides, currently the information that is available concerning the chronology of Holocene marine transgression is mainly that emerging from studies related to coastal occupation by early human communities. Briefly, at North Carn on the northeast coast of Jura (Fig. 5), raised (storm) beach gravels overlie charcoal dated 7414 $\pm$80 years B.P. (SRR–161, Harkness and Wilson 1974, p. 250) suggesting that the local maximum of the Holocene marine transgression was c. 7400 years B.P. Similarly, on the island of Oronsay, shells of *Arctica islandica* (Linné) from sites on the raised beaches adjacent to Mesolithic shell mounds yielded radiocarbon dates ranging between 7020$\pm$140 and 7610$\pm$150 years B.P. (Jardine 1974), suggesting that the Holocene marine transgression was in progress there approximately contemporaneously with the transgression in Jura. In Oronsay there is evidence also that occasional storms at c. 5800 years B.P. continued to cover parts of occupation sites (yielding charcoal that corroborates the shell-derived dates), implying that, although the sea may have begun to recede by that time, it had not receded far from or dropped much in height below its maximal position.

Both on the mainland and on the islands detailed information concerning the recession of the sea to its present position after the maximum of the main Holocene transgression is scanty. On Mull, Gray (1972) identified at least two shore-lines (of unknown age) below that of the main Holocene transgression, and in Dumfriesshire, Galloway and Ayrshire a shore-line at c. +5 to +6 m (compared with the main Holocene shore-line at c. +7 to +10 m) can be identified at numerous locations (Jardine 1975, p. 187). From radiocarbon-dated shell material from sites near the head of Wigtown Bay, it appears that the lower shore-line was formed c. 2000 years B.P. during a short pause in the regression of the sea to its present position.

6.   Conclusion

The Quaternary marine record in southwest Scotland and the Scottish Hebrides is exceedingly complex, and the areas for which information is available are rather unevenly distributed. Published data on the Holocene coastal deposits of the Oban area, Knapdale and Kintyre, together with data collected from the northern seaboard of the Solway Firth and the south Ayrshire coast are sufficient now to allow the history of coastal evolution in these areas during the last 10,000 years to be reconstructed with some confidence. Published information relating to the Holocene history of other parts of the coast still is inadequate, however, and only in a few areas, *e.g.* around Oban, is the late-Devensian marine record well known. The coastal history during earlier parts of the Quaternary sub-era requires intensive investigation, but unless hitherto undiscovered biogenic interglacial and/or interstadial deposits are found, the chronology that may be established is unlikely to be convincing.

Recent offshore investigations have added greatly to information concerning the stratigraphy of the Quaternary sediments on the continental shelf around and beyond the Hebridean islands and within the Firth of Clyde and Solway Firth. The detailed chronology of the submarine deposits, however, requires to be soundly established before the history of Quaternary marine transgression and regression across the shelf areas can be reconstructed in detail.

Relevant current research in the landward parts of the area is concentrated in the Hebridean islands of Jura, Colonsay and Oronsay and, on the mainland, in the Oban area, Knapdale, Kintyre, Cowal, the environs of Glasgow, Ayrshire and the Solway Firth. Publication of the results of this research will provide important details concerning the Quaternary history of the area. Broadly, however, it is clear on the basis of information presently available that the two main factors that influenced the evolution of the coast of western Scotland during the Quaternary sub-era were global changes in sea-level occasioned by growth and decay of ice masses, and the interaction of eustatic sea-level movements and regional movements of the land. In combination with local depositional features resulting from glacial and marine sedimentary processes, the two major factors mentioned above produced the intricate coastal changes that have been briefly discussed in this essay. It is possible that a third factor, intense cold in a periglacial regime, also played an important part in coastal development at the time of the Loch Lomond Readvance. This, however, remains to be proved.

Acknowledgments. Grateful thanks are accorded to Professor T. Neville George and Dr. R. J. Price who read the manuscript and offered helpful advice towards the improvement of the text. Dr. J. von Weymarn kindly made available a copy of his thesis for consultation.

References

BAILEY, E. B., CLOUGH, C. T., WRIGHT, W. B., RICHEY, J. E. and WILSON, G. V. 1924. Tertiary and post-Tertiary geology of Mull, Loch Aline and Oban. *Mem. geol. Surv. U.K.*

BINNS, P. E., HARLAND, R. and HUGHES, M. J. 1974. Glacial and post-glacial sedimentation in the Sea of the Hebrides. *Nature, Lond.* **248**, 751–4.

——, MCQUILLAN, R. and KENOLTY, N. 1973. The geology of the Sea of the Hebrides. *Rep. Inst. geol. Sci.* No. 73/14.

BISHOP, W. W. and DICKSON, J. H. 1970. Radiocarbon dates related to the Scottish Late-glacial sea in the Firth of Clyde. *Nature, Lond.* **227**, 480–2.

BRADY, G. S., CROSSKEY, H. W. and ROBERTSON, D. 1874. Post-Tertiary Entomostraca of Scotland. *Palaeontogr. Soc. [Monogr.]*.

COOPE, G. R. 1975. Climatic fluctuations in northwest Europe since the Last Interglacial, indicated by fossil assemblages of Coleoptera. *In* Wright, A. E. and Moseley, F. (Editors). *Ice Ages: Ancient and Modern*, 153–68. *Geological Journal Special Issue* No. 6.

CUNNINGHAM CRAIG, E. H., WRIGHT, W. B. and BAILEY, E. B. 1911. The geology of Colonsay and Oronsay, with part of the Ross of Mull. *Mem. geol. Surv. U.K.*

CURRAY, J. R. 1965. Late Quaternary history, continental shelves of the United States. *In* Wright, H. E. and Frey, D. G. (Editors). *The Quaternary of the United States*, 723–35. Princeton University Press, Princeton.

DEEGAN, C. E., KIRBY, R., RAE, I. and FLOYD, R. 1973. The superficial deposits of the Firth of Clyde and its sea lochs. *Rep. Inst. geol. Sci.* No. 73/9, 42 pp.

DOBSON, M. R. and EVANS, D. 1975. Glacial and post-glacial sedimentation in the Malin Sea. Unpublished contribution, *Geol. Soc. Lond.* 22 October 1975.

DREIMANIS, A. and KARROW, P. F. 1972. Glacial history of the Great Lakes—St. Lawrence Region, the classification of the Wisconsin(an) Stage, and its correlatives. *Proc. Int. geol. Congr.* **24** (12), 5–15.

GEMMELL, A. M. D. 1973. The deglaciation of the island of Arran, Scotland. *Trans. Inst. Br. Geogr.* **59**, 25–39.

GODWIN, H. 1963. Isostatic recovery in Scotland. *Nature, Lond.* **199**, 1277–8.

GRAY, J. M. 1972. *The inter-, late- and post-glacial shorelines, and ice-limits of Lorn and eastern Mull.* Unpublished Ph.D. thesis, University of Edinburgh.

—— 1974. The main rock platform of the Firth of Lorn, western Scotland. *Trans. Inst. Br. Geogr.* **61**, 81–99.

—— 1975. The Loch Lomond Readvance and contemporaneous sea-levels in Loch Etive and neighbouring areas of western Scotland. *Proc. Geol. Ass.* **86**, 227–38.

—— and BROOKS, C. L. 1972. The Loch Lomond Readvance moraines of Mull and Menteith. *Scott. J. Geol.* **8**, 95–103.

HARKNESS, D. D. and WILSON, H. W. 1974. Scottish Universities Research and Reactor Centre Radiocarbon Measurements II. *Radiocarbon* **16**, 238–51.

JARDINE, W. G. 1966. Landscape evolution in Galloway. *Trans. J. Proc. Dumfries. Galloway nat. Hist. Antiq. Soc.* **43**, 1–13.

—— 1971. Form and age of late Quaternary shore-lines and coastal deposits of south-west Scotland: critical data. *Quaternaria* **14**, 103–14.

—— 1973. The Quaternary geology of the Glasgow district. *In* Bluck B. J. (Editor). *Excursion guide to the geology of the Glasgow district*, 156–69. Geological Society of Glasgow.

—— 1974. Oronsay shell mounds. *In*, Stewart, M. E. C. and Lythe, C. M. (Editors). *Discovery and Excavation in Scotland 1973*, 9–10. C.B.A. Scottish Regional Group.

—— 1975. Chronology of Holocene marine transgression and regression in south-western Scotland. *Boreas* **4**, 173–96.

MCCALLIEN, W. J. 1937. Late-glacial and early post-glacial Scotland. *Proc. Soc. Antiq. Scot.* **71**, 174–206.

MCCANN, S. B. 1964. The raised beaches of north-east Islay and western Jura. *Trans. Inst. Br. Geogr.* **35**, 1–16.

—— 1968. Raised shore platforms in the western isles of Scotland. *In*, Bowen, E. G., Carter, H. and Taylor, J. A. (Editors). *Geography at Aberystwyth*, 22–34. University of Wales Press, Cardiff.

—— and RICHARDS, A. 1969. The coastal features of the island of Rhum in the Inner Hebrides. *Scott. J. Geol.* **5**, 15–25.

MACGREGOR, M. 1941. The Leven valley, Dumbartonshire. *Trans. geol. Soc. Glasg.* **20**, 121–35.

MILLIMAN, J. D. and EMERY, K. O. 1968. Sea levels during the past 35,000 years. *Science* **162**, 1121–3.

MITCHELL, G. F. 1972. The Pleistocene history of the Irish Sea: second approximation. *Scientific Proc. R. Dublin Soc.* **4**, 181–99.

PANTIN, H. M. 1975. Quaternary sediments of the northeastern Irish Sea. *Quaternary Newsletter* **17**, 7–9.

PEACOCK, J. D. 1971. Marine shell radiocarbon dates and the chronology of deglaciation in western Scotland. *Nature phys. Sci., Lond.* **230**, 43–5.

PENNY, L. F., COOPE, G. R. and CATT, J. A. 1969. Age and insect fauna of the Dimlington Silts, East Yorkshire. *Nature, Lond.* **224**, 65–7.

RICHARDS, A. 1969. Some aspects of the evolution of the coastline of north east Skye. *Scott. geogr. Mag.* **85**, 122–31.

RITCHIE, W. 1966. The post-glacial rise in sea-level and coastal changes in the Uists. *Trans. Inst. Br. Geogr.* **39**, 79–86.

—— 1972. The evolution of coastal sand dunes. *Scott. geogr. Mag.* **88**, 19–35.

SISSONS, J. B. 1967. *The evolution of Scotland's scenery.* Oliver and Boyd, Edinburgh.

—— 1974a. Late-Glacial marine erosion in Scotland. *Boreas* **3**, 41–8.

—— 1974b. The Quaternary in Scotland: a review. *Scott. J. Geol.* **10**, 311–37.

VON WEYMARN, J. 1974. *Coastline development in Lewis and Harris, Outer Hebrides, with particular reference to the effects of glaciation.* Unpublished Ph.D. thesis, University of Aberdeen.

WELIN, E., ENGSTRAND, L. and VACZY, S. 1975. Institute of Geological Sciences Radiocarbon Dates VI. *Radiocarbon* **17**, 157–9.

WRIGHT, W. B. 1911. On a preglacial shoreline in the western isles of Scotland. *Geol. Mag.* **48**, 97–109.

—— 1928. The raised beaches of the British Isles. In: *First Report of the Commission on Pliocene and Pleistocene Terraces*, 99–106. International Geographical Union.

—— 1937. *The Quaternary Ice Age* 2nd Edition. MacMillan and Co. Ltd., London.

W. G. Jardine, Department of Geology, The University, Glasgow, G12 8QQ.

The coasts of northwest England

D. Huddart, M. J. Tooley and P. A. Carter

The evidence for and nature of the glacigenic sediments is presented and four phases of glaciation during the Late Devensian are recognised in Cumbria. Flandrian marine sediments from Lancashire and Cumbria are briefly described and marine transgression sequences are correlated. A comparison is made with the correlation table for the Quaternary in northwest England and modifications are recommended.

1. Introduction

The coasts of northwest England can be subdivided into a Cumbrian unit and a Lancashire unit (Fig. 1). In Cumbria, the coast is backed by lowland, which is broadest in the north, where it is drained by rivers flowing to the Solway Firth, and decreases in width as it extends south and west around the edge of the Lake District fells. In the south, it is discontinuous and is confined to the river valleys discharging into Morecambe Bay. Further south, in Lancashire, the relief is dominated again by coastal lowland, extensive in the Fylde and southwest Lancashire.

This area contains few topographic features that have not been the result of Quaternary drift deposition, although most belong to the Devensian glaciation. There are occasional exceptions such as the Triassic Sandstone St. Bees Head, the modest sandstone scar at Cockersand Abbey and the Carboniferous moorlands and scars of west and south Cumbria. However, even these exceptions have been modified by glacial processes. The mostly glacigenic drift deposition often attains

J

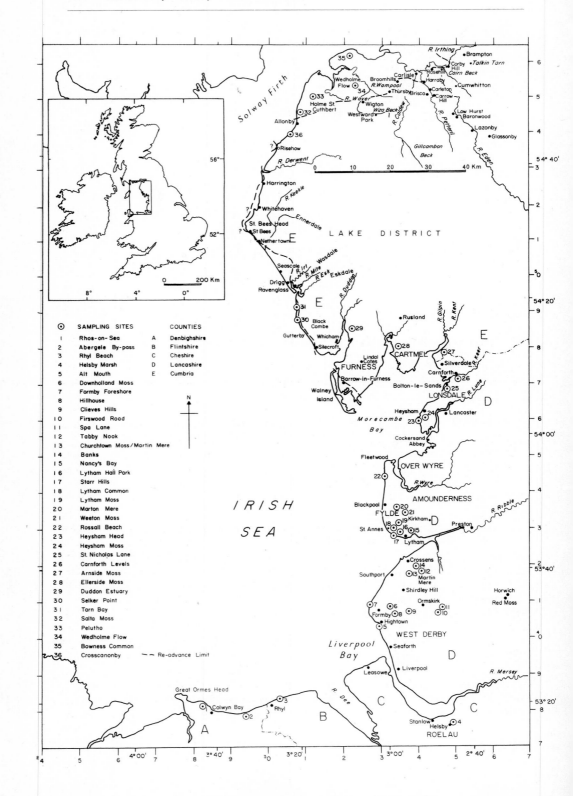

SAMPLING SITES

COUNTIES

1	Rhos-on-Sea
2	Abergele By-pass
3	Rhyl Beach
4	Helsby Marsh
5	Alt Mouth
6	Downholland Moss
7	Formby Foreshore
8	Hillhouse
9	Clieves Hills
10	Firswood Road
11	Spa Lane
12	Tabby Nook
13	Churchtown Moss/Martin Mere
14	Banks
15	Nancy's Bay
16	Lytham Hall Park
17	Starr Hills
18	Lytham Common
19	Lytham Moss
20	Marton Mere
21	Weeton Moss
22	Rossall Beach
23	Heysham Head
24	Heysham Moss
25	St. Nicholas Lane
26	Carnforth Levels
27	Arnside Moss
28	Ellerside Moss
29	Duddon Estuary
30	Selker Point
31	Tarn Bay
32	Salta Moss
33	Pelutho
34	Wedholme Flow
35	Bowness Common
36	Crosscanonby

A   Denbighshire
B   Flintshire
C   Cheshire
D   Lancashire
E   Cumbria

— — Re-advance Limit

considerable thicknesses: in Lower Wasdale at Haggs Wood there is 86·6 m of drift; at Blackpool the till reaches maximum thicknesses of 77·0 m and at Southport 32·3 m. Adjacent to the coast, thick accumulations of Flandrian minerogenic and biogenic sediments have been recorded: in Morecambe Bay, marine sands, silts and clays attain thicknesses of 25·3 m whereas tidal flat and lagoonal deposits on Lytham Common attain thicknesses of 16·9 m and on Downholland Moss 19·0 m. Enclosed basins in the till, isolated from any marine effects during the Flandrian Stage, such as Marton Mere, record more than 9·0 m of gyttja of late Devensian and Flandrian age.

There have been periods of intensive study of the Quaternary deposits in this area, such as the late Victorian period, the 1920's and 30's and from the 1960's onwards. Hence the literature of this area is characterised by a few relevant works, such as Goodchild (1875, 1887), Hollingworth (1931), Huddart (1970), Huddart and Tooley (1972), Oldfield (1960, 1963), de Rance (1871a, 1871b), Reade (1871) Tooley (1974, 1976), Trotter (1929) and Walker (1966) and the Geological Survey Memoirs.

Early studies of the drifts of northwest Lancashire and parts of Cumbria were made by Mackintosh (1878), de Rance (1871a, 1871b), Smith (1877) and Kendall (1879, 1881). All these workers incorporated the idea of a marine origin for at least part of the tripartite sequence that was generally recognised: a Lower and Upper Boulder Clay separated by Middle Drifts. However, the work of Goodchild (*op cit.*) in Edenside was far in advance of its time. He stressed the complexity of the deposits and considered that practically all were formed subglacially or englacially during the melting of a stagnant ice sheet.

Smith (1912, 1931) corroborated the threefold drift division but stressed both the lateral and vertical variability in character and height of the deposits. He described the Whicham valley, Duddon, Eskdale, Miterdale and Wasdale glacial lakes and the marginal channels on the slopes of Black Combe. The Lower Boulder Clay was described as the ground moraine of the Main Glaciation; sands and gravels accumulated in embayments in the ice margin and the Upper Boulder Clay was the result of a limited readvance of the Irish Sea ice.

The glaciation of Eastern and Western Edenside, the Alston Block and the Carlisle plain, including evidence of retreat stages and glacial lakes both during the Main and Scottish Readvance glaciation was discussed by Trotter (1929) and Hollingworth (1931). Both workers were members of the Geological Survey team and the Memoirs (1926, 1930, 1931, 1932a, 1937 and 1968) added much local detail. However, the glacial model into which the landforms and sediments were fitted was one of a continuously retreating ice front, with extraglacial deposition and little importance was attached to subglacial or englacial processes. For example, glacial drainage channels were mapped as lake overflows and many glacial lakes were suggested on little evidence.

Since the Geological Survey Memoirs have been published, little further systematic work was undertaken in Cumbria until Huddart (1970, 1971a, 1971b, 1973 and in the press), using sedimentological and geomorphological techniques established the ice marginal limits of a readvance stage, identified as the Scottish

Fig. 1. Map of northwest England to show the main sampling sites and locations referred to in the text.

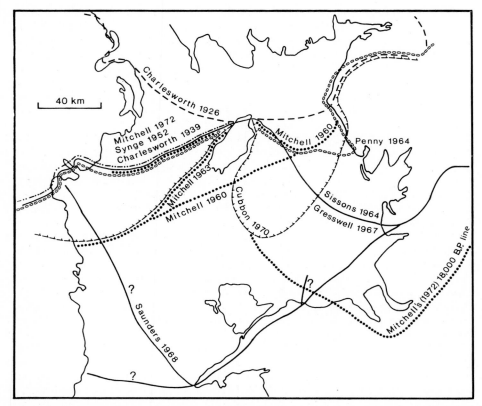

Fig. 2. Scottish Re-advance Limits in northwest England, the Isle of Man and Ireland according to several authors.

Readvance, and reconstructed the glacigenic environments associated with this re-advance and those associated with the decay of the Main Glaciation ice sheet.

Work on sea-level changes, vegetational history, climatic change and the influence of man on plant communities since the end of the last glaciation has been undertaken in Cumbria by Walker (1956, 1966). This work was extended south to the Duddon estuary (Huddart and Tooley 1972; Pennington 1970, 1975) and was complemented in southern Cumbria by the contributions of Smith (1958, 1959), Oldfield (1960, 1963), Oldfield and Statham (1963) and Dickinson (1973, 1975) on sites around Morecambe Bay.

In the Lancashire coastal plain work on glacigenic sediments has been confined both in extent and scope (de Rance 1877; Wray and Cope 1948; Gresswell 1967). A tripartite division of glacigenic sediments into an Upper Boulder Clay, Middle Sands and Lower Boulder Clay was recognised in the nineteenth century. Whether they constitute evidence for different glaciations or stages of the same glaciation is open to speculation, but recently Evans and Arthurton (1973) have suggested that the Lower Boulder Clay is a lodgement till and that the Upper Boulder Clay being less compact and lacking lodgement till at its base may represent the same phase of deglaciation. The age, extent and nature of the landforms associated with glacial deposition are not well-known in this area. Gresswell (1967) has suggested that the till sheet in southwest Lancashire is attributable to the Main Glaciation, whereas the till ridge bisecting the Fylde from east to west and culminating in the 33 metres high cliffs at Blackpool is a terminal feature associated with the culminating stages of the last glaciation. He tentatively correlates this Kirkham End Moraine with the Bride Hills Moraine in the Isle of Man, and suggests that it is contemporaneous with the Scottish Readvance.

Evidence of sea-level changes pre-dating the Flandrian in northwest England is limited. At Wigton, in the Solway Lowlands, a borehole proved lenses of drab clay, containing *Turritella communis* Risso, foraminifera and ostracods, beneath gravel and till. This marine clay is placed provisionally in the Ipswichian but, unfortunately, there were gaps in the core, both above the clay and between the clay and the underlying Stanwix Shales: it is possible that the clay was not *in situ* and was an erratic within the till. In southwest Lancashire, inferential evidence of early Quaternary sea-levels may be derived from the network of river valleys in the sub-drift surface of southwest Lancashire and north Cheshire (Howell 1973). However, the long profiles of the valleys are ungraded and Howell suggests that over-deepened hollows may have been eroded sub-glacially during one or more of the Quaternary glaciations. Most of these valleys have become filled with glacigenic sediments, but along the coasts of the Fylde and southwest Lancashire a sufficiently accented relief existed at the end of the Devensian Glaciation for marine sedimentation to proceed from the early part of the Flandrian Stage onwards. From these areas, a graph of sea-level movements has been derived and a Flandrian marine chronology established (Tooley 1969, 1970a, 1970b, 1971, 1973, 1974, 1976).

The geometry of the northeast quadrant of the Irish Sea Basin and the alignment of the present coasts of northwest England is explained by structure, inheritance, glaciation, marine sedimentation, sea-level movements and contemporary coastal processes. In this contribution a summary of the evidence for and nature of the glacigenic and marine sediments is considered and regional correlations suggested. A comparison is made with the correlation table for the Quaternary of north-west England (Evans and Arthurton 1973) and modifications are recommended.

2. Glacigenic Sediments

In Cumbria four glacial phases have been recognised, each producing a till or till/fluvioglacial complex. They have been referred to local stages: Early Scottish; Main Glaciation; Furness readvance and Scottish Readvance (Trotter and Hollingworth 1932b; Huddart 1971b). In Lancashire the North British or Kirkham readvance was recognised by Gresswell (1967).

2a. Deposits pre-dating the Late Devensian

i. Glacial sediments older than the Devensian. There are scattered basal till units from the Early Scottish glacial recognised at Willowford, Glassonby, Gillcambon Beck, Westward Park, the Derwent valley and below organic sediments in Low Furness (Trotter 1929; Eastwood *et al.* 1968; Huddart 1971b). In Edenside a small percentage of Scottish granites has been observed in the Main Glaciation drift at Carrow Hill, Baronwood and Lazonby. As it is considered that ice move ment was from south to north at this time these erratics must be derived from an earlier Scottish ice advance (Huddart 1970, 1971b).

This till unit could have been deposited by Wolstonian Scottish ice, although there does not seem to be evidence of complementary activity from the Lake District. An alternative could be that during the Late Devensian ice built up in Scotland much more quickly than in the Lake District and enabled an ice sheet to penetrate the Lake District foothills before ice moved out from the local mountain core. However, evidence from west Cumbria presented later suggests that, at least in that area, this hypothesis is not true. However, it seems that the basal till of Low Furness must be Wolstonian in age but the chronology of the other sites is speculative and can only be placed justifiably as pre-Main Glaciation.

ii. Ipswichian sediments. Three sites with organic sediments, and a marine clay beneath Devensian glacigenic sediments are recorded from Cumbria. The Wigton marine clay has been referred to already. At Lindal Cotes and Crossgates in Low Furness thicknesses of up to 8 m of organic sediments have been recorded between two tills (Bolton 1862; Kendall 1881). This unit is regarded tentatively as Ipswichian but an attempt to relocate it by an IGS borehole proved negative (Anon. 1972). However, the pollen from organic layers in silts beneath a till at Scandal Beck seem to be Ipswichian (Carter, unpublished). At first a finite date of 36300^{+2160}_{-1700} B.P. (Birm. 161a) was obtained but later samples gave infinite dates of over 32500 and over 42000 B.P. (Birm. 245) (Shotton *et al.* 1970; Shotton and Williams 1971, 1973).

iii. Mid-Devensian interstadial. At Low Hurst an undated organic sand underlies the main glacial complex (Evans and Arthurton 1973) and its associated sands and gravels are presumed synchronous.

2b. Late Devensian sediments

i. Main Glaciation tills. These have been discussed in detail by Huddart (1970). Away from Edenside two tills in the same stratigraphic sequence are rare, except for the St. Bees and Black Combe coastal sections, and have only been recorded in the Ehen and Keekle valleys. In these valleys a basal grey Lake District till is overlain by a red till derived from the north and northwest. It seems that in these western foothill fringes the valley glaciers reached far into the lowland before they were incorporated into an ice sheet composed of the northern Lake District valley glaciers and the Scottish ice sheet. Both till units are basal tills from the Main Glaciation maxima which can be traced over much of the Cumbrian lowland from Edenside to West Cumbria by the drumlin belt. The orientation of this belt together with till fabric orientations and erratic transport studies show that ice flow in Edenside was from SE–NW, with a gradual orientation change to E–W north of the Lake District, to NE–SW in the west coastal lowland (Huddart 1970). Below the Lake District till in the Derwent valley is a thick sequence of proglacial sandur sediments laid down as the valley glacier was advancing down the pre-glacial valley.

ii. Landforms created by the Main Glaciation ice decay. In Edenside it has been shown that this ice stagnated *in situ* during deglaciation in the Eden, Petteril and Caldew valleys (Huddart 1970, in press). Compressive flow would cause basal ice layers to rise to counteract the Pennine obstruction and even though the ice was not cut off from its source, shear planes would bring debris to the ice surface during the early stages of deglaciation. At any one time period a narrow (c. 7·5 km) marginal belt seems to have stagnated *in situ* with common examples of ice marginal fluvial and lacustrine; subglacial; open crevasse and ice-walled lacustrine fluvioglacial depositional environments.

The Brampton 'kame' belt illustrates the mode of deglaciation in this part of Cumbria. The highest major landform is the Hallbankgate esker system (200–220 m) which trends from Castle Carrock reservoir to the South Tyne drainage. Between this stage and Talkin Tarn to the west there is a series of kame terraces marking marginal positions of the downwasting ice. The major depositional stage is at 133–143 m where there are flat-topped kames, hummocky kame belts, moulin kames and both linear and sinuous ridges. A further lower depositional tract occurs west of this at 66–100 m where there are many linear ridges which have been interpreted as open crevasse fluvial deposits.

Drainage associated with this progressive decay can be divided into several landform systems: at the 200–220 m stage drainage was from Edenside to the South Tyne; at the 133–200 m stages it was through the esker system from Kirkcambeck through the Gilsland col into the Tyne drainage; all later meltwater flowed to the west eroding the channel systems focussing on the Wampool, Wiza Beck, Waver, Caldew, Petteril, Eden, Cairn Beck and Irthing megachannels.

In the Eden, Petteril and Caldew valleys below 66 m the ice sheet stagnated in deep valleys between the Pennines and the Triassic ridge north-west from Penrith; between this ridge and the drumlinised higher ground west of the Petteril and again between drumlinised higher ground between the two valleys.

Similar deglaciation stages can be seen in the Ennerdale, Wasdale and Black Combe/Waberthwaite Fell areas where many overflow channels and lake beaches have been reinterpreted as subglacial meltwater channels and ice-marginal terraces (Smith 1967; Huddart 1967). However, in the Carlisle plain there is evidence that the ice sheet retreated westwards with an active ice-front depositing a series of proglacial, lacustrine sediments between 66–33 m (Huddart 1970).

2c. Later Readvances

i. The problem of the Scottish Readvance. This Devensian phase was first recognised by the Geological Survey in the 1920's as a distinct readvance of ice from Southern Scotland onto the Cumbrian lowland. The ice was thought to have dammed up lakes and deposited an upper till sheet, although Trotter (1929) stated that unless there were underlying sands the 'Upper Boulder Clay' could not be recognised from the 'Lower Boulder Clay'. Since the 1920's many workers have drawn much differing, generalised lines for the marginal limits of this readvance in Northern Britain as can be seen from Figure 2. (Trotter and Hollingworth 1932; Charlesworth 1926, 1939; Synge 1952; Mitchell 1960, 1963; Penny 1964; Sissons 1964; Gresswell 1967; Saunders 1968 and Walker 1966) whilst others do not recognise such a phase at all (Pennington 1970; Evans and Arthurton 1973; Sissons 1974).

Mitchell (1972) suggests that many of our much fought after 'advances' and 'retreat stages' are probably largely illusory and depend as much on personal whim as on field evidence. He correlated the Bar Hill-Wrexham moraine with the Bride Hills moraine but there is no indication of a readvance limit in Cumbria.

With the background of uncertainty as to the existence or limits of such a readvance phase the criteria used to establish the validity of this glacial readvance are presented. The best approach is a study of both landforms and stratigraphy in order to establish the depositional environment of any particular sedimentary association. This was attempted in coastal Cumbria, in sand and gravel pits and in motorway excavations in the Carlisle plain. This work has established well defined proglacial depositional environments. If these proglacial morpho-statigraphic units can be shown to have been associated with an advancing ice sheet rather than a retreating one and their marginal limits mapped, then the validity and extent of a readvance can be established. The problem is both strati-graphic and genetic and proof depends both on the validity of distinguishing criteria and the origin of glacial sediments defined as either Main Glaciation or Scottish Readvance in age.

ii. Depositional environments associated with the Scottish Readvance. Evidence has been presented in a series of papers by Huddart (1971a, 1971b, 1973, in press) and Huddart and Tooley (1972) that readvance of Irish Sea ice did occur in the Cumbria lowland. The following depositional environments were associated with the readvance ice:

1. proglacial lacustrine with overlying till in the eastern Carlisle plain
2. subglacial till in the western Carlisle plain
3. proglacial sandur at Broomhills in the eastern Solway lowlands
4. subglacial esker at Thursby in the eastern Solway lowlands
5. proglacial lacustrine at Holme St. Cuthbert in the Solway lowland
6. proglacial sandur at Harrington in the West Cumbria coastal plain
7. terminal morainic at St. Bees and in the West and South Cumbrian coastal plain from Nethertown to Seascale
8. proglacial lacustrine in lower Wasdale
9. terminal morainic along the Black Combe section of the South Cumbria coastal plain.

The limit for the readvance has been established by mapping the morphological expression of these sedimentary associations. At each locality the stratigraphy and lithology of the units was recorded and the depositional environment deter-mined by analysing the grain size, fabric and the lateral and vertical changes in the sediment.

iii. The Validity of the Upper Till as a Basal Readvance Till. The stratigraphic position of the upper till in the Cumberland lowland was the fundamental difference between the Main Glaciation basal till and the Scottish Readvance basal till of the early Geological Survey workers and was the initial basis for subdividing the two tills into two distinct time periods of formation. However, upper tills are not always automatically basal in origin as was thought by these workers in the 1920's. Recently it has been appreciated that upper tills could be ablation tills, flow tills or englacial tills (Boulton 1972) and because two till units are separated by 'Middle Sands' it does not mean that the sequence denotes a threefold ice advance-retreat-advance sequence. All can be deposited from the decay of a complex ice sheet and the fact that there is an upper till in the Cumberland lowland will not stand by itself as evidence for an ice readvance. However, the upper till does have certain characteristics which indicate that it is of basal origin. The till, especially in the Carlisle plain, shows incorporation of the underlying lake sediments which are disturbed, as at Brunstock, the new Eden Bridge, Tower Farm, Scotby and the Rosehill interchange (Huddart 1971b). The till matrix is composed of silt and clay and there is an almost indistinguishable junction between the till and the clay/silt, although the latter is complexly con-torted in the upper 50 cm. West of the new Eden Bridge, where the upper till overlies bedrock, the till shows an irregular, erosional contact with the underlying white, Kirklinton sandstone and the lowest 60 cm of till has a sand matrix with white sandstone clasts.

The fact that the upper till seems to be of regional significance with a relatively uniform thickness, as in the Carlisle plain and along the Black Combe coastal plain, indicates that the processes which led to its deposition are not just of local importance. The Black Combe upper till units have a length thickness ratio of 1000:1 which indicates a tabular to blanket geometry in Krynine's (1948) shape categories. Also in local areas the till fabric orientations give consistently similar results.

The dominant orientation is W–E or NW–SE, with an imbrication indicating deposition from the west. These two facts—the till's regional importance and consistent till fabric orintations suggest that the upper till is of basal origin Other characteristics which together suggest that the till is of basal origin are that it has a consistently fine grain size which gives much lower values of sand/matrix ratio than the Main Glaciation till unit (Huddart 1971a); its stratigraphic position, especially in south and west Cumberland, where it is above a proglacial fluvial sequence which indicates an advancing ice front and its lithological suite which is composed dominantly of southern Scottish erratics, together with lithologies which indicate some erosion of Main Glaciation sediments by an advancing ice sheet. It is realised that flow tills could be associated with an advancing ice front and could overlie proglacial sediments but it has already been noted that thin till units (up to 78 cm thick) are occasionally associated with proglacial sandur sediments along the Black Combe coast. These gravels are overlain by much thicker till units which are interpreted as basal in origin.

iv. Readvance Marginal Limits in Cumbria. The precise ice-marginal limits of the readvance of the Irish Sea basin ice sheet are difficult to establish accurately in many parts of Cumbria but are approximately indicated in Figure 1. This is especially the case in the Carlisle plain where the readvance ice deposited only a thin till sheet and did not form a terminal morainic landform. In the Carlisle area the readvance ice is thought to have advanced into the topographically low area, spreading out as a relatively thin lobe. The ice seems to have reworked the marginal slopes but did not override the Rosehill deltas which reach a maximum height of 106 feet (31·8 m) O.D. An upper till has been located as far east as Greenholme, near Corby Hill (Huddart 1971b) which is probably the maximum distance east that the effects of the readvance ice can be traced in the Cumberland lowland. Thus, Trotter's (1929) readvance margin between Lanercost, Brampton and Cumwhitton is thought to be too far east. Similarly in the Eden, Petteril and Caldew valleys, outside the Carlisle plain, no evidence of an upper till unit can be found. In the Petteril valley, where the best exposures were available, the upper till has not been traced further south than Harraby. Within the valley proper the glacigenic sediments and their interpreted depositional environments at Carlton (ice-marginal fluvial), Brisco (ice-marginal deltaic) and Carrow Hill (ice-walled lacustrine) all suggest that they were formed during the decay of the 'Main Glaciation' icesheet (Huddart 1970). Hence, no upper, readvance till has been located in these valleys which join together in the Carlisle plain where perhaps topographically it might have been expected. The most likely reason for its absence is that these valleys were still occupied by stagnant 'Main Glaciation' ice during the time period when the ice readvanced into the Carlisle plain.

Southwest of the Carlisle plain the readvance limit is difficult to trace, although

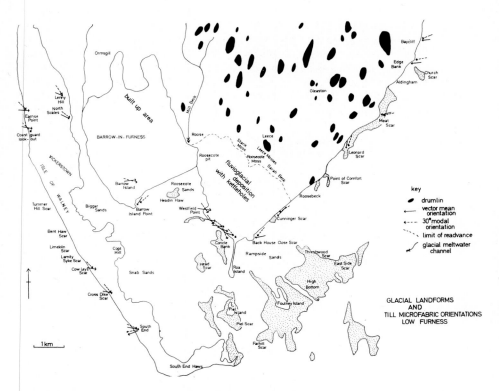

Fig. 3. Glacial landforms and till microfabric orientations in Low Furness.

proglacial braided river sediments exposed at Broomhills were certainly associated with the readvance, as was the esker system around Thursby (Huddart 1973). The deltas around Wigton, discussed in Eastwood *et al.* (1968) could be associated with a lake ponded up by the readvance ice front. The Holme St. Cuthbert fluvio-glacial complex was formed at the outer margin of the readvance ice but south of this depositional zone the ice probably did not reach the present day coast, except perhaps at Risehow, where there is a high percentage of Criffel granite in the till and there is a till fabric orientation indicating ice movement from the northwest (Huddart 1971), until the region of Workington. In this area there is the well-defined sandur at Harrington which was deposited in association with an advancing ice sheet.

South of Harrington the readvance ice did not override the present day coast because the Whitehaven and St. Bees Sandstone cliffs reach between 200–462 feet (60–138·6 m) O.D. The next positive readvance ice marginal position is at St. Bees, at the southern end of the Whitehaven—St. Bees valley. Here there is the proglacial sequence and upper till which form the terminal landform of the readvance ice sheet.

South of St. Bees the readvance ice probably did not reach inland until south of Coulderton where a belt of 'kame and kettle' topography is the principal feature between Nethertown and Drigg. The stratigraphy reveals proglacial

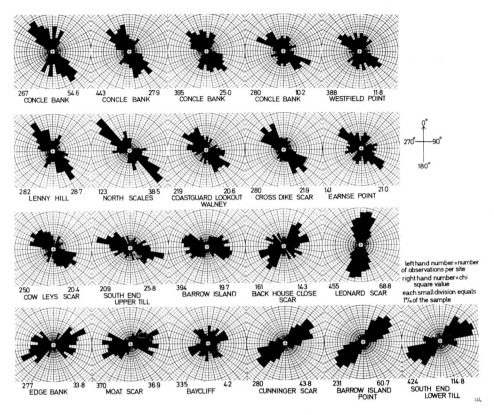

Fig. 4. Till microfabric distributions in Low Furness.

outwash sediments overlain, in cases, by an upper till unit. The marginal position of the readvance ice is thought to be represented by the deltas in lower Wasdale. Around and south of Ravenglass the readvance ice is not considered to have reached as far as the Lake District foothills but a terminal morainic position can be established along the narrow, coastal plain between the River Annas and Silecroft.

The Geological Survey workers in the 1920's and Eastwood *et al.* (1968) considered that the 'Scottish Readvance' ice reached as far as the 400 feet (120 m) contour but this contour limit is considered much too high and too far east in the Cumbria lowland. Maximum heights of landforms associated with the readvance sediments are from north to south around 90 feet (27 m) in the Carlisle plain, 191 feet (57·3 m) at Thursby, 157 feet (47·9 m) in the Holme St. Cuthbert area, just over 100 feet (30 m) at Harrington and St. Bees, 214 feet (68·3 m) in lower Wasdale and 181 feet (55·2 m) at Gutterby Spa along the Black Combe coastal plain. This height difference depended on local conditions and how far east the readvance ice had reached. It is not considered that the whole northern Irish Sea basin was deglaciated during the interval between the 'Main Glaciation' and the readvance as was suggested by Trotter (1929). It is thought that a climatic change resulted in a greater snow accumulation over a time period in the Southern

Table 1. Sand/matrix ratios for the Low Furness tills

Lower Till			Upper Till		
n = 8	S/M	Total	n = 13	S/M	Total
Baycliff	0·43	1041	Lenny Hill	0·195	1375
Moat Scar	0·865	854	North Scales	0·571	1301
Barrow Island Point	0·748	804	Earnse Point	0·516	549
Back House Close Scar	0·872	807	Concle Bank (18)	0·41	862
Edge Bank	0·649	861	Concle Bank (15)	0·614	807
Leonard Scar	0·57	697	Concle Bank (9)	0·256	1154
Cunninger Scar	1·303	896	Barrow Island	0·653	858
South End	0·899	1094	Cross Dike Scar	0·54	829
			Cow Leys Scar	0·595	777
			Coastguard Station, Walney	0·407	788
			South End	0·581	939
			Westfield Point	0·704	1184
			Concle Bank (8)	0·63	1074

mean 0·792, range 0·43–1·303
S.D. 0·246
S.E. of the mean 0·087

mean 0·513, range 0·195–0·704
S.D. 0·148
S.E. of the mean 0·041

$t = 2·79$, therefore the difference in means is significant (1%).

Uplands which was still ice covered and the Irish Sea basin ice sheet was reactivated and advanced an unknown distance on to the Cumbrian Lowland. This readvance was a major readvance stage in the general deglaciation of the Irish Sea basin.

v. *Furness readvance.* Recent work on the Low Furness glacigenic sediments suggests that there is evidence for two distinct glacial phases during the Late Devensian. Two tills are present on Walney and the mainland as far east as a line from Roose to Roosebeck (see Fig. 3). In the latter area the upper till is underlain by a thick series of fluvioglacial sediments.

The sand/matrix ratio illustrates a significant difference between the tills. In Table 1 the lower till has a mean sand/matix ratio of 0·792 whilst the upper till has a mean of 0·513. The lower till lithologically is composed of a high percentage of Borrowdale Volcanic Series rocks, Lower Palaeozoic grits and Carboniferous Limestone (see Table 2). In Figure 4 a southern Lake District source is also indicated by till fabric distributions and drumlin orientations (NE–SW). The

upper till contains erratics from the Western Lake District and Irish Sea basin
and till fabric distribtions indicate a NW–SE moving ice.

The fluvioglacial sediments are composed of fining upwards cycles of horizontal
stratification, trough cross stratification and small scale ripple cross stratification.
This lower sand sequence passes vertically into horizontal pebble gravel bars and
channels, with occasional thin, laterally impersistent till units. This gravel facies
is capped by a thin, but extensive till unit. This succession indicates a proximal
sandur with braid bars channel and flow tills from the nearby ice front, which
eventually passed over the succession.

3. Flandrian marine transgression sequences

In 1913, Clement Reid recommended the low-lying coasts of northwest England
for detailed study of sea-level changes: he noted that in the strongly marked alternations of peat and clay, "the geologist should be able to study ancient changes of
sea-level under such favourable conditions as to leave no doubt as to the reality
and exact amount of these changes".

Pollen, diatom and sedimentological analyses on biogenic and minerogenic
material from critical sites identified by extensive stratigraphic surveys in these
low-lying areas has borne out this promise. The use of radiocarbon dating has
allowed the establishment of the ages of the marine transgressions identified and the
construction of a sea-level curve (Tooley 1974, 1976, 1977a). Eleven marine transgressions have been identified and are named after the type area at Lytham:

Lytham I	9270–8575 radiocarbon years B.P.
Lytham II	8390–7800
Lytham III	7605–7200
Lytham IV	6710–6157
Lytham V	5947–5775
Lytham VI	5570–4897
Lytham VIa	4800–4545
Lytham VII	3700–3150
Lytham VIII	3090–2270
Lytham IX	1795–1370
Lytham X	~817

A summary of the evidence for these transgressions is given from Cheshire,
Lancashire, Morecambe Bay and Cumbria. Detailed evidence for Lancashire is
found elsewhere (Tooley 1977a).

3a. Cheshire

There is evidence of two marine transgressions affecting areas adjacent to
the estuaries of the Mersey and Dee: the first occurred towards the end of Flandrian
II on Helsby Marsh and the second is recorded at Leasowe, Wirral during
Flandrian III.

Along the southern margin of Helsby Marsh, till is overlain by a coarse white
sand, locally a metre thick. Overlying this sand is a ten centimetres thick bed of
compact woody detrital peat overlain in turn at an altitude of +0·73 m O.D. by
a layer of blue-grey silty marine clay with rhizomes of *Phragmites,* some 53 cm
thick. Above the marine clay are layers of turfa with *Phragmites* and *Eriophorum*
and woody detrital peat with *Alnus* and *Betula* up to the surface at an altitude of

Table 2. Distribution of lithologies in the Low Furness glacial sediments

Location	No. of stones	Carboniferous Limestone	Grits	Borrowdale Volcanic Series	Eskdale granite	St. Bees Sandstone	Mudstones/ Siltstones	Others
Earnse Point	303	1·7	34·7	20·8	3·6	20·8	1·0	16·5
Lime Kiln Scar	158	3·8	42·4	28·9	7·7	3·2	1·3	13·4
Lamity Syke Scar	161	—	30·4	37·3	2·8	8·1	0·6	20·5
Cross Dike Scar	196	3·1	29·6	21·5	5·1	5·6	1·0	15·6
South End Upper Till	269	—	35·3	34·5	3·7	4·1	0·7	21·6
South End Lower Till	300	19·0	37·3	19·0	3·7	2·3	2·7	16·0
Ormsgill	289	4·2	33·6	25·6	4·5	17·0	2·3	12·7
Barrow Island Point	316	12·3	34·2	31·3	3·8	2·2	3·4	12·6
Roosecote	318	10·4	22·6	28·9	18·6	7·2	2·5	9·8
Westfield Point	311	3·5	29·9	33·7	12·9	1·0	7·3	11·1
Back House Close Scar	311	27·0	17·7	28·6	6·8	9·6	4·9	4·1
Leonard Scar	311	50·2	20·6	11·6	—	—	12·2	5·4
Moat Scar	300	76·0	7·0	11·0	0·3	—	0·3	2·9
Wadhead Scar	322	50·6	37·3	4·3	—	—	5·2	2·5

+9·12 m O.D. The beginning of marine conditions is dated at 5470±155 (HV. 2686) at an altitude of +0·73 m O.D., and ended at 5250±385 (HV. 2685) at an altitude of +1·29 m O.D.

Morton (1888) has recorded a lower peat, a grey estuarine clay and an upper peat in the Stanlow embayment and these are equivalent to the deposits recorded at Helsby Marsh, but he shows a second estuarine clay intercalating the upper peat. This clay is evidence of a second transgression, that penetrated a short distance southwards in the Stanlow, Helsby and Frodsham areas, and may correlate with the transgression proved at Leasowe.

At Leasowe, wood has been dated from a peat bed overlain by silty estuarine clay containing valves of *Scrobicularia plana* da Costa (Godwin and Willis 1962). Two radiocarbon assays on wood gave dates of 3695 ± 110 and 3680 ± 110 (Q.620). These dates refer to the age of the wood sampled and give a maximum date for the beginning of a marine transgression during Flandrian III at an altitude a little above $+3.05$ m O.D. and up to at least $+4.26$ m O.D. This deposit is a landward extension of a similar deposit at Dove Point, Leasowe on the Wirral Coast described by Erdtman (1928) and Travis (1929) which is at a higher altitude and is more recent than the intertidal peat bed recorded at the Alt Mouth in Lancashire (Tooley 1969, 1970a, 1977b).

The stratigraphic evidence from Helsby Marsh (Tooley 1969) particularly from the buried valley of the Hornsmill Brook, points to possible earlier marine transgressions in this area at an altitude of -5.0 m O.D., but these lack corroboration. There is unequivocal evidence for two marine transgressions, which occurred whilst Lytham VI and VII were underway in west Lancashire.

3b. Lancashire

Along the coast of west Lancashire, where the relief is modest and valleys in the till open to the coast, marine transgressions persistently penetrated landward at progressively higher altitudes during the Flandrian Stage. Catchments were restricted in area and stream discharge from the valleys during the Flandrian was low. In every case the stream basin was small in area and drained peat mosses. Southwest Lancashire and the southwest and west Fylde are open coasts bordered by the Irish Sea and mosslands such as Downholland, Altcar, Halsall and Lytham Moss discharged their water directly to the sea. Nancy's Bay, however, on the north side of the Ribble estuary may have been affected by periods of high river discharge from the River Ribble.

The evidence for all but one of the marine transgressions in northwest England comes from four sites in the south Fylde–the Starr Hills, Lytham Common, Lytham Hall Hall Park and Nancy's Bay—all of which lie within the former township of Lytham (Tooley 1969, 1974, 1977a).

The culminating stages of the first transgression, Lytham I, are recorded from Lytham Common where a grey clay with sandy partings gives way to a gyttja in which the pollen of open habitat, coastal taxa are recorded at an altitude of -9.75 m O.D. some 8575 ± 105 radiocarbon years ago. Lytham II is recorded from the Starr Hills as a grey fine clay, and transgressed the present coast ending biogenic sedimentation in basins in the till at an altitude -11.13 m O.D. shortly after 8390 ± 105. The transgressive phase is recorded well landward at Heyhouses Lane, but at a higher altitude (-9.62 m O.D.) and during Flandrian Id. The end of Lytham II is registered in Nancy's Bay at a mean altitude of -2.58 m O.D. shortly before 7800 years B.P. Lytham II includes the very rapid rise of sea-level registered in other parts of the world; in the Fylde, relative sea-level rose from -9.6 to -2.5 m O.D., and records the final disintegration of Laurentide ice sheet and attenuation of Antarctic shelf ice.

Lytham III is recorded exclusively from Nancy's Bay. It has altitudinal limits of -2.51 to -1.35 m O.D. and comprises a blue grey silt.

In southwest Lancashire, only a single marine transgression appears to have affected this area during Lytham I, II and III. The present coast was first transgressed at an altitude of −10·21 m O.D. some 8000 years ago. The transgression facies comprise a compact grey silt coarsening upwards to a grey-brown fine medium sand.

Lytham IV comprises a complex of small, short-lived transgressions with slight altitudinal variation recorded in Nancy's Bay, Lytham and its northern extension into the Lytham-Skippool valley. The early stages of the transgression are characterised by grey sand and silt whereas the later stages are fine silt and clay with sheets of *Phragmites* peat containing pollen both of coastal taxa—*Plantago maritima* and *Armeria maritima* and freshwater taxa—*Cladium, Typha angustifolia* and *Nuphar*. In southwest Lancashire Lytham IV is represented by two lithologic units: the first has altitudinal limits of −0·72 to −0·19 m O.D. and the second +0·33 to +1·07 m O.D. At DM−15 sampling site, the first unit is represented by a 76 cm thick layer of blue silty clay. There are significant changes in particle size distribution through the unit: coarse silts dominate the middle of the unit, and above and below as the boundaries are approached the frequency of medium and fine silts increases and the clay fraction is dominant at the boundaries. Across the upper boundary there is a significant decline in the frequency of both marine and marine-brackish diatoms such as *Melosira sulcata* (Ehr) Kützing and an increase in fresh-brackish water forms, such as *Diploneis ovalis* (Hilse) Cleve.

Shortly after the end of Lytham IV, and simultaneously throughout Nancy's Bay, Lytham V is recorded: it is dated at NB−6 at 5950±80 and at NB−10 at 5945±50, and ended at NB−6 at 5775±85. The mean altitude of the transgressive phase is +1·30 m O.D. and of the regressive phase +1·59 m O.D. Although the facies characteristics varied, at NB−6, it was a blue clay with black organic partings.

The transgressive phase of Lytham VI is also recorded from Nancy's Bay. The transgression has a mean altitude of +2·88 m O.D. and comprises a blue-grey clay with rounded pebble lenses, some 30–80 cm thick. The end of Lytham VI is recorded throughout Lytham at the Flandrian II/III chronozone boundary, that is about 5000 years B.P., at a mean altitude of +3·03 m O.D.

In southwest Lancashire Lytham V was underway at a slightly lower altitude: it comprised a silty clay with sandy and ferruginous partings. The diatom assemblage is dominated by marine and marine-brackish water taxa such as *Melosira sulcata* (Ehr.) Kützing, *Podosira stelliger* (Bail.) Mann and *Diploneis didyma* Ehrenberg.

There is evidence for a post-elm decline (early Flandrian III) transgression in southwest Lancashire and in Lonsdale, but not in the type area. Lytham VIa is recognised at the Alt Mouth and at Heysham Moss: at the former, the regressive phase occurred at 4545±90, at an altitude of +3·11 m O.D., whereas at the latter it occurred at 4190±150 at an altitude of +4·49 m O.D. In the type area, there is indirect evidence of this transgression, for simultaneously there is a regional elevation of the freshwater table after the elm decline.

Lytham VII is recorded at the northern end of the Lytham-Skippool valley as a layer of blue clay with iron partings, and in Lytham Hall Park where its culminating phase occurred 3150±150 radiocarbon years ago at an altitude of +3·51 m O.D. In Lytham Hall Park, Lytham VII is very close to its marine limit, and a short period separates it from Lytham VIII.

Fig. 5. Pollen diagrams from Morecambe Bay Boreholes. A key to the stratigraphic signatures is shown on Figure 11, and the stratigraphic and geographical relationships in Tooley (1974). Pollen frequences are calculated according to the formula shown for each taxonomic group.

MORECAMBE BAY - POLLEN DIAGRAMS

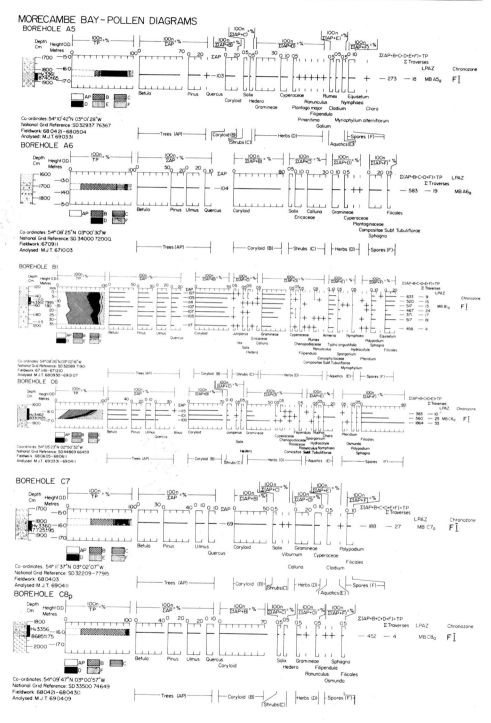

Table 3. Correlation of Quaternary deposits in northwest England

Stage	Ch.	Lancashire	Age	Cumbria	Age
		Starr Hills' turfa. (Lytham X)	830–805		
		Grey-blue marine clay (Lytham IX)	1800–1370	Arnside Moss estuarine sand	1545 ±35
	FIII	Marine silty clay (Lytham VIII)	3090–2270	Selker Point raised shingle beach	2820 ±55
		Blue marine clay (Lytham VII)	3700–3150		
				Heysham Moss marine clay	4190 ±150
		Blue marine clay (Lytham VIa)	4800–4545	Pelutho marine clay	4845 ±100
		Blue grey marine clay (Lytham VI)	5570–4897	Arnside Moss marine clay	5015 ±100
				Duddon Estuary marine clay and silt	4960 ±150
		Blue marine clay (Lytham V)	5947–5775	Silverdale, Helsington and Ellerside marine clays	5865 ±115 to 5277 ±120
	FII			Williamson's Moss marine sand, silt, clay	6236 ±85
				Crosscanonby marine clay	6810 ±130 to 6495 ±95
				Bowness Common marine clay	6850 ±60 to 5875 ±220
Flandrian		Gray marine sand and silt (Lytham IV)	6710–6157	Wedholme Flow marine clay	6870 ±95 to 5385 ±280
	FId	Blue-grey marine silt (Lytham III)	7605–7200	Rusland Pool marine clay	
	FIc	Marine clay (Lytham II)	8390–7800	Black Dub peat	8480 ±205
	FIb	Grey marine clay (Lytham I)	9270–8575	Morecambe Bay marine clay, silt, shells	9270 ±200 to 8330 ±125
	FIa	Gytja Red Moss, Marton Mere			

Stage		Date	Event / Deposit
Devensian	Late	10455 ±110	Shirdley Hill Sand. Basal Peat
		11170 ±260	Coarse detritus mud at Skitham
		12200 ±160 to	Coarse detritus mud at High Furlong
		11665 ±140	
		12320 ±155	Gyttja at Rossall Beach
			High altitude valley moraines in the Lake District
		12810 ±810 to	Peat at St. Bees, Scaleby and Oulton
		12560 ±170	
		14305 ±230	Sapropel mud Blelham Bog
			PHASE 3. SCOTTISH RE-ADVANCE Readvance of Irish Sea ice in coastal lowlands: deposition of proglacial fluvial and lacustrine sediments and till. Gyttja at Moorthwaite and Abbot Mosses.
			PHASE 2. FURNESS RE-ADVANCE Readvance of Irish Sea ice in Low Furness: deposition of proglacial fluvial sediments and till.
			PHASE 1. MAIN GLACIATION Till-fluvio-glacial complex in Edenside during ice decay. Drumlins of Edenside, Coastal Lowland and Furness Lake District till: Derwent, Eden, Keekle valleys Lake District Sandur sediments: Derwent valley. Low Hurst organic sands and gravels
	Middle		Kirkham moraine (?)
	Early		Till-fluvio-glacial complex
Ipswichian		>32,500	Scandal Beck organic silts
		>42,000	Lindal Cotes and Crossgate organic beds. Wigton marine clay
Wolstonian			EARLY SCOTTISH Tills at Willowford, Glassonby, Gillcambon Beck, Gillcambon Beck, Wiza Beck Valley, Low Furness.

Ch = Chronozone (Hibbert *et al.* 1971)

Lytham VIII is recorded in Lytham Hall Park and is a silty clay, although further seaward the transgression is represented by a silty sand. The mean altitude of the transgressive phase is +3·68 m O.D., and +4·34 m O.D. for the regressive phase. Marine sedimentation lasted from 3090±135 until 2270±65.

The end of Lytham IX is recorded from the southwest corner of Lytham Hall Park and between the Park and the present coast. The mean altitude of the regressive phase is +4·46 m O.D., although it attains a maximum altitude of +5·39 m O.D. The maximum altitude appears to have occurred about 1795±240 radiocarbon years ago, but marine sedimentation was being sustained nearer the coast at a lower altitude some 1370±88 radiocarbon years ago.

A final transgression can be inferred from Lytham. Dates from biogenic strata intercalating the coastal dune sand ranged from 805±70 to 830±50. On the basis that marine transgression episodes are closely related to periods of dune stability, increased precipitation and biogenic sedimentation (Jelgersma *et al.* 1970) then the dates recorded from the biogenic strata do indicate that a marine transgression of limited extent was underway in this area. Documentary evidence from the accounts and chartulary of Lytham Monastery from the mid-fifteenth to the early sixteenth century (Fishwick 1907; Piper *pers. comm.*) indicates a period of dune instability, from which a period of relatively low sea-level can be inferred.

3c. Morecambe Bay

In Morecambe Bay, and the coasts flanking it in north Lancashire and south Cumbria, there is evidence of five marine transgressions.

The first transgression inundated the Bay at an altitude of −17·60 m O.D. shortly after 9270±200, which is an assay on peat from Heysham Harbour (Shotton and Williams 1971). By 8330±125, the transgression is recorded at an altitude of −16·03 m O.D., and the lowest reaches of the Kent and Leven estuaries had been inundated. Peat growth on a bench in the Leven estuary ended about 7995±80 at an altitude of −11·16 m O.D., and the pollen diagram (Fig. 5 Borehole B.1) shows rising frequencies of aquatic taxa, particularly *Typha angustifolia, Nymphaea* and *Hydrocotyle,* and low sporadic frequencies of open habitat, coastal taxa, such as Chenopodiaceae and *Armeria* heralding the inception of full marine conditions at the site. The marine facies comprise silty clay and fine sand with occasional pebble beds, lenses of re-worked biogenic material and beds of *Scrobicularia plana* da Costa, *Mytilus edulis* Linné and *Cerastoderma edule* Linné.

North of the Leven estuary, in the valley of Rusland Pool, Dickinson (1973) has recorded a marine clay at an altitude of from −0·75 to +1·35 m O.D. and laid down during Flandrian Id. (8196 to 7107 radiocarbon years ago Hibbert *et al.* 1971). This transgression may represent the maximum landward penetration of the transgression recorded in Morecambe Bay, but at a much higher altitude. In the Rusland Pool valley, as far as Crooks Bridge, the marine facies is a grey-blue clay containing *Globigerina* sp., fragments of Radiolaria and sponge spicules.

The evidence for a second transgression comes from Silverdale, Helsington and Ellerside Mosses. At Silverdale, marine sedimentation began at an altitude of +2·93 m O.D. at 5734±129 (Q.256) (Oldfield 1960 and *pers. comm.*; Godwin and Willis 1961). The end of marine conditions occurred at an altitude of +3·85 m O.D., and the radiocarbon assay on *Alnus* wood 30 cm above the clay/peat contact of 5865±115 (Q.261), may testify to the very short duration and rapid rate of sedimentation in the basin. Further north, in the Leven estuary at Ellerside Moss marine sedimentation ended at an altitude of +3·72 m O.D. shortly before

5435±105 (Hv. 3844) which is the date on *Phragmites* peat immediately above the marine clay. At a higher altitude, on Helsington Moss, north of Silverdale Moss and the Kent estuary, in the Gilpin valley, Smith (1959) has recorded the end of marine conditions at +4·88 m O.D. some 5277±120 (Q.85) years ago (Godwin and Willis *op. cit.*). On the basis of altitude and age, the data from Helsington Moss probably indicate the culminating stage of a third transgression. Substantiating evidence comes from Arnside Moss, where, shortly before 5015±100 (Hv. 3460) marine conditions ended at an altitude of +4·98 m O.D., and on Carnforth Levels minerogenic sedimentation ended at an altitude of +4·65 m O.D.

Further south, at Heysham Moss, the culminating stages of a post-elm decline transgression are recorded at an altitude of +4·49 m O.D. The transgression facies comprise a grey-blue, clayey silt. An assay on organic material immediately above the marine silt yielded a date of 4190±150 (Hv. 2920).

A fifth marine transgression may be recorded on Arnside Moss, where a ten centimetre thick layer of yellow silty fine sand with iron concretions is recorded at an altitude of +5·78 m O.D. The layer overlies woody detrital peat and an assay on this material yielded a date of 1545±35 (Hv. 3461). This marine episode may be represented on Carnforth Levels by a layer of stiff grey clay from +5·46 to +5·61 m O.D.

A correlation of these five transgressions with those recorded at Lytham is given in Table 3.

3d. Cumbria

The oldest marine deposit on the west Cumbrian coast has been recorded adjacent to the Black Dub north of Allonby. Here interdigitating bands of clay, peat and sands are exposed, beneath the surface of the 25 ft raised beach identified by Eastwood *et al.* (1968). Pollen analysis of the peat demonstrated an assemblage rich in halophiles, such as Chenopodiaceae, *Artemisia* and *Plantago* cf. *maritima,* and indicated a Flandrian Ic age, which was confirmed by the radiocarbon date of 8480±205. It is probable that the sedimentary complex at Black Dub is an early Flandrian sand-dune system, and that the organic lenses are fossil dune slacks initiated by a relatively high sea-level, an elevation of the coast freshwater table and an increase in the humidity of the coastal zone (Jelgersma *et al.* 1970).

Walker (1966) described two marine transgressions from Bowness Common, but a recent stratigraphic survey has proved only a single marine episode, represented by a clay facies. Marine sedimentation began shortly after 6850±60 (Hv. 6208) at an altitude of +4·73 m O.D. and ended at 5875±220 (Hv. 6207) at an altitude of +5·92 m O.D.

Further south on Wedholme Flow, the same transgression is recorded from 6870±95 (Hv. 5228) to 5385±280 (Hv. 4713) at an altitude of from +4·80 to +6·18 m O.D. The transgression facies comprised grey tenacious clay and grey sand and were laid down during a period of relatively high sea-level, during which Wedholme Flow was inundated by way of the Rivers Waver and Wampool.

At Crosscanonby (Fig. 6), this transgression is represented by a clay bed 150 cm thick, rich in brackish water and marine diatoms, such as *Diploneis fusca* (Gregory), *Podosira stelliger* (Ehr.) Kützing, and *Scoliopleura tumida* (de Brebisson). Full marine conditions are replaced as the lithologic boundaries are approached by brackish water and freshwater taxa. In the gyttja and detrital peat below and above the marine clay, the proximity of open habitat coastal taxa is indicated

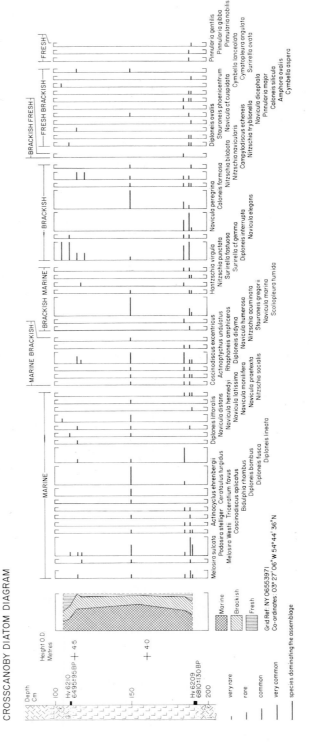

Fig. 6. Diatom diagram from Crosscanonby. The assignment and proportion of marine, brackish and freshwater diatoms is shown according to the conventions of van der Werff and Huls (1958–1966).

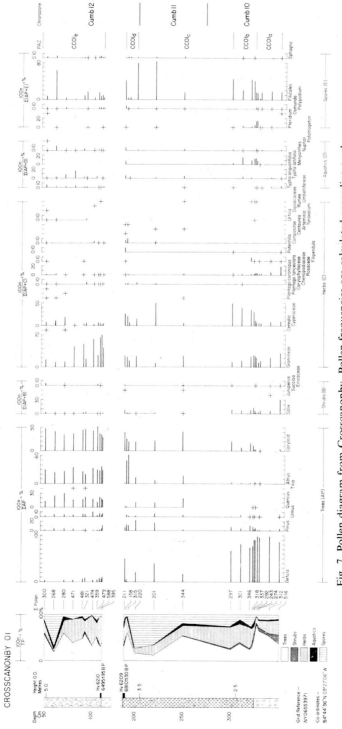

Fig. 7. Pollen diagram from Crosscanonby. Pollen frequencies are calculated according to the formula shown for each taxonomic group.

in the pollen diagram (Fig. 7) by Chenopodiaceae. The transgression occurred between 6810 ± 130 (Hv. 6209) and 6495 ± 95 (Hv. 5210) at altitudes from $+3\cdot71$ to $+4\cdot44$ metres O.D. It is probable that during a period of relatively high sea-level and during the erosional stage of the "25 ft" raised beach, the seaward side of the drumlin (Swarthy Hill) protecting the inter-drumlin depression was eroded and the lowest point breached. The sea inundated the freshwater lagoon and the period of marine influence is represented by silts, clays, sands and shingle in the borehole records. The early end of marine sedimentation here compared to Wedholme Flow may be explained by the accumulation of sand and shingle on the raised beach, effectively sealing the breach.

A similar explanation may be advanced for Williamson's Moss where a basin in the raised beach has been isolated from the sea by a complex of shingle ridges. The basin bottoms in grey clay, which contains brackish water and halophile diatoms (Fig. 8) such as *Melosira sulcata* (Ehr). Kützing, *Podosira stelliger* (Bail.) Mann and *Actinoptychus undulatus* (Bail.) Ralfs. The freshwater element, represented particularly by *Gomphonema acuminatum* (Ehrenberg) and *Stauroneis anceps* (Ehrenberg), increases towards the lithologic boundary, and the base of the gyttja is characterised by few marine forms. The clay is overlain by gyttja, and an assay on the basal gyttja at $+3\cdot18$ m O.D. yielded a date of 6320 ± 85 (Hv. 5227). This date is corroborated by pollen analytical evidence (Fig. 9) and the proximity of open habitat coastal taxa is weakly shown by the presence of Chenopodiaceae at the base of pollen assemblage zone WMO1–1. (See also Pennington 1975.)

In the valley of the Kirkby Pool and the mosses adjacent to the Duddon estuary (Fig. 10), there is evidence for the culminating stages of a Flandrian II/III transgression that is correlated with Lytham VI. The transgression facies comprise a grey clay or silt or brown coarse sand. Locally, discrete iron partings are apparent (e.g. DE–11 and DE–16 on Fig. 11). The marine clays, silts and sands pass up into biogenic sediments—woody detrital peats, monocotyledonous turfas and bryophyte turfas. The altitude of the top of the marine facies ranges from $+4\cdot92$ to $+4\cdot01$ m O.D. Two pollen diagrams (Figs 12 and 13) from Waitham Common and Bank End Moss indicate an autogenic succession from salt marsh communities with Chenopodiaceae, *Plantago maritima*, *Artemisia* and *Armeria* to freshwater reed-swamp communities of *Phragmites, Typha angustifolia, T. latifolia, Lythrum salicaria, Filipendula, Hydrocotyle, Lemna* and *Potamogeton*, replaced by a fen of *Alnus* and *Quercus* with shrubs of *Salix* and *Frangula* and woody climbers, such as *Solanum*. In both pollen diagrams, salt marsh communities thrive immediately prior to the elm decline, and a radiocarbon assay on woody detrital peat with *Betula* from DM–16 yielded a date of 4960 ± 150. An assay on *Phragmites* peat from a stratum above marine silt at DE–11 yielded a younger date of 4760 ± 45, and, although at a higher altitude, the tree pollen assemblage indicates a pre-elm decline date for this stratum and the date is in error by about 250 years. The culminating stage of the same transgression is also recorded from Pelutho, south of Silloth and is dated to 4845 ± 100 (Hv. 4418).

There is some evidence for a final mid-Flandrian III transgression at Selker Point where a raised shingle beach overlies freshwater clays at an altitude of $+6\cdot69$ m O.D., and, in turn, monocotyledonous turfa and gyttja. A radiocarbon assay on material immediately subjacent to the freshwater clays yielded a date of 2820 ± 55 (Hv. 3842), and is thus maximal for the onset of marine conditions at the site (see also Huddart and Tooley 1972).

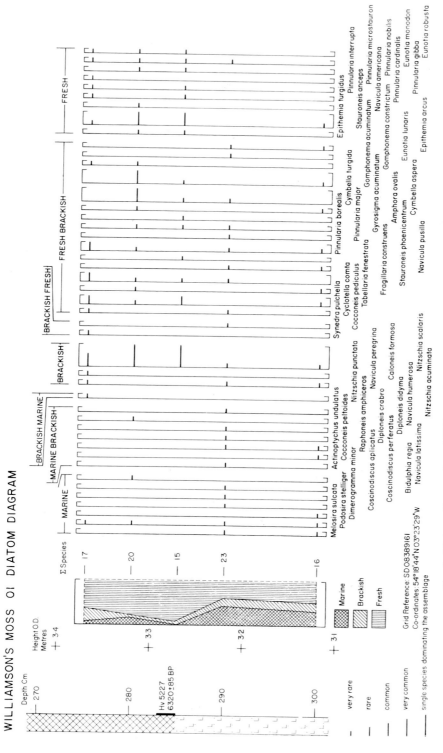

Fig. 8. Diatom diagram from Williamson's Moss. The assignment and proportion of marine brackish and freshwater diatoms is shown according to the conventions of van der Werff and Huls (1958–1966).

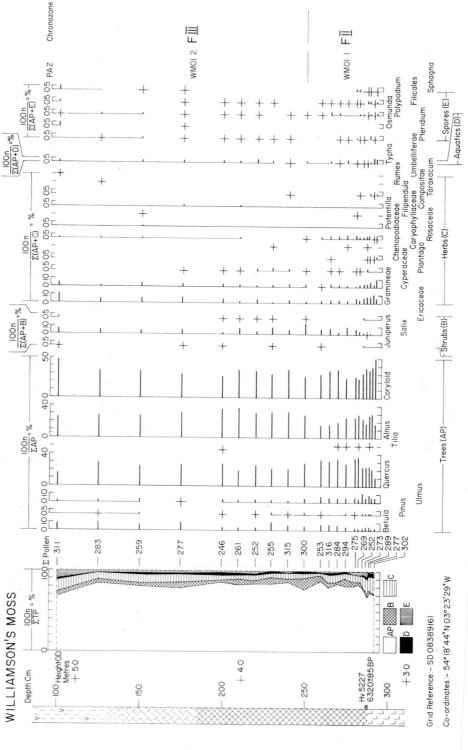

Fig. 9. Pollen diagram from Williamson's Moss. Pollen frequencies are calculated according to the formula shown for each taxonomic group.

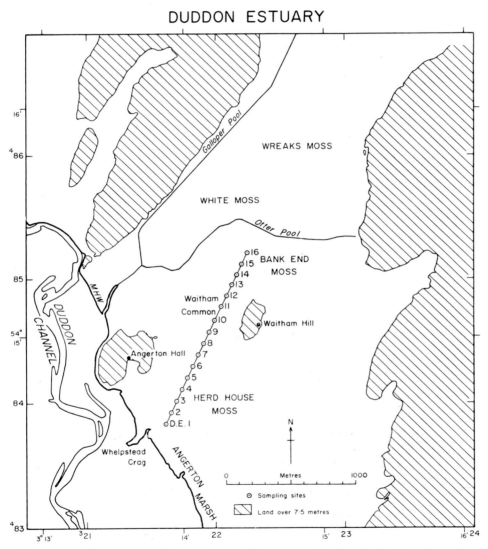

Fig. 10. Map of the mosses adjacent to the Duddon estuary showing the location of sampling sites on Angerton Marsh, Herd House Moss, Waitham Common and Bank End Moss.

4. Conclusions, including a correlation of deposits in northwest England

(a) Little is known about pre-Late Devensian events and sediments in this region. In Cumbria there are examples of till which have been placed in the Wolstonian (see Table 3) but only in Low Furness would there seem to be conclusive proof of this age for the thin, remnant till.

(b) Ipswichian sediments are represented by the Scandal Beck organic silts and much more tentatively the Lindal Cotes and Crossgates organic sediments in Low Furness and the marine clay at Wigton. The former were placed in the Middle Devensian by Evans and Arthurton (1973) but infinite C^{14} dates and the pollen analytical and stratigraphic evidence of Carter (unpublished) proves that they are older.

(c) The undated Low Hurst organic sands (Evans and Arthurton 1973) need more investigation before they can be conclusively placed in the Mid-Devensian interstadial.

(d) The Late Devensian glacigenic sediments are complex in Cumbria and have been interpreted in more detail than in Lancashire. Although Lancashire was the type area for the tripartite sequence of Upper Boulder Clay, Middle Sands and Lower Boulder Clay, the exact significance of this succession is uncertain. Evans and Arthurton (1973) assume that the units are "integral parts of the same melt but this is not necessarily true for all occurences".

It seems likely that the maximum glacial phase in the Devensian in Britain occurred between 25000 and 20000 B.P. During the retreat from this maximum, readvances or stillstands in the Irish Sea basin have been claimed by various authors in mid-Cheshire, the Isle of Man, Wexford, Lleyn and in the area under consideration in this contribution.

During Phase 1 of this glacial in Cumbria (Table 3) thick, proglacial valley sandur sediments were aggraded in the Derwent valley. These underlie a till derived from the Lake District. The valley glaciers which deposited these sediments were then incorporated in an ice sheet composed of Scottish, northern, western and southern Lake District ice. Johnson, Tallis and Pearson (1972) suggest that part of this ice sheet, which deposited the Edenside, coastal lowland and Furness drumlin fields, was formed after the maximum advance to Wolverhampton, about 18000 B.P. The Lake District ice was thought by them to have extended at this time to the Kirkham moraine and the Irish Sea ice to North Wales and the Ellesmere-Whitchurch moraine (Johnson 1971). However, dateable organic sediments have not been found associated with these readvance/stillstand positions so it is difficult at present to fit the sequence of stages into an absolute chronology. The Kirkham moraine, part of the North British Readvance (Gresswell 1967), has not been described sufficiently well so that its exact mode of formation cannot be established.

The final stage in Phase 1 was the decay *in situ* of this ice sheet in suitable topographic localities, for example in Edenside and some of the western Lake District valleys. Here complex interdigitations of till, fluvioglacial and glacio-lacustrine sediments were deposited in ice-contact environments.

The Irish Sea ice is considered to have readvanced during the deglaciation (Phase 2) to produce the Low Furness proglacial and subglacial sediments. It is tempting to correlate this ice frontal limit with the terminal moraine west of Black Combe, only 12 km to the north. However, in the latter area the upper tills have

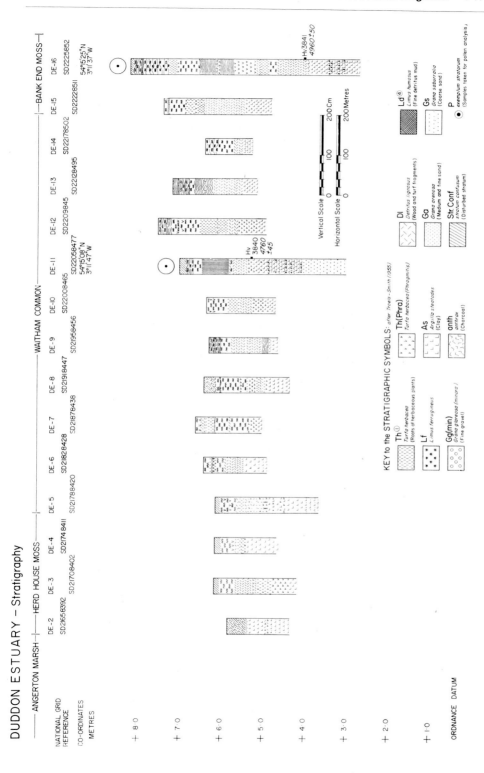

Fig. 11. Stratigraphy across the mosses adjacent to the Duddon estuary.

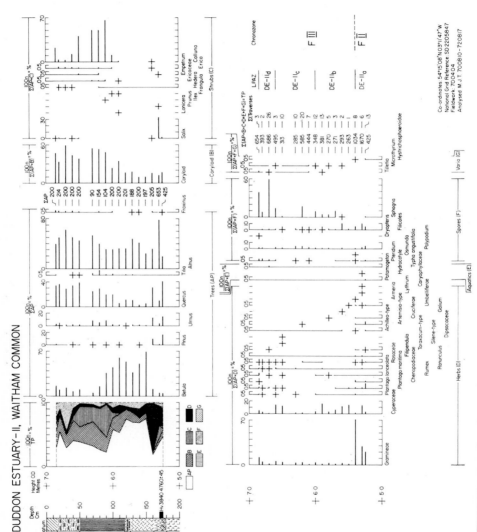

Fig. 12. Pollen diagram from Duddon Estuary – 11. Pollen frequencies are calculated according to the formula shown for each taxonomic group.

a low value for the sand/matrix ratio compared with the Low Furness readvance tills. Nevertheless this oscillation of the Irish Sea basin ice sheet produced a similar sequence and landform assemblage to the Scottish Readvance farther north in Cumbria. It probably indicates a slightly earlier readvance phase.

The Irish Sea ice again readvanced to produce the Scottish Readvance (Phase 3) sequence of sediments and landforms in the Cumbrian coastal lowlands. Whether all the environments referred to this phase were formed during a single synchronous advance is unknown and there is no evidence to date this readvance from extra-glacial, biogenic sequences. However, Walker's (1966) evidence from Moorthwaite and Abbot Mosses, well outside the reinvestigated readvance limits, shows a Cumbrian Oscillation (his zones C1–C3) which he equated with the Main Glaciation–Scottish Readvance interstadial. His zone C4 is equated with the period of maximum effect of this readvance, although we must be aware of Pennington's (1970) criticism of these sites. More data from both sites inside and outside the readvance limit are needed to check Walker's conclusions.

This readvance must be older than the oldest C^{14} date that is available from a kettle sequence at St. Bees on the terminal moraine. This is $12,810\pm180$ B.P. (Godwin and Willis 1959) although this date's validity has been debated (Huddart and Tooley 1972). It must also be older than the Perth Readvance of Sissons (1967) which is dated at around 13000 B.P.

In many papers the Scottish Readvance in Cumbria has been correlated with the Bride Hills moraine in the Isle of Man and this correlation is favoured, rather than Mitchell's (1972) line linking the Bride Hills with the mid-Cheshire moraine. In the Isle of Man C^{14} dates of 18900 ± 330 (Birm. 213), 18700 ± 500 (Birm. 270a), 18550 ± 85 (Birm. 270b) and 18400 ± 500 B.P. (Birm. 270c) (Dickson *et al.* 1970; Shotton and Williams 1971, 1973) from a moss layer in silts at the base of a kettle hole sequence does not appear to have been overridden by ice and, therefore, the implication is that the area south of the Bride Hills has been ice free since around 19000 B.P. (Thomas 1971). This has resulted in Mitchell's attempts to show a by-passing of the Isle of Man by Late Devensian ice, with a lobe extending down the eastern side of the Irish Sea at 18000 B.P. On this basis the age of the readvance in Cumbria is younger than 18000 B.P. on the other hand, if one accepts the Bride Hills moraine and the readvance landforms in Cumbria as broad correlatives then the latter must be older than around 19000 B.P.

A major problem in the Irish Sea basin then is that there are coastal lowlands with evidence of Irish Sea ice sheet end moraine complexes which are composed of similar stratigraphic sequences and landform assemblages. These occur for example at the Screen Hills (Wexford), northern Lleyn, the Bride Hills and the Cumbrian readvances described here. At present there seems to be no method of establishing their age or relationships with one another. What is needed first is a close comparison of the sedimentary associations found in these sequences and a continued search for organic sediments that can be linked with them. Nevertheless it seems clear that the Irish Sea ice did oscillate and readvance during the Late Devensian contrary to the views of Mitchell (1972) and Evans and Arthurton (1973). Possibly three such stages have been recognised in northwest England.

(e) In view of the stratigraphic, micropalaeontological and radiometric evidence presented here and elsewhere (Huddart and Tooley *op cit.*; Tooley 1969, 1970a, 1974, 1976, 1977a, b), it is necessary to reconsider the chronology and nature of sediments in coastal Lancashire and Cumbria proposed by Evans and Arthurton (*op cit.*) for the Flandrian Stage. In Table 3, a sub-division of the Flandrian Stage is

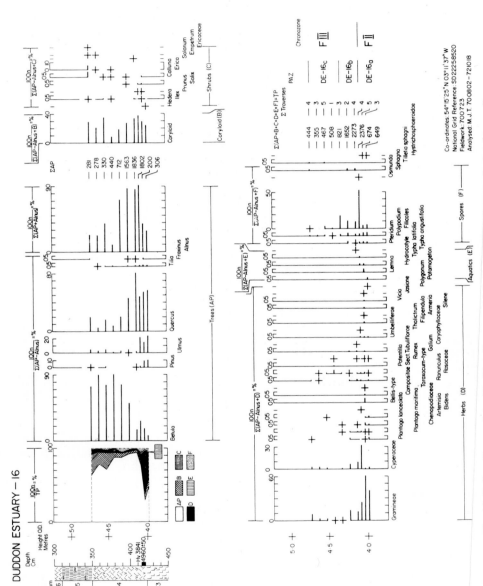

Fig. 13. Pollen diagram from Duddon Estuary – 16. Pollen frequencies are calculated according to the formula shown for each taxonomic group.

proposed on the basis of the marine episodes recognised in Lancashire and their correlatives in Cumbria. Both the Basal Peat and the Shirdley Hill Sand formation are reassigned to the Late Devensian Stage. A sample of Basal Peat from near Clieves Hills in southwest Lancashire yielded a pollen assemblage dominated by herb and shrub taxa, such as Gramineae, Cyperaceae, *Filipendula, Potentilla* and *Selaginella* with shrubs of *Hippophae, Empetrum* and *Salix*. Of the tree taxa recorded, *Betula* was dominant. This late Devensian assemblage has been confirmed by a radiocarbon assay on the same peat of 10455 ± 100 (Hv. 4710). At Clieves Hills, the Basal Peat is overlain by Shirdley Hill Sand, which is interpretted as a late Devensian cover sand—a conclusion reached by Godwin (1959).

It is proposed that the terms Preesall Shingle, Older marine alluvium and coastal peat mosses used by Evans and Arthurton (*op cit.*) are replaced by the sequence of marine sediments recorded and dated radiometrically from the tidal flat and lagoonal zone of Lancashire and Cumbria, and named after the type area of Lytham. Persistent beds of monocotyledonous turfa, detritus lignosa and limus intercalate the marine sediments, and would permit further subdivision into freshwater and terrestrial stages.

Acknowledgments. Dr. M. A. Geyh Niedersächsisches Landesamt für Bodenforschung Hannover carried out most of the radiocarbon dates cited in this paper. Mr. G. Brown, Department of Geography, University of Durham, drew the Figures. M. J. T. acknowledges with gratitude a grant from Durham University Staff Research Fund to carry out a series of deep borings along the Lancashire coast and to undertake a reconnaissance survey in the Isle of Man. P. A. C. was in receipt of an NERC Studentship whilst working on sites along the Cumbrian coast.

References

Anon. 1972. Borehole Records. *In*, Annual Report of the Institute of Geological Sciences 1971. p. 119.

Bolton, J. 1862. On a deposit with insects, leaves, *etc.*, near Ulverston. *Q. J. geol. Soc. Lond.* **18**, 274–277.

Boulton, G. S. 1972. Modern Arctic glaciers as depositional models. *J. geol. Soc. Lond.* **128**, 361–393.

Charlesworth, J. K. 1926. The Readvance marginal kame moraine of the south of Scotland and some later stages of the retreat. *Trans. R. Soc. Edinb.* **67**, 25–50.

—— 1939. Some observations on the glaciation of northeast Ireland. *Proc. R. J. Acad.* B. **45**, 255–295.

de Rance, C. E. 1871a. On the two glaciations of the Lake District. *Geol. Mag.* **8**, 107–117.

—— 1871 b. On the glaciation of the North of England. *Geol. Mag.* **8**, 412–418.

—— 1877. The superficial geology of the country adjoining the coast of southwest Lancashire. *Mem. geol. surv. U.K.*

L

DICKINSON, W. 1973. The development of the raised bog complex near Rusland in the Furness District of North Lancashire. *J. Ecol.* **61**, 871–886.
—— 1975. Recurrence surfaces in Rusland Moss, Cumbria (formerly north Lancashire). *J. Ecol.* **63**, 912–936.
DICKSON, C., DICKSON, J. H. and MITCHELL, G. F. 1970. The Late-Weichselian flora of the Isle of Man. *Phil. Trans. R. Soc. Lond.* **B.258**, 31–79.
DIXON, E. E. L., MADEN, J., TROTTER, F. M., HOLLINGWORTH, S. E. and TONKS, L. H. 1926. The geology of the Carlisle, Longtown and Silloth districts. *Mem. geol. Surv. U.K.*
——, EASTWOOD, T. HOLLINGWORTH, S. E. and SMITH, B.1931. The geology of the Whitehaven and Workington district. *Mem. geol. Surv. U.K.*
EASTWOOD, T. 1930. The geology of the Maryport district. *Mem. geol. Surv. U.K.*
——, HOLLINGWORTH, S. E., ROSE, W. C. C. and TROTTER, F. M. 1968. The geology of the country around Cockermouth and Caldbeck. *Mem. geol. Surv. U.K.*
ERDTMAN, O. G. E. 1928. Studies in the post-arctic history of the forests of North West Europe, I. Investigations in the British Isles. *Geol. För. Stockh. Förh.* **50**, 123–192.
EVANS, W. B. and ARTHURTON, R. S. 1973. Northwest England. *In* Mitchell, G. F. *et al.* (Editors). A correlation of Quaternary deposits in the British Isles. *Geol. Soc. Lond. Special Report* No. 4. 99 pp.
FISHWICK, H. 1907. The history of the parish of Lytham. *Chetham Soc.* N.S.60.
GODWIN, H. 1959. Studies of the post-glacial history of British vegetation. XIV. Late-glacial deposits at Moss Lake, Liverpool. *Phil Trans. R. Soc.* **B.242**, 127–149.
—— and WILLIS, E. H. 1961. Cambridge University Natural Radiocarbon Measurements. III. *Radiocarbon* **3**, 60–76.
—— 1959. Cambridge University Natural Radiocarbon Measurements VI. *Radiocarbon* **4**, 116–137.
GOODCHILD, J. G. 1875. The glacial phenomena of the Eden valley and the west part of the Yorkshire Dale district. *Q. J. geol. Soc. Lond.* **31**, 55–99.
—— 1887. Ice work in Edenside and some of the adjoining parts of northwestern England. *Trans. Cumb. West. Ass. Adv. Lit. Sci.* **12**, 111–167.
GRESSWELL, R. K. 1967. *Physical Geography.* Longmans, London. 504 pp.
HIBBERT, F. A., SWITSUR, V. R. and WEST, R. G. 1971. Radiocarbon dating of Flandrian pollen zones at Red Moss, Lancashire. *Proc. R. Soc. Lond.* **B.177**, 161–176.
HOLLINGWORTH, S. E. 1931. Glaciation of western Edenside and adjoining areas, and the drumlins of Edenside and the Solway Basin. *Q. J. geol. Soc. Lond.* **87**, 281-359.
HOWELL, F. T. 1973. The sub-drift surface of the Mersey and Weaver catchment and adjacent areas. *Geol. J.* **8**, 285–296.
HUDDART, D. 1967. Deglaciation in the Ennerdale area: a re-interpretation. *Proc. Cumb. geol. Soc.* **2**, 63–75.
—— 1970. *Aspects of glacial sedimentation in the Cumberland Lowland.* Unpublished Ph.D. Thesis, University of Reading. 340 pp.
—— 1971a. Textural distinction between Main Glaciation and Scottish Readvance tills in the Cumberland Lowland. *Geol. Mag.* **108**, 317–324.
—— 1971b. A relative glacial chronology from the tills of the Cumberland Lowland. *Proc. Cumb. geol. Soc.* **3**, 21–32.
—— 1973. The origin of Esker sediments, Thursby, Cumberland. *Proc. Cumb. geol. Soc.* **4**, 59–69.
—— in press. The origin of readvance glacigenic sediments in the Cumberland Lowland.
—— in press. Ice walled lacustrine sediments and the decay of the Devensian ice in Edenside.
—— and TOOLEY, M. J. 1972. *The Cumberland Lowland Handbook.* Quaternary Research Association. 96 pp.
JELGERSMA, S., DE JONG, J., ZAGWIJN, W. H. and VON REGTEREN ALTENA, J. F. 1970. The coastal dunes of the western Netherlands; geology, vegetational history and archaeology. *Med. Rijks. geol. Dienst.* NS. **21**, 93–167.
JOHNSON, R. H. 1971. The last glaciation in northwest England. *Amateur geol.* **5**, 18–37.
——, TALLIS, J. H. and PEARSON, M. 1972. A temporary section through late Devensian sediments at Green Lane, Dalton-in-Furness, Lancashire. *New Phytol.* **71**, 533–544.
KENDALL, J. D. 1879. Distribution of boulders in west Cumberland. *Trans. Cumb. Ass.* **5**, 151–157.
—— 1881. Interglacial deposits of west Cumberland and north Lancashire *Q. J. geol. Soc. Lond.* **37**, 29–39.
KRYNINE, P. D. 1948. The megascopic study and field classification of sedimentary rocks. *J. geol.* **56**, 130–165.
MACKINTOSH, D. 1878. Tripartite origin of the boulder clays of the north-west of England. *Geol. Mag.* **14**, 575–576.
MITCHELL, G. F. 1960. The Pleistocene history of the Irish Sea. *Advant. Sci., Lond.* **17**, 313–325.
—— 1963. Morainic ridges on the floor of the Irish Sea. *Ir. geog.* **4**, 335–344.
—— 1972. The Pleistocene history of the Irish Sea: second approximation *Sci. Proc. R. Dublin Soc. Ser. A.* **4**, 181–199.

MORTON, G. H. 1888. Further notes on the Stanlow, Ince and Frodsham Marshes. *Proc. Lpool geol. Soc.* **6**, 50–55.

OLDFIELD, F. 1960. Late Quaternary changes in climate, vegetation and sea-level in Lowland Lonsdale. *Trans. Inst. Br. Geogr.* **28**, 99–117.

—— 1963. Pollen analysis and man's role in the ecological history of the south-east Lake District. *Geogr. Annlr.* **45**, 23–40.

—— and STATHAM, D. C. 1963. Pollen analytical data from Urswick Tarn and Ellerside Moss. *New Phytol.* **62**, 53–66.

PENNINGTON, W. 1970. Vegetation history in the north-west of England: a regional synthesis. *In* Walker, D. and West, R. G. (Editors). *Studies in the vegetational history of the British Isles*. Cambridge University Press, 97–116.

—— 1975. The effect of Neolithic man on the environment in northwest England: the use of absolute pollen diagrams. *In* Evans, J. G., Limbrey, S., and Cleere, H. (Editors). *The effect of man on the landscape: the Highland Zone*. Council for British Archaeology. Research Report No. 11, 74–86.

PENNY, L. F. 1964. A review of the last glaciation in Great Britain. *Proc. Yorks. geol. Soc.* **34**, 387–411.

READE, T. M. 1871. The geology and physics of the post-glacial period as shown in deposits and organic remains in Lancashire and Cheshire. *Proc. Lpool geol. Soc.* **2**, 36–88.

REID, C. 1913. *Submerged Forests*. Cambridge University Press. 129 pp.

SAUNDERS, G. E. 1968. Glaciation of possible Scottish Readvance age in northwest Wales. *Nature Lond.* **218**, 76–78.

SHOTTON, F. W., BLUNDELL, D. J. and WILLIAMS, R. E. G. 1970. Birmingham University Radiocarbon Dates. IV. *Radiocarbon* **12**, 385–399.

—— and WILLIAMS, R. E. G. 1971. Birmingham University Radiocarbon Dates. V. *Radiocarbon* **13**, 141–156.

—— and —— 1973. Birmingham University Radiocarbon Dates. VII. *Radiocarbon* **15**, 451–468.

SISSONS, J. B. 1964. The glacial period. *In* Wreford Watson J. and Sissons J. B. (Editors). *The British Isles, a systematic geography*. Nelson, Edinburgh and London, 131–152.

—— 1967. *The evolution of Scotland's Scenery*. Oliver and Boyd, Edinburgh.

—— 1974. The Quaternary in Scotland: a review. *Scott. J. Geol.* **10**, 311–337.

SMITH, A. G. 1958. Two lacustrine deposits in the south of the English Lake District. *New Phytol.* **57**, 363–386.

—— 1959. The mires of south-western Westmorland: stratigraphy and pollen analysis. *New Phytol.* **58**, 105–127.

SMITH, B. 1912. The glaciation of the Black Combe district. *Q. J. geol. Soc. London.* **68**, 402–448.

—— 1931. The glacial lakes of Eskdale, Mitredale and Wasdale, Cumberland and the retreat of the ice during the Main Glaciation. *Q. J. geol. Soc. London.* **88**, 57–83.

SMITH, C. 1877. Boulder clay. *Trans. Cumb. Ass.* **3**, 91–108.

SMITH, R. A. 1967. The deglaciation of southwest Cumberland: a reappraisal of some features in the Eskdale and Bootle areas. *Proc. Cumb. geol. Soc.* **2**, 76–83.

SYNGE, F. M. 1952. Retreat stages of the last ice sheet in the British Isles. *Ir. Geog.* **2**, 168–171.

THOMAS, G. S. P. (Editor) 1971. *Isle of Man Field Guide*. Quaternary Research Association. 55 pp.

TOOLEY, M. J. 1969. *Sea-level changes and the development of coastal plant communities during the Flandrian in Lancashire and adjacent areas.* Unpublished Ph.D. Thesis. University of Lancaster. 160 pp.

—— 1970a. The peat beds of the south-west Lancashire coast. *Nature in Lancashire* **1**, 19–26.

—— 1970b. Sea-level changes during the Flandrian. *Brit. Ass. Advant Sci.* Paper given to Section E at the Annual Meeting, Durham.

—— 1971. Changes in sea-level and the implications for coastal development. *Association of River Authorities Yearbook and Directory* 220–225.

—— 1973. Flandrian sea-level changes in northwest England and pan-northwest European correlations. *Ninth Congress of the International Union for Quaternary Research, Abstracts.* 373–4.

—— 1974. Sea-level changes during the last 9000 years in northwest England. *Geogr. J.* **140**, 18–42.

—— 1976. Flandrian sea-level changes in west Lancashire and their implications for the "Hillhouse Coastline" *Geol. J.* **11**, 37–52.

—— 1977a. *Sea-level changes: the coast of northwest England during the Flandrian Stage.* Oxford University Press.

—— 1977b. *INQUA Excursion Guide to North West England and the Isle of Man.*

TRAVIS, C. B. 1929. The peat and forest beds of Leasowe, Cheshire. *Proc. Lpool geol. Soc.* **15**, 157–178.

TROTTER, F. M. 1929. The glaciation of the eastern Edenside, Alston Block and the Carlisle Plain. *Q. J. geol. Soc. Lond.* **88**, 549–607.

—— and HOLLINGWORTH, S. E. 1932a. The geology of the Brampton District. *Mem. geol. Surv. U.K.*

TROTTER and HOLLINGWORTH 1932b. The glacial sequence in the North of England. *Geol.*
—, —— *Mag.* **69,** 374–380. EASTWOOD, T. and ROSE, W. C. C. 1937. The geology of the
Gosforth district. *Mem. geol. Surv. U.K.*
WALKER, D. 1956. A late-glacial deposit at St. Bees, Cumberland. *Q. J. geol. Soc. Lond.* **112,**
93–101.
—— 1966. Late Quaternary history of the Cumberland lowland. *Phil. Trans. R. Soc.* **B.251,**
1–210.
WERFF, A. VAN DER and HULS, H. 1958–1966. *Diatomeeenflora van Nederland.*
WRAY, D. A. and COPE, F. W. 1948. Geology of Southport and Formby. *Mem. geol. Surv.
U.K.*

D. Huddart, University of Dublin, Department of Geography, Trinity College, Dublin 2,
Ireland.
M. J. Tooley, University of Durham, Department of Geography, South Road, Durham,
DH1 3LE.
P. A. Carter, University of Durham, Department of Geography, South Road, Durham,
DH1 3LE.

The Quaternary of the Isle of Man

G. S. P. Thomas

The Quaternary deposits of the Isle of Man are divided into two suites: one the product of local periglacial processes, the other of successive foreign ice penetration. Four major stratigraphic groups are distinguished and a tentative chronology suggests that Anglian, Hoxnian, Wolstonian and Ipswichian deposits may lie deeply buried. The Devensian is represented by a multiple sequence of both local and foreign formations. The island is considered not to have been overridden during this period. Attention is directed to some fundamental problems of chronology, ice gradient, multiple readvance and marine penetration, as they affect current concepts concerning the Devensian of the Irish Sea.

1. Introduction

Situated athwart successive ice advances from the major source areas of western Scotland, Lamplugh (1903 p. 332) considered that the Isle of Man offered ". . . an unrivalled field for the study of the conditions that have ruled in the northern part of the basin of the Irish Sea during the Glacial Period".

The greater part of the island has a relatively thin drift cover, but the northern third (Fig. 1) shows Quaternary strata to at least 145 m below O.D.* (Lamplugh 1903). The maximum thickness, including that above sea-level, is in the order of

---

* Isle of Man zero datum was established in 1865. The datum is mean sea-level on a tide pole fixed to a low water jetty north of Pollock Rocks and southwest of St. Mary's Rocks, Douglas. A bench mark cut in the coping stone of the jetty is 5·69 feet below the Mean Water Datum. Isle of Man Ordnance Datum is 0·04 m above Newlyn Ordnance Datum.

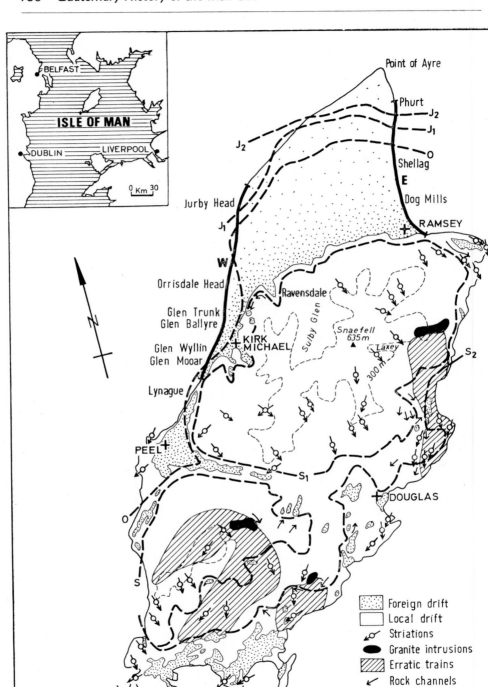

Fig. 1. Elements of the Quaternary geology of the Isle of Man.

250 m. The 30 km of almost continuous coastal cliff section bounding the northern area displays a sequence of deposits of such extraordinary diverse character that Kendall (1894 p. 397) considered to be unsurpassed in the United Kingdom.

Traditionally, the Quaternary deposits have been divided into two great and mutually exclusive suites (Cumming 1846; Horne 1874; Kendall 1894; Lamplugh 1903). A high-level suite of local composition is restricted to the area underlain by the Cambrian Manx Slate, while a low-level suite composed of material foreign to the island occupies the north and the island margin (Fig. 1). Whilst it is convenient for descriptive purposes to treat the two suites separately, the succession is best divided using a formal system of stratigraphic classification (Am. Comm. Strat. Nomenclature 1961). Hence, Thomas (1976) has divided the succession into four major groups (Table 1). The lowest two, the Basement and Sub-Surface Groups, are known only from deep bore-holes (Lamplugh 1903; Smith 1930). The third, or exposed Surface Group, is divided into five formations. Three are of foreign glacigenic origin with the other two of local, but glacially related, origin. One of these, the Upland Formation, is entirely synonomous with the local or high-level suite of earlier workers. The fourth, or Ayres Group, is of post-Pleistocene age.

A schematic representation of the Quaternary succession of the island is given in Figure 2. Individual members are identified in this figure, and in the text by abbreviations as, for example, Shellag Till (ST). A key is given in Table 1 and formal stratigraphic definition may be found in Thomas (1976). In this paper the detail of the succession is not given. In its place we shall consider an interpretation of the succession and the establishment of an event sequence. We shall also direct attention to the many fundamental problems that this provides for the Quaternary history of the surrounding Irish Sea area.

2. The local deposits

The local suite of deposits were considered by Kendall (1894) and Lamplugh (1903) to represent the products of a great glaciation, "one and indivisible" (Kendall 1894 p. 424), that swept over the island to its summit. Both workers explained the entire and enigmatic absence of foreign erratics from amongst the deposits by introducing the concept of a clean-ice shear across the island at heights in excess of 180 m, the limit of foreign penetration. Other workers have interpreted the suite as head, developed during the last glaciation when the island existed as a nunatak (Wirtz 1953; Cubbon 1957; Thomas 1971; Mitchell 1972). Evans and Arthurton (1973) and Bowen (1973a), however, have suggested that the deposits are Devensian tills, modified and re-worked by periglacial processes during the Late-Glacial. Some dispute therefore exists on the question of the origin of the local deposits, and of whether the Isle of Man either nurtured its own glaciers or was overridden by Irish Sea ice during the last glaciation.

From an examination of the morphology, stratigraphy and sedimentology of the deposits, the present author is of the view that they represent the products of a severe periglacial environment. This operated during the major part of the Devensian period, when the island existed as an ice-free nunatak, above the levels of successive advances of Irish Sea ice around its margins.

Lamplugh (1903) and his predecessors made many pertinent comments concerning the entire absence from the uplands of moraine forms, cirques or other

Table 1. Summary of the stratigraphic nomenclature of the Quaternary of the Isle of Man

Group	Formation	Member
Ayre	Point of Ayre	Cranstal Silts (CSS) Ayre Beach (AB)
	Curragh	Curragh Peat (CP)
	Moorland	Upland Peat (UP) Sulby Gravel (SRG)
Surface	Ballaugh	Ballaleigh Debris Fan (BDF) Wyllin Debris Fan (WDF) Ballyre Debris Fan (BYDF) Ballaugh Debris Fan (BHDF) Ballure Debris Fan (BLDF) Crawyn Sand (CS) Jurby Kettles (JK) Wyllin Kettles (WK)
	Upland	Upper Stratified Head (USH) Upper Gravel (UG) Mid Stratified Head (MSH) Massive Head (MH) Lower Gravel (LG) Lower Stratified Head (LSH) Rock Rubble (RR) Upper Blue Head (UBH) Lower Blue Head (LBH) Ballure Slope Wash (BSW) Brown Head (BH) Mooar Head (MRH) Ballure Scree (BS) Mooar Scree (MS)
	Jurby	Andreas Platform Gravel (APG) Trunk Till (TT) Trunk Gravel (TG) Cranstal Till (CT) Ballaquark Till (BQT) Ballaquark Sand (BQS) Jurby Till (JT) Jurby Sand (JS)
	Orrisdale	Dog Mills Series (DMS) Ballure Clays (BC) Kionlough Till (KT) Ballavarkish Sand (BVS) Ballavarkish Till (BVT) Ballavarkish Marginal Series (BVMS) Orrisdale Gravel (OG) Orrisdale Sand (OS) Orrisdale Till (OT)
	Shellag	Kionlough Gravel (KG) Shellag Gravel (SG) Shellag Sand (SS) Shellag Till (ST) Wyllin Sand (WS) Wyllin Till (WT) Ballure Till (BT)
Sub-Surface	Sub-Surface	Middle Sands (MSS) Middle Boulder-Clay (MBC)
Basement	Basement	Ayre Marine Silts (AMS) Lower Sand (LSS) Lower Boulder-Clay (LBC)

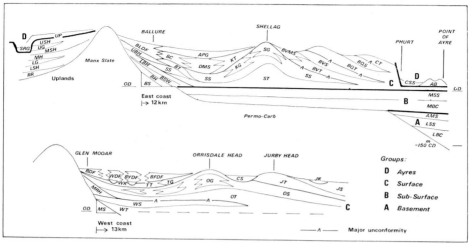

Fig. 2. Schematic representation of the Quaternary succession in the Isle of Man. For identification of members see Table 1 and text.

geomorphological evidence of local glaciation. The almost ubiquitous scenery of the Manx uplands is one of gently rolling hills, heavily drift covered and with little rock exposure. Slopes display a zonal character reflecting weathering, transportation and deposition in separate and successive down-slope elements. Accordingly, there is a general tendency for the deposits to thicken down-slope. This thickening, however, is preferentially developed across slopes of particular facing directions and gives rise to pronounced drift terraces. The best developed of these, both in width and thickness, are almost exclusively in the sector from northwest, through north to northeast. This produces a marked asymmetry in the drift distribution of the upland valleys, a characteristic noted by many as an indicator of former periglacial environments (Gloriad and Tricart 1952; Ollier and Thomasson 1957; Tricart 1970; French 1971 and 1972). Overall, the assemblage of slope forms in the uplands accords well with a stage between the mature and ultimate phases of an evolutionary sequence of periglacial slope conditions (Tricart 1970).

On interfluves and summits the local drift is a thin, angular rubble with little fines. On lower slopes and terraces it is thicker, more compact and silty, and displays much evidence of a rude stratification parallel to slope. The contact between the drift and the underlying bed-rock is everywhere gradational, and smooth, polished or striated rock surfaces are unknown. The fabric of the deposits everywhere shows rock fragments orientated down-slope (Fig. 3A), throughout the vertical range (Fig. 3B).

On the more extensive terrace slopes a number of distinct sedimentary types can be distinguished, and a crude stratigraphy is displayed. Thick sequences of stony clay or head occur between thinner units of coarser, often sorted washed gravel. An outline stratigraphy, derived from a type-site in the upper part of the Sulby River (Fig. 3A), displays the following members. Above the bed-rock is a coarse, angular Rock Rubble (RR, Fig. 2 and Fig. 3B), that passes upwards into a finer, more clayey, Lower Stratified Head (LSH). This is succeeded by a coarse,

Table 2. Origin of the Upland Formation

Cycle	Phase	Member	Process
	3	Upland Peat (UP)	Soil-formation
3	2	Sulby Gravel (SRG)	Slope-wash
	1	Upper Stratified Head (USH)	Gelifluction
	2	Upper Gravel (UG)	Slope-wash
2	1	Mid Stratified Head (MSH)	Gelifluction
		Massive Head (MH)	
	2	Lower Gravel (LG)	Slope-wash
1	1	Lower Stratified Head (LSH)	Gelifluction

edge-rounded, often open-work Lower Gravel (LG); an unstratified Massive Head (MH); and a Middle Stratified Head (MSH). Above is a distinctive, iron-stained, often indurated Upper Gravel (UG), that is succeeded by an Upper Stratified Head (USH). These members form part of the Upland Formation of the Surface Group (Table 1). They are locally overlain by extensive Upland Peat (UP) and Sulby Gravel (SRG), that together comprise the Moorland Formation of the post-Pleistocene Ayre Group (Table 1).

From one section to another the two gravel members retain their stratigraphic position, suggesting that they may have some palaeo-climatic significance. If this is the case, we can resolve the sequence of heads and gravels into a series of periglacial climatic cycles, similar to those proposed by Alexandre (1960). Ignoring the Rock Rubble, which would represent initial frost break-down caused by the onset of cold conditions, the first phase in a cycle is that of gelifluction, represented here by the Lower, Middle and Upper Stratified Heads (Table 2). The second phase is that of slope-wash, represented by the Lower, Upper and Sulby Gravels. The third phase is that of soil formation, represented only by the thick Upland Peat and modern soil. We therefore have two truncated cycles, and one full cycle (Table 2), representing two partial and one full transition from periglacial to temperate conditions.

Whilst the slope form, fabric and deposit type of the local suite point strongly to a periglacial origin, this does not imply that they are derived exclusively from frost break-down of the bed-rock. They may in part be derived from pre-existing

Fig. 3. A. Fabric characteristics of the local deposits in the Sulby River basin. Coarse stipple: solifluction terraces; fine stipple: local drift.

B. Fabric characteristics through the succession at Lhergyrhenny (X on Fig. 3A). For identification of members see Table 1 and text.

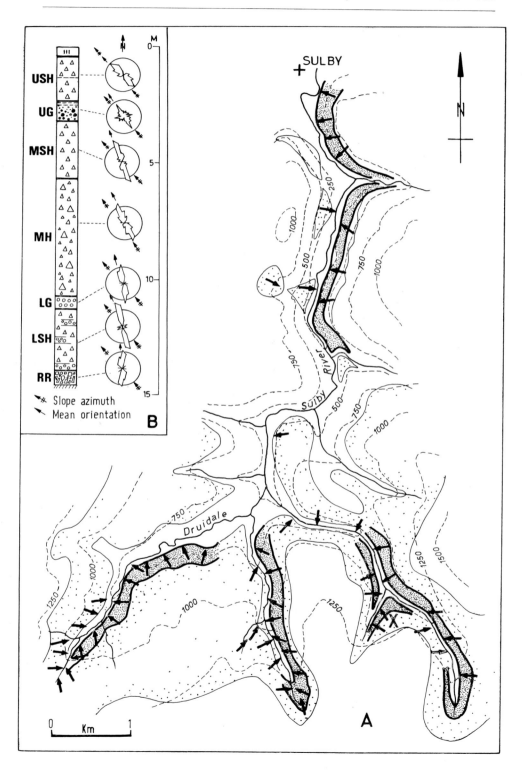

till. That a glaciation of the uplands has occurred is substantiated by evidence from the south and east of the island, where local granite erratics (Fig. 1) can be traced to heights of 483 m, some 280 m above the highest bed-rock source (Darwin 1848; Lamplugh 1903). There is no direct evidence of the age for this glaciation, however, and no unaltered till from it is recognised. At the margin of the uplands, head deposits of identical character to those of the Upland Formation are found below the lowest members of the complex Surface Group foreign formations. This implies that periglacial conditions were operative on the island before, during, and after the last major glacial episode. In this case, the granite erratics, and possibly the faint striae (Fig. 1), must represent all that is recognisable of a major glaciation that precedes this episode.

3. The foreign deposits

The foreign deposits of the island are best displayed in a broad triangle north of the probably pre-glacial Manx Slate scarp that runs west from Ramsey (Fig. 1). The triangle is broken across its apex by the ridges of the Bride Moraine. South of the moraine, the topography is essentially that of a large, flat sandur. To the north, the moraine is truncated by a post-Pleistocene raised cliff and associated complex of beach ridges and dunes (Ward 1970).

3a. East coast

The stratigraphy displayed on the east coast is shown in Figure 4. Three stratigraphic formations are exposed. The lowest, the Shellag, comprises four distinct lithostratigraphic members (Table 1). The lowest member, the Shellag Till (ST), is exposed discontinuously along the base of the cliffs throughout the 13 km length of section. This till is a stiff, over-consolidated, calcareous, red clay. It is succeeded by the Shellag Sand (SS), most thickly developed immediately south of the moraine. In the tectonically complex area adjacent to the moraine, the Shellag Sand passes locally into a coarse, cemented gravel, the Shellag Gravel (SG). A further passage into the Kionlough Gravel (KG) is seen south of the moraine. All the members of the Shellag Formation are highly disturbed tectonically north of the moraine. South of it the degree of disturbance diminishes rapidly.

Immediately to the rear of the moraine, the Shellag Formation is succeeded unconformably by the Orrisdale Formation. This formation, whose type-site is found on the west coast, is here divided into five members. Overlying the contorted Shellag Till and Sand is a basal Ballavarkish Till (BVT). This sandy and stony till terminates against the rear of the moraine where it passes into a series of sands, gravels and clays, as the Ballavarkish Marginal Series (BVMS). Northwards, the till is succeeded by a Ballavarkish Sand (BVS). To the south of the moraine, the

Fig. 4. East coast succession. 1: Shellag Till (ST); 2: Shellag Sand (SS); 3: Shellag Gravel (SG); 4: Kionlough Gravel (KG); 5: Ballavarkish Marginal Series (BVMS); 6: Ballavarkish Till (BVT); 7: Ballavarkish Sand (BVS); 8: Kionlough Till (KT); 9: Dog Mills Series (DMS); 10: Ballaquark Till (BQT); 11: Ballaquark Sand (BQS); 12: Cranstal Till (CT); 13: Crosby Channel Deposits (CCD); 14: Andreas Platform Gravels (APG); 15: Ayre Beach (AB); 16: Cranstal Silts (CSS). For identification of members at Ballure see Fig. 6.

BALLURE

DOG MILLS

KIONLOUGH

SHELLAG POINT

CROSBY CHANNEL

PHURT

2 km Gap

90 m Gap

240 m Gap

Marx Scale

Metres
100
50
0
Horizontal and Vertical scale

Surveyed Summer 1969-71

Shellag Sand and Kionlough Gravel are succeeded conformably by a well-bedded, sandy till, the Kionlough Till (KT). This thickens rapidly off the moraine face to pass into a series of laminated silts, sands and clays, as the Dog Mills Series (DMS).

Some 2 km north of the moraine the members of the Orrisdale Formation are succeeded, again unconformably, by deposits of the Jurby Formation. This formation, whose type site is also found on the west coast, is here divided into a sandy Ballaquark Till (BQT) and a succeeding Ballaquark Sand (BQS) (Table 1). Towards Phurt a further member of this formation, the Cranstal Till (CT), rests unconformably upon these members.

The three foreign glacigenic formations seen on this coast form part of the Surface Group (Table 1). North of Phurt, the Cranstal Till Member of the upper-most Jurby Formation is truncated by a raised marine cliff. Northwards, the till is overlain by the lacustrine Cranstal Silts (CSS) and the gravel ridges of the Ayre Beach (AB). These two deposits form part of the post-Pleistocene Ayre Group (Table 1).

3b. West coast

The stratigraphy displayed on the west coast is shown in Figure 5. The sections between Glen Mooar and Jurby Head have been described by Mitchell (1965). He distinguished two major formations, the Ballateare and Orrisdale, but re-mapping suggests a revision.

The basal stratigraphic member exposed is the Wyllin Till (WT). This is a tectonically disturbed, clay-rich till, similar in all respects to the Shellag Till of the east coast. It is succeeded by a series of sands, the Wyllin Sands (WS), that contain seams of rich shelly gravel (Lamplugh 1903; Mitchell 1965). Both these members are placed in the Shellag Formation by virtue of their lithological similarity, structural deformation and position beneath a major unconformity.

Above the unconformity along this coast lies the widely exposed Orrisdale Formation, the three west coast members of which, the Orrisdale Till (OT), the Orrisdale Sand (OS) and Orrisdale Gravel (OG), dominate the coastal sections (Fig. 5). The Orrisdale Till, up to 25 m in thickness, is of similar lithological character to its east coast equivalent, the Ballavarkish Till, being sandy, stony, well-bedded and relatively undeformed.

The type-site for the succeeding Jurby Formation occurs at Jurby Head (Fig. 5). Resting conformably upon the Orrisdale Formation below, it here comprises a lower Jurby Till (JT), and an upper Jurby Sand (JS). South of the major morainic ridge of Orrisdale Head these two members are replaced stratigraphically by a thinly bedded and partially laminated till, the Trunk Till (TT). This till thickens rapidly off the southern face of the Orrisdale ridge, attains its maximum thickness north of Glen Wyllin and thins towards Glen Mooar (Fig. 5). It is locally succeeded by the Trunk Gravel (TG). The Trunk Till is deformed, and on its surface have developed a series of five basins of organic sediment, the Wyllin Kettles (WK). Further shallow basins containing organic sediment, the Jurby Kettles (JK), occur above the Jurby Sand just north of Jurby Head.

The Wyllin Kettles are succeeded by a series of cryoturbated debris fans of local origin. They include the Ballaleigh Debris Fan (BDF) (Mitchell 1965), the Wyllin Debris Fan (WDF) and the Ballyre Debris Fan (BYDF). The heavily cryoturbated Crawyn Sand (CS), that occurs between the Orrisdale and Jurby ridges, forms the distal portion of a similar fan, the Ballaugh Debris Fan (BHDF), that issues from the mouth of Glen Dhoo (Watson 1971). The debris fans, and the kettles that underlie them, are grouped together as the Ballaugh Formation (Table 1). This formation is overlain by the Curragh Peat (CP), which is part of the Ayre Group.

3c. The sub-surface succession

Underlying the majority of the northern drift plain is a remarkably level rock platform at between 41 and 52 m below O.D. (Smith 1930). North of the Bride Moraine it drops rapidly away to at least 145 m below O.D. at Point of Ayre. From bore-hole records, Smith (1930) recognised three major units. The lowest, recorded only in the deepest bores at Point of Ayre, comprised a Lower Boulder Clay (LBC) and a Lower Sand (LSS). Above this, Lamplugh (1903) described a series of marine deposits, the Ayre Marine Silts (AMS), that include richly fossiliferous sands and silts. These three deposits are grouped together here as a Basement Group (Table 1). The second of Smith's units included a Middle Boulder Clay (MBC) and Middle Sand (MSS), identified here as a Sub-Surface Group (Table 1). The third unit, comprising an Upper Boulder Clay and Upper Sand, includes all the foreign deposits seen above sea-level, and is equivalent to the Surface Group.

4. The relationship between the local and foreign deposits

At Ballure on the east coast, and south of Glen Mooar on the west (Fig. 1), the foreign suite is banked against the Manx Slate core of the island, and the relationship between the local and foreign suites is well seen.

At Ballure, a fossil slate cliff is buried by a coarse, angular local scree, the Ballure Scree (BS, Fig. 1 and Fig. 6A). Rounded pebbles in the lower part of the scree suggest derivation from a degraded beach, but none is seen. Above the scree are three local head members, all lithologically identical to those of the Upland Formation. Between the lower two, the Brown Head (BH) and the Lower Blue Head (LBH), are a thin series of sands and clays identified as the Ballure Slope Wash (BSW). Between the Lower Blue Head and the Upper Blue Head (UBH) is a thick sequence of foreign sands and gravels that are probably equivalent to the widespread Shellag Sand (SS) exposed further north. Within these sands, Kendall (1894 p. 414) noted a "boss of clay" of foreign origin. No longer visible, this clay is probably equivalent to the Shellag Till.

Above the Upper Blue Head is a till, the Ballure Till (BT), of mixed but predominantly local composition. On the foreshore it is overlain by further Shellag Till. In this area, therefore, two Shellag Tills can be recognised, one below and one above the Upper Blue Head. Above the Ballure Till is a repeated sequence of stoneless, foreign red clay, the Ballure Clays (BC) and washed and sorted local flood gravel. The latter form part of the Ballure Debris Fan (BLDF).

At and south of Glen Mooar, a broadly similar succession is seen (Fig 6B). Banked against the slate is a coarse scree, the Mooar Scree (MS). Above are irregular patches of Wyllin Till (WT) and Wyllin Sand (WS) capped by thick Mooar Head (MRH). Orrisdale Till (OT) lies above, followed by a thick succession of Orrisdale Sand (OS) and a thin remnant of Trunk Till (TT). The sequence is capped by cryoturbated gravel of the Ballaleigh Debris Fan (BDF).

The local deposits at both these sites are heads by their fabric and other sedi-

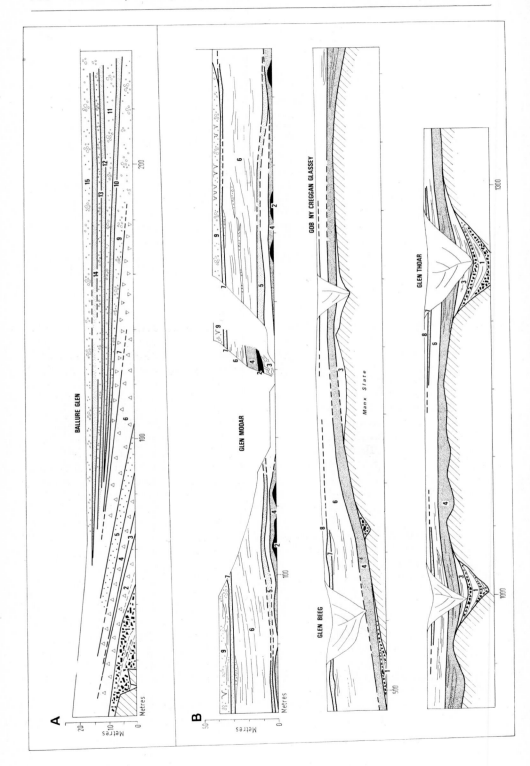

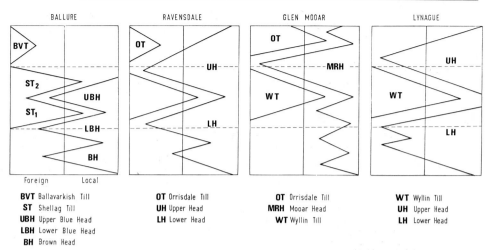

BVT Ballavarkish Till
ST Shellag Till
UBH Upper Blue Head
LBH Lower Blue Head
BH Brown Head

OT Orrisdale Till
UH Upper Head
LH Lower Head

OT Orrisdale Till
MRH Mooar Head
WT Wyllin Till

WT Wyllin Till
UH Upper Head
LH Lower Head

Fig. 7. Schematic representation of the relationship between the local and foreign deposits. For location of sections see Fig. 1.

mentological properties. They are assigned, therefore, on this and general litho-logical grounds, to the Upland Formation (Table 1). The essential relationship between the members of this formation and those of the foreign formation is that they occur both above and below the most significant and widespread foreign one, the Shellag. Hence, three major stratigraphic episodes can be recognised (Fig. 7). Prior to the foreign ice invasion that deposited the members of the Shellag Formation, unrestricted head deposition occurred around the island margin and by inference in the island interior. During the Shellag advance a competitive situation existed at the island margin. At Glen Mooar, a location more open to full foreign ice assault, foreign deposition dominates. At Ballure, a more sheltered locality, local head deposition predominates. After the Shellag advance, the less extensive readvances that deposited the succeeding Orrisdale and Jurby Formations allowed the re-occupation of the margin by further head deposition.

The relationship between the local and foreign formations of the Surface Group can therefore be seen as part of a continuum between full periglacial deposition in the uplands, intercalating facies competition at the margin, and full foreign glacial deposition to the north. This implies that some members of the Upland Formation in the uplands are equivalent chronologically to members of the same

Fig. 6. The relationship between the local and foreign deposits.

A. The succession at Ballure. 1: Ballure Scree (BS); 2: Brown Head (BH); 3: Ballure Slope Wash (BSW); 4: Lower Brown Head (LBH); 5: Shellag Sand (SS); 6: Upper Blue Head (UBH); 7: Ballure Till (BT); 8: Shellag Till (ST) (On foreshore); 9, 11, 13 and 15: Ballure Debris Fan (BLDF); 10, 12 and 14: Ballure Clays (BC).

B. The succession around Glen Mooar. 1: Mooar Scree (MS); 2: Wyllin Till (WT); 3: Wyllin Sand (WS); 4: Mooar Head (MH); 5: Orrisdale Till (OT); 6: Orrisdale Sand (OS); 7: Trunk Till (TT); 8: Loam; 9: Ballaleigh Debris Fan (BDF).

formation at the margin and to members of the foreign formations to the north.

5. The sequence of events

We begin the record of the events of the Quaternary of the Isle of Man at the base of the exposed Surface Group, for it is at this point that the only possible evidence for an interglacial datum can be established. An age determination for the underlying Sub-Surface and Basement Groups is therefore dependant upon that made for the Surface Group and will be considered subsequently.

The first major event is represented by the fossil cliff at Ballure. This is tentatively assigned to the Ipswichian interglacial, for nowhere in the complex of cold-climate deposits of the overlying Surface Group is there any indication of a substantive climatic break. The climatic deterioration at the beginning of the Devensian saw the frost shattering of the cliff and its burial by scree. Subsequent Lower and Middle Devensian time is represented at Ballure by the Brown Head, the Ballure Slope Wash, which may be assigned to the Upton Warren interstadial complex, and the Lower Blue Head.

Subsequent to the main expansion of Devensian ice around 25000 yrs B.P. (Mitchell 1972), a foreign glaciation advanced onto the island and laid down the deposits of the Shellag Formation. With a well-defined limit of advance to an absolute altitudinal maximum of 180 m around the island margin (Line S1 Fig. 1) (Lamplugh 1903; Temple 1960; Mitchell 1965), there is no evidence that ice overrode the island. Instead, it by-passed most of the solid core, which emerged as an ice-free nunatak. On the east coast at least, the Shellag ice suffered a phase of minor retreat, for the Upper Blue Head locally intercalates the lower and upper Shellag Tills. The readvance associated with the upper Shellag Till may have some expression in the foreign limit mapped by Temple (1960) (Line S2 Fig. 1).

The Shellag episode was terminated by a retreat to a position running approximately from Orrisdale Head towards the Point of Ayre. The Shellag and Wyllin Sands were laid out before the retreating margin and the dominantly sandy outwash was diversified by the Shellag and Kionlough Gravel spreads.

Following the general retreat a powerful but short-distance readvance, the Orrisdale, took place. Its limits were less extensive than those of the Shellag. On the western side of the island it just penetrated the rock margin to low elevation in the area of Kirk Michael and deposited the Orrisdale Till. From here the limit looped northeast (Line O Fig. 1) along the line of the Bride Hills to Shellag Point. To the rear of the moraine it deposited the Ballavarkish Till. Extensive tectonic activity accompanied the readvance and is responsible for both the form and tectonic structure of the Bride Moraine (Slater 1931) and the general unconformity that separates the Shellag Formation from the Orrisdale.

South-east of the Bride Moraine the ice limit probably swept across Ramsey Bay to a position a little seaward of Maughold Head. This is supported by the deposits of the Dog Mills Series, which are in part lacustrine and which represent the infilling of Lake Andreas (Lamplugh 1903). This lake was trapped between the ice margin to the north and east and the solid core of the island to the south. Partial escape for the ponded lake water is found in the wide melt-water channel that separates Maughold Head from the island proper.

With the decline of the Orrisdale readvance, and the tectonic break-up of its margin at Shellag, large volumes of flow-till were let down the long distal slope of the moraine to impinge into the north shore of the lake. A progressive passage is therefore seen north of the Dog Mills where this till, the Kionlough, passes into the Dog Mills Series. On the southern margin of the lake, at Ballure, the Dog Mills Series passes into alternations of Ballure Debris Fan and partially laminated Ballure Clay, suggesting a large delta issuing from the mouth of Ballure Glen.

Following the Orrisdale readvance, the ice retreated to a position a little north of the island, and large sandur sedimentation took place beyond the margin. On the east this was directed westwards by the topographic obstruction of the Bride Moraine, in a series of marginal channels running parallel to the ice-front. At Jurby on the west the discharge was mainly south-east across a large delta that abutted into the western margin of Lake Andreas. Around Orrisdale Head the discharge was concentrated in a series of large marginal channels running south-west parallel to the declining ice margin. These led to the build-up of large thicknesses of Orrisdale Sand and Gravel.

After the Orrisdale retreat the ice again readvanced. On the east it overrode and mildly contorted the Ballavarkish members of the Orrisdale Formation to deposit the Ballaquark Till Member of the Jurby Formation to the rear of the Bride Moraine (Line J1 Fig. 1). The ice limit then passed west to cross the present coast at Jurby Head where it deposited the Jurby Till. From the margin of the ice in this area, large volumes of flow-till slumped south-east, as the Trunk Till, to bury the Orrisdale Sand and Gravel.

The Jurby ice retreated briefly and its outwash pattern broadly resembles that of the preceding Orrisdale retreat. Some outwash, however, broke through the Bride Moraine barrier, via the Lhen Trench (Lamplugh 1903), to deposit the Andreas Platform Gravels. After a short interval the retreating Jurby ice locally readvanced to deposit the Cranstal Till over the rear of the Bride Moraine towards Phurt. From here the limit ran west (Line J2 Fig. 1), off the present coast, for no equivalent till is seen on the west coast. After this time ice permanently left the island.

In hollows on the collapsed surface of the Trunk Till south of Orrisdale Head, and on the Jurby Till at Jurby Head kettle basins developed. With the return of more genial climatic conditions, fluvial action in the uplands was resumed. The large amounts of glacial and periglacial deposit stored in the lower parts of the upland glens were re-worked and issued as large debris fans around the island rock margin. In the Kirk Michael district much of the morainic topography and the kettle basins were buried. Many of the fans show cryoturbation structures and Watson (1971) has interpreted some of the depressions that occur on the margins of the Ballaugh Debris Fan as the remains of pingos.

One published view detracts from the suggestion that the above sequence of events is of Devensian age. In the upper part of the Upper Blue Head at Ballure, Mitchell (1972) identified a de-calcified horizon that he referred to the Ipswichian interglacial. This view would assign at least part of the Shellag Formation to the Wolstonian and the Ballure Cliff to the Hoxnian. The present author finds no evidence of progressive or systematic de-calcification in this head, for like those below it, it is composed of local material, and is therefore non-calcareous. As no other evidence of a substantial climatic break is recorded throughout the Surface Group, the safest assumption is that the Ballure Cliff marks the Ipswichian.

An upper age bound for the succession of foreign advances and readvances in the Devensian is provided on the west coast. A moss band at the base of one of the

Ballaugh Formation kettle basins, near Glen Ballyre, has yielded five ^{14}C dates in the range 18900 (Birm–213) to 18400 yrs B.P. (Birm–270) (Shotton and Williams 1971 and 1973). These dates overlie all the foreign formations of the Surface Group and suggest that the Shellag advance, and the succeeding Orrisdale and Jurby readvances, occurred in the period 25000 to c. 18400 yrs B.P. Some 30 cm above the basal moss at Ballyre, early Zone I deposits yield ^{14}C dates of 12645 (Birm–214) (Shotton and Williams 1971) and 12210 yrs B.P. (GRO–1610) (Mitchell 1965). Further dates in the upper portions of other basins in the Wyllin Kettle series range from 12150 to 10250 yrs B.P. (Mitchell 1965; Dickson, Dickson and Mitchell 1970).

No direct evidence of age is available from the members of the local Upland Formation. The fact that head and scree underlie the Shellag Formation at Ballure and Glen Mooar, and overlie the Ipswichian Ballure Cliff, provides strong evidence that the Upland Formation represents the whole of Devensian time. Some of the upland members must therefore be equivalent to those seen at the margin. Although no direct correlation can be established, it appears likely that the Rock Rubble is equivalent to the Ballure and Mooar Scree, and that the Lower Stratified Head is equivalent to the Brown Head. The Lower Gravel may be correlated with the Ballure Slope Wash, and may together represent the Upton Warren interstadial complex. The Massive Head, finer and more comminuted than the others, may suggest the more severe conditions pertaining in the uplands during the succession of foreign ice advances and readvances around the island margin. The Upper Gravel may then represent the milder conditions of the Allerød interstadial.

The Flandrian in the Isle of Man is represented by the beach gravels at Point of Ayre, the peat and alluvium of the Curragh basin, and the hill-peat and flood gravel within the uplands. The marked Flandrian cliff-line that truncates the rear of the Bride Moraine was probably cut around 9500 yrs B.P. (Phillips 1967), for Boreal (Pollen Zone V) deposits are banked against it at Phurt. To c. 5000 yrs B.P. a developing barrier beach created brackish water ponds or lagoons against the cliff. After this date the beach accreted rapidly as the sea-level regressed from its Flandrian maximum. The removal of immediate marine influence reduced the brackish water element in the ponds and they became dominantly fresh.

It remains only to consider the age of the buried Quaternary strata. The Surface Group lies stratigraphically above the Ballure Cliff, which we have suggested is best considered Ipswichian. The Sub-Surface Group may hence be logically assigned to the Wolstonian, the Ayre Marine Silts to the Hoxnian, and the Lower Boulder Clay and Lower Sand of the Basement Group to the Anglian. It is possible that the Ayre Marine Silts represent a deep water facies of the same Ipswichian sea that cut the Ballure Cliff. In this case, the Sub-Surface Group would become Devensian, and the Lower Boulder Clay of the Basement Group, Wolstonian (Wirtz 1953). It is also possible that the Ayre Marine Silts represent some torn-up and transported mass of very much older sea-floor, not *in situ*. It is significant, however, that they lie on the same horizon as the remarkably level platform that underlies the majority of the Sub-Surface Group. The level, trimmed nature of this platform, cut across a variable basement geology, may indicate a marine origin, although no marine deposits have actually been recognised upon it. If this is the case, the platform, and the Ayre Marine Silts, may be related to one sea-level quite distinct in time from the very much higher sea-level recorded at Ballure.

6. Discussion

Arising from its strategic location and complex stratigraphic succession, the Isle of Man carries obvious significance for the Quaternary history of the Irish Sea basin. Indeed, in the context of the current controversy surrounding the chronology and limits of the last glaciation of the basin (Mitchell 1972; Bowen 1973a and b), the proposed sequence of events in the Isle raises some rather intractable problems of chronology, ice gradient and rate and condition of retreat. In addition, the excellence of the sections generates some points of significance for general stratigraphic principle. It is not opportune to discuss these topics in detail here and we shall suffice with a general discussion.

6a. Till facies and glacial stratigraphy

The exposed Surface Group of deposits essentially shows a multiple till sequence developed over an apparently short phase of the Devensian glaciation. Between and partly within each widely traceable till member are displayed considerable facies variations. Three major till types can be identified.

The first type includes the Shellag and Wyllin Tills. Both are plastic, clay-rich, over-consolidated and highly deformed. They appear to have been deposited as lodgement tills under thick, active ice, at a point well to the rear of the margin. The second type includes the Ballavarkish, Ballaquark, Cranstal, Orrisdale and Jurby Tills. All are sandy, stony and crudely bedded. By their proximal thinning and termination they appear to result from the marginal melt of debris-rich ice. They accord with the "melt-out" tills of Boulton (1970). The third type, including the Kionlough and Trunk Tills, are flow-tills (Boulton 1968). They demonstrate by their proximal and distal thinning, intra-formational flow and slump structures, frequent lamination, and position beyond major moraine limits, deposition by normal mass-movement processes beyond the active ice margin.

The essential stratigraphic relationship between the three till types is shown schematically in Figure 8, and is illustrated by the sequence of Shellag, Ballavarkish and Kionlough Tills on the east coast of the island (Fig. 4). The relationship demonstrates that during one short phase of one glacial event a number of very different till facies may be deposited as a function of position relative to the ice margin. This implies that till lithostratigraphy does not provide a good basis for event correlation. The point is illustrated in Figure 8B. Two major glacial events are shown separated in time. During each event three till types are deposited. If these tills are correlated by lithostratigraphy, and each correlated till is taken to mark a glacial event, three major events, not two, are distinguished. The heavy pecked line in Figure 8B clearly indicates that such a correlation is entirely time-transgressive. Devoid of a firm chronological base, long-distance correlation of tills by lithostratigraphy appears to be an entirely inappropriate procedure.

6b. Chronology

The Glen Ballyre dating sequence, between 18900 and 18400 yrs B.P., occurs above all the Manx glacigenic formations. No ice can, therefore, have passed further south than the island after this date. This concept conflicts with later dates much further to the south, for many of these require the wide presence of Irish Sea ice to considerable elevation against North Wales in Late Devensian times (Rowlands 1970). It also conflicts with the almost equivalent date of 18500 yrs B.P. at Dimlington in East Yorkshire (Penny, Coope and Catt 1969). This date

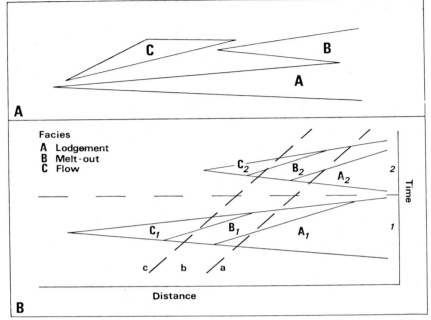

Fig. 8. A. Schematic relationship between lodgement, melt-out and flow till facies generated in one glacial episode.
  B. Till facies and time.

derives from material beneath a complex sequence of tills, whereas the Ballyre dates are derived from material above. The contrast implies that the major phase of Devensian glaciation in western Britain pre-dates 18500 years B.P., while in eastern Britain it post-dates 18500 years B.P. Andrews (1973) has pointed out, in the context of similar disparities between sections of the Wisconsin Laurentide cap, that the maximum of glaciation may be exceedingly time-transgressive. His explanation is that glacier changes are not so much a function of shifts in world climate, but are a response to small, and often climatologically insignificant, changes in glacier mass balance. Paterson (1969) has demonstrated that such changes may have a disproportionate effect upon snout response. Glacier advances may, therefore, be neither chrono-correlative nor synchronous with world climatic change.

6c. Ice gradient

We have suggested that Devensian Irish Sea ice rose no higher than 180 m onto the margins of the Isle of Man. A very low ice gradient of approximately 100 m in 300 km is therefore indicated for the limits of Devensian Irish Sea ice in Pembrokeshire and south-east Ireland proposed by Bowen (1973a and b). Even if the limit is placed at the Lleyn Peninsula (Mitchell 1960; Synge 1964), the gradient would still remain low, at about 100 m in 150 km. Both these gradient estimates ignore a relative reduction in current altitudinal differences caused by a normal, but unknown, amount of isostatic depression at the time of maximum glaciation, an effect that might flatten the gradients considerably. They are an order of magnitude or more lower than the observed gradients of the Greenland ice-cap (Paterson 1969) and the reconstructions of both the Scandinavian (Flint 1970) and the Laurentide (Bryson *et al.* 1969; Andrews 1973). They are also below the anomalously low gradients reported by Mathews (1974) on some margins of the Laurentide.

Given that the surface of an ice-cap approximates to a parabolic curve (Paterson 1969); making the assumption, implicit in the work of Bowen (1973a), that a line from central Scotland to the southern Irish Sea is a radial flow-line; and ignoring isostatic effects, it is possible to make estimates of the ice thicknesses required over the Isle of Man to sustain a snout in either Pembrokeshire or the Lleyn. Using the coefficients derived by Hollin (1962) for the Antarctic, and by Mathews (1974) for Greenland (both with $A = 4\cdot7$ $m^{\frac{1}{2}}$), a limit in Pembrokeshire would require an ice thickness of approximately 2500 m in the Isle of Man and 3300 m in central Scotland. To the Lleyn, ice thicknesses would be 1800 and 2750 m respectively. If the coefficients are adjusted to accommodate a maximum Devensian ice thickness of some 1500 m over central Scotland (Sissons 1967), the required ice thicknesses in the Isle of Man would be 1150 m ($A = 2\cdot17$ $m^{\frac{1}{2}}$) and 975 m ($A = 2\cdot53$ $m^{\frac{1}{2}}$), respectively, for limits in Pembrokeshire and the Lleyn. There is no evidence that even the lowest of these ice thickness estimates occurred in the Isle of Man during the Devensian.

6d. Devensian marine activity

The somewhat unique form of moraines like that of the Bride suggest that special conditions may be necessary for their creation. Andrews (1973) has drawn attention to the fact that similar moraines around the former margins of the Wisconsin Laurentide ice-cap commonly occur immediately landward of, or just within, glacio-lacustrine or glacio-marine boundaries. This suggests that they may be a response to a change from a water-based to a land-based margin. Ablation rates in water are substantially higher than land rates, due to calving. If an ice-sheet retreats so that its margin becomes land-based, the glacier system will attempt to compensate for the lower rates of loss at the snout by lower rates of flow. There will, however, be a lag in the response of the glacier system. During this lag, the unablated ice may build up at the snout to give a pronounced still-stand, a series of minor readvances, or a tectonic response. This in consequence may account for the topographic form of the Bride Moraine, its rearward complex of readvance tills, and its tectonic structure.

After the Shellag advance there was certainly a large pro-glacial lake trapped between the ice-front and the island rock margin. It is doubtful, however, if this were large enough to induce the necessary changes in the glacier. This invites the possibility that the Dog Mills Series might, in part, be not glacio-lacustrine, but glacio-marine. Wright (1902) and Wright and Reade (1906) provide lists of foraminifera obtained from the series, and a very tentative interpretation suggests cold-water accumulation, although no definite arctic indicators are present (J. R. Haynes, *personal communication*). The majority of the fauna is open-water marine forms, but two estuarine and intertidal species occur in abundance. If the fauna is in place, deposition was either cold-water marine, with an introduced brackish water element, or more likely, considering the laminated nature of much of the Dog Mills Series, estuarine-intertidal, with a current-swept, open-water element. Partial corroboration of a marine episode is provided on the east, south and southeast coasts of the island where two distinct marine limits can be traced at 23 to 29 m and 12 to 16 m O.D. (F. M. Synge, *personal communication*). Both occur outside the Orrisdale readvance limit, and the lower is at approximately the same level as the Dog Mills Series.

The Dog Mills Series form part of the Orrisdale Formation. This lies stratigraphically below the ^{14}C dating series at Glen Ballyre, and is hence older than 18900 yrs B.P. If the foraminifera are *in situ* they would represent a marine incursion after 25000 yrs B.P., the maximum age for the underlying Shellag Formation, but before 18900 yrs B.P. Stephens *et al.* (1975) have recently summarised Late Devensian shoreline evidence in northeast Ireland, and although the situation is by no means resolved, it is becoming increasingly clear that arctic seas followed the withdrawal of the Late Devensian ice from at least as far back as 14000 yrs B.P. The evidence from the Dog Mills Series suggests that this date may be pushed back to c. 19000 yrs B.P. Although there are difficulties with the heights of the associated marine features, this view would be consistent with the mapping by Synge (*personal communication*) of a marine level outside the limit of the Drumlin readvance phase of eastern Ireland. Stephens *et al.* (1975) correlate this phase with a limit no further south than Ballyre at about 18–19000 yrs B.P. This extension of marine accompaniment raises the possibility that marine conditions may have been associated with Devensian ice right back to the time of its maximum extension.

6e. Multiple readvance and stratigraphic classification

The study of glacial readvance has figured significantly in many recent Quaternary investigations. Emphasis has been placed on tracing the limits of readvance, the correlation of such limits from area to area, and their placement in a chronological sequence. Implicit in many of these studies are two underlying assumptions. The first is that glacial readvance is a direct response to climatic change. The second is that the response is synchronous with the climatic change and that both are chrono-correlative regionally or even world-wide. These assumptions are enshrined in a recent paper by Jardine (1972), in which it is considered that readvance provides substance for a division of the Quaternary based on climato-stratigraphic units.

Jardine (1972) has proposed that evidence of readvance be established by reference to both geomorphic and stratigraphic criteria. Even though many workers have defined both morpho-stratigraphic units (Frye and Willman 1962) and morpho-stratigraphic classification (Mangerud *et al.* 1974), the use of morphological criteria is of doubtful validity (Richmond 1959; Flint 1970). A large morainic structure, for instance, may mark a readvance, a recessional still-stand, or even a terminal maximum. Despite this difficulty, many workers have attempted to correlate the Bride Moraine with other limits of readvance around the margins of the northern Irish Sea basin (Charlesworth 1926 and 1939; Cubbon 1957; Mitchell 1960; Penny 1964; Sissons 1964; Gresswell 1967; Stephens *et al.* 1975). It would appear essential that before morpho-stratigraphic criteria are used to correlate readvances and to trace geographic limits, the very existence of a readvance should be established by other, less ambiguous, criteria first.

The stratigraphic criterion most commonly adopted is the presence of one unit of sub-glacier deposit, usually till, above another. This itself is not, however, unambiguous for Boulton (1967) has shown that complex till sequences may have been formed during one single glacier retreat phase rather than by multiple advance and retreat. A more exact criterion would be the presence of a strong unconformity, indicating override, separating pro-glacial deposits below from sub-glacial deposits above.

In this work we have established considerable stratigraphic evidence for at least three episodes of readvance after the major Shellag advance. On the east coast, for instance, the Ballavarkish, Ballaquark and Cranstal Tills all unconformably off-lap one another to the north. Accordingly, if this evidence of readvance is used to define a series of stades, as Jardine (1972) proposes, we must recognise at least three complementary interstadial episodes between them. Moreover, given the chronology that we have established, these three interstades must be fitted into the short time span between 25000 and 18400 yrs B.P. There is no support for this from work elsewhere.

We have already proposed in the preceding section that the morpho-tectonic structure of the Bride Moraine may be a function of a change from a water-based to a land-based margin. A similar explanation may account for the sequence of readvances. This is supported by the fact that although the evidence for readvance is striking, the forward extent is, in each case, exceedingly limited, and can be measured in terms of single rather than tens or hundreds of kilometres. Because of the continuity of section it can be demonstrated that the Orrisdale readvance, for example, has a forward movement of less than 5 km, and the others considerably less. It would seem unlikely that the cause of such limited readvance lies in climatic change, and certainly not climatic change of sufficient magnitude to justify the identification of a significant interstade. At least from evidence in the Isle of Man we may conclude that many readvances are of local significance, are caused by local not secular conditions, have limited palaeo-climatic implication, and are time-independent events that cannot be correlated one with the other. If this is so, it implies that whilst a stade may indeed be distinguished by a readvance in many cases, the identification of a readvance does not, of necessity, distinguish a stade.

6f. A tentative resolution

The problems of chronology, ice gradient and marine accompaniment can be resolved if the basic geometry of the Devensian Irish Sea ice is conceived to be somewhat different from that implicit in the views of Bowen (1973a). Instead of one large cap filling the whole of the Irish Sea basin and coeval with Irish, Welsh and Scottish ice, it is possible to conceive two separate masses essentially bisecting about the pivot of the Isle of Man. This conception is little different from that pictured by Mitchell (1972) and Stephens *et al.* (1975), except that their geometry relates to a stage early in the retreat of the Devensian. Our conception is that the geometry they outline was broadly the same at the Devensian maximum itself. The western mass formed the outer margin of an Irish cap, fed in part by ice from western Scotland and the North Channel. The eastern mass formed a similar outer margin to a Southern Uplands cap, assisted by ice from the Solway and the Lake District. Under this concept, the low rise of foreign ice onto the Isle of Man would be a function of the low altitude margins of the two flanking ice masses, and not the thick ice centre of one. Any chronological limit in the Isle of Man would thence become quite compatible with limits of similar age further south in the basin. The narrow area between the two lobes to the south of the Isle of Man might also provide an opening for marine penetration.

Within its insular boundaries, the Isle of Man presents a complex, yet easily traceable sequence of deposits and events. Externally however, it provides some very considerable and seemingly intractable difficulties for some of the currently held interpretations of the Devensian glaciation of the Irish Sea basin. In particular, it leads to the speculative conclusions that the Devensian maximum may have been considerably earlier in western Britain than in eastern; that its magnitude, gradient and form were somewhat different from that currently conceived; that readvance within it may have more to do with local rather than secular causes; and that the post-maximum marine history of the Irish Sea is more complex than hitherto considered. We may, in consequence, agree with Lamplugh's comment that the Isle of Man " . . . is pre-eminently an area wherein the various theories by which the drift phenomena of the Irish Sea basin have been explained, may be put to the test" (1903 p. 332).

References

ALEXANDRE, J. 1960. La succession probable des phases morphologiques au cours d'un cycle climatique quaternaire en haute Belgique. *Biul. peryglac.* **9**, 63–72.

AMERICAN COMMISSION ON STRATIGRAPHIC NOMENCLATURE 1961. Code of stratigraphic nomenclature. *Bull. Am. Ass. Petrol. Geol.* **45**, 645–665.

ANDREWS, J. T. 1973. The Wisconsin Laurentide Ice Sheet: Dispersal centers, problems of retreat and climatic implications. *Arct. Alp. Res.* **5**, 185–199.

BOULTON, G. S. 1967. The development of a complex supraglacial moraine at the margin of Sørbreen, Ny Friesland, Vestspitzbergen, *J. Glaciol.* **6**, 717–735.

—— 1968. Flow tills and related deposits on some Vestspitzbergen glaciers, *J. Glaciol.* **7**, 391–412.

—— 1970. On the deposition of subglacial and melt-out tills at the margins of certain Svalbard glaciers. *J. Glaciol.* **9**, 231–245.

BOWEN, D. Q. 1973a. The Pleistocene succession of the Irish Sea. *Proc. Geol. Ass.* **84**, 249–273.

—— 1973b. The Pleistocene history of Wales and the borderland. *Geol. J.* **8**, 207–224.

BRYSON, R. A., WENDLAND, W. M., IVES, J. D. and ANDREWS, J. T. 1969. Radiocarbon isochrones on the disintegration of the Laurentide Ice Sheet. *Arct. Alp. Res.* **1**, 1–14.

CHARLESWORTH, J. K. 1926. The Readvance marginal Kame Moraine of the south of Scotland, and some later stages of retreat. *Trans. R. Soc. Edinb.* **55**, 25–56.

CHARLESWORTH, J. K. 1939. Some observations on the glaciation of north-east Ireland. *Proc. R. Ir. Acad.* **45B,** 255–295.

CUBBON, A. M. 1957. The Ice Age in the Isle of Man. *Proc. Isle of Man nat. Hist. antiq. Soc.* **5,** 499–512.

CUMMING, J. G. 1846. On the geology of the Isle of Man. Part 2. The Tertiary formations. *Q. J. geol. Soc. Lond.* **2,** 335–348.

DARWIN, C. 1848. On the transport of erratic blocks from a lower to a higher level. *Q. J. geol. Soc. Lond.* **4,** 315–329.

DICKSON, C. A., DICKSON, J. H. and MITCHELL, G. F. 1970. The late-Weichselian flora of the Isle of Man. *Phil. Trans. R. Soc. B,* **258,** 31–79.

EVANS, W. B. and ARTHURTON, R. S. 1973. Northwest England, in Mitchell *et al.* A correlation of Quaternary deposits in the British Isles. *Geol. Soc. Lond. Special Report* No. 4, 28–35.

FLINT, R. F. 1970. *Glacial and Quaternary Geology.* New York. 892 pp.

FRENCH, H. M. 1971. Slope asymmetry of the Beaufort Plain, Northwest Bank Island, NWT, Canada. *Can. J. Earth Sci.* **8,** 717–734.

—— 1972. Asymmetrical slope development in the Chiltern Hills. *Biul. peryglac.* **21,** 51–73.

FRYE, J. C. and WILLMAN, H. R. 1962. Morphostratigraphic units in Pleistocene stratigraphy. *Bull. Am. Ass. Petrol. Geol.* **46,** 112–113.

GLORIAD, A. and TRICART, J. 1952. Etudes statistique des vallees asymmertique de la feuille St. Pol. *Revue Géomorph. dyn.* **3,** 88–92.

GRESSWELL, R. K. 1967. *Physical Geography.* Longmans, London. 504 pp.

HOLLIN, J. T. 1962. On the glacial history of Antarctica. *J. Glaciol.* **4,** 173–195.

HORNE, J. 1874. A sketch of the geology of the Isle of Man. *Trans. Edinb. geol. Soc.* **2,** 323–347.

JARDINE, W. G. 1972. Glacial readvances in the context of Quaternary classification. *Proc. Int. Geol. Congr.* **24(12),** 48–54.

KENDALL, P. F. 1894. On the glacial geology of the Isle of Man. *Yn Lioar Manninagh.* **1,** 397–437.

LAMPLUGH, G. W. 1903. The Geology of the Isle of Man. *Mem. Geol. Surv. U.K.* 620 pp.

MANGERUD, J., ANDERSEN, S. T., BERGLUND, B. E. and DONNER, J. J. 1974. Quaternary stratigraphy of Norden, a proposal for terminology and classification. *Boreas.* **3,** 109–128.

MATHEWS, W. H. 1974. Surface profiles of the Laurentide ice sheet in its marginal areas. *J. Glaciol.* **13,** 37–43.

MITCHELL, G. F. 1960. The Pleistocene history of the Irish Sea. *Advmt Sci.* **17,** 197–213.

—— 1965. The Quaternary deposits of the Ballaugh and Kirkmichael districts, Isle of Man. *Q. J. geol. Soc. Lond.* **121,** 359–381.

—— 1972. The Pleistocene history of the Irish Sea: Second Approximation. *Sci. Proc. R. Dubl. Soc. A,* **4,** 181–199.

OLLIER, C. D. and THOMASSON, A. J. 1957. Asymmetrical valleys in the Chiltern Hills. *Geog. J.* **73,** 71–98.

PATERSON, W. S. B. 1969. *The Physics of Glaciers.* Oxford. 250 pp.

PENNY, L. F. 1964. A review of the last glaciation in Great Britain. *Proc. Yorks. geol. Soc.* **34,** 387–411.

——, COOPE, G. R. and CATT, J. A. 1969. Age and insect fauna of the Dimlington silts, East Yorkshire. *Nature, Lond.* **224,** 65–67.

PHILLIPS, B. A. M. 1967. The post-glacial raised shoreline around the northern plain of the Isle of Man. *Nth Univ. geogr. J.* **8,** 56–63.

RICHMOND, G. M. (Ed.) 1959. Application of stratigraphic classification and nomenclature to the Quaternary. *Bull. Am. Ass. Petrol. Geol.* **43,** 663–675.

ROWLANDS, B. M. 1970. Radiocarbon evidence of the age of an Irish Sea glaciation in the Vale of Clwyd. *Nature, Lond.* **230,** 9–10.

SHOTTON, F. W. and WILLIAMS, R. E. G. 1971. Birmingham University Radiocarbon dates III. *Radiocarbon* **13,** 141–156.

—— and —— 1973. Birmingham University Radiocarbon dates IV. *Radiocarbon* **15,** 1–12.

SISSONS, J. B. 1964. The Perth Readvance in central Scotland. *Scott. geogr. Mag.* **80,** 28–36.

—— 1967. *The Evolution of Scotland's Scenery.* Edinburgh 259 pp.

SLATER, G. 1931. The structure of the Bride Moraine, Isle of Man. *Proc. Lpool geol. Soc.* **14,** 184–196.

SMITH, B. 1930. Borings through the glacial drifts of the northern plain of the Isle of Man. *Mem. geol. Surv. U.K. Summ. Prog.* **3,** 14–23.

STEPHENS, N., CREIGHTON, J. R. and HANNON, M. A. 1975. The late-Pleistocene period in north-eastern Ireland: an assessment 1975. *Ir. Geogr.* **8,** 1–23.

SYNGE, F. M. 1964. The glacial succession in Caernarvonshire. *Proc. Geol. Ass.* **75,** 431–444.

TEMPLE, P. H. 1960. Some aspects of the geomorphology of the Isle of Man. M.A. Thesis, University of Liverpool. (Unpublished).

THOMAS, G. S. P. 1971. *Field Guide to the Isle of Man.* Quaternary Research Association, Liverpool.

—— 1976. The Quaternary stratigraphy of the Isle of Man. *Proc. Geol. Ass.* (In the press).

TRICART, J. 1970. Geomorphology of cold environments. London. 320 pp.

WARD, C. 1970. The Ayre raised beach, Isle of Man. *Geol. J.* **7,** 217–220.

WATSON, E. 1971. Remains of pingos in Wales and the Isle of Man. *Geol. J.* **7,** 381–387.

WIRTZ, D. 1953. Zur Stratigraphie des Pleistocans in Westen der Britisch Inseln. *Neues Jb. Geol. Palläont. Abh.* **96,** 267–303.

WRIGHT, J. 1902. The foraminifera of the Pleistocene clay, Shellag. *Yn Lior Manninagh.* **3,** 627–629.

—— and READE, T. M. 1906. The Pleistocene clays and sands of the Isle of Man. *Proc. Lpool geol. Soc.* **17,** 103–117.

G. S. P. Thomas, Department of Geography, Roxby Building, The University, P.O. Box 147, Liverpool L69 3BX, England.

Late-Pleistocene ice movements and patterns of Late- and Post-Glacial shorelines on the coast of Ulster, Ireland

N. Stephens and A. M. McCabe

A series of late-glacial shorelines were formed on the Ulster coastline during the Late Midlandian Cold Period, which is regarded as approximately equivalent to the Late Devensian of Great Britain. The earliest of these late-glacial shorelines were contemporaneous with a major readvance of the inland Irish ice to the coastlands. This Drumlin Readvance occurred sometime between 18000 and 14000 radiocarbon years B.P.

A period of relatively low sea-level followed the late-glacial marine transgressions, when organic deposits accumulated in coastal situations. These peats and forest beds are overlain by post-glacial beach and estuarine deposits representing marine transgressions younger than 8000 years B.P. and in some cases younger than 5000 years B.P.

1. Introduction

The coastal environment of Ulster reflects combinations of various geological and geomorphological events from Pre-Cambrian times to the present day. The sea is in contact with resistant quartzites, gneisses and granites aligned along the dominant northeast to southwest Caledonian trend in Co. Donegal. In Co. Londonderry and Co. Antrim Tertiary (post-basaltic) faulting and late-Pleistocene landslips strongly control the general coastal alignment and morphology. Active mass movement is changing many coastal profiles along the North Channel coast of Co. Antrim (Prior *et al.* 1968; Hutchinson *et al.* 1974). In Co. Down the Palaeozoic basement declines in height towards the Irish Sea, but is mainly hidden by a thick cover of glacial drifts, which, if removed, would reveal an archipelago of rocky promontaries, rock platforms and islands between Belfast Lough and Dundrum Bay.

The presence of countless till drumlins in eastern Co. Down has facilitated the development and preservation of many late- and post-glacial shoreline notches and raised beaches with systems of 'dead' cliffs. The age of the sub-drift rock platforms is not known, and no beach deposits of the type found on the south coast of Ireland in association with such platforms have been detected (Stephens 1957, 1958, 1970). In southeastern Co. Down the former positions of moraines can sometimes be traced by lines of large boulders stretching seaward from the present coastline. The extensive drift deposits have everywhere provided copious quantities of sediments for the construction of beach and dune systems, such as those seen in Dundrum Bay. There is therefore a marked contrast between the basic physical conditions found on the drift-clad Co. Down coast, and the generally rocky, cliffed coast of northern Ulster, where elevated shorelines are more fragmentary and difficult to identify.

2.   The general pattern of glaciation

Lowland Ulster was glaciated during at least two separate cold periods in the Late-Pleistocene (Mitchell *et al.* 1973). During the earlier or Munsterian Cold Stage, ice moved outwards from centres of dispersal in north-central Ireland to reach the Foyle estuary in the north and the lower Boyne estuary in the southeast (Colhoun 1970, 1971; Colhoun & McCabe 1973; McCabe 1973). A strong outflow of Scottish ice crossed the northeast coast and extended to Lough Foyle, the Lough Neagh lowland, and probably encircled the Mourne – Carlingford massif (Stephens *et al.* 1975).

Environmental conditions during the early and middle parts of the Last Cold Stage (Midlandian Glaciation in Ireland) cannot be accurately reconstructed. Two type-sites near Maguiresbridge, Co. Fermanagh indicate that no major ice sheets existed in lowland Ulster until after 30000 years B.P. Drumlin-forming till of late-Midlandian age overlies organic-rich silts which have been dated to greater than 41500 years B.P. (Birm. 309) (Hollymount Silts), and $30500 \, {}^{+1170}_{-1030}$ years B.P. (Birm. 166) (Derryvree Silts). The pollen content of the silts indicates that a long period of periglacial, tundra conditions must have persisted throughout Middle Midlandian times (Colhoun *et al.* 1972; Shotton and Williams 1973; McCabe *et al.* in prep.).

During the Late Midlandian a main phase of glaciation involved the expansion of ice masses from various centres of accumulation within Ulster and from Scotland (Fig. 1). Ice streams from the Irish centres covered much of the coastlands although the extent of ice in the northwest is not known (Synge 1970). But there is no evidence to support the views of Dwerryhouse (1923) and Charlesworth (1939, 1963, 1973) that Scottish ice penetrated deeply into northern Ulster.

Withdrawal of the Main Midlandian ice sheet involved large masses of ice occupying the northern Irish Sea Basin, as well as much of Ulster. In east-central Ireland the ice retreat was generally towards the northwest (McCabe 1971, 1972), and for a brief period ice centred in the northern Irish Sea expanded into the coastal lowlands of Co. Louth and the lower Boyne Valley. It is not certain if a

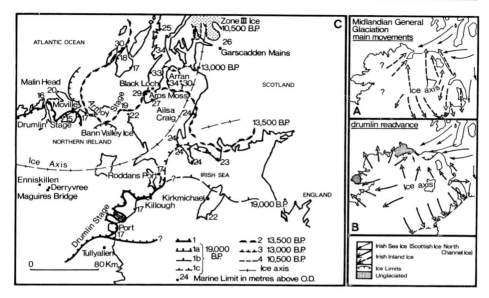

Fig. 1. General directions of ice movement are shown for the Midlandian (Last) Glaciation in Ulster (A and B).

The approximate extent of ice during the Drumlin Readvance stage, together with some important sites where the late-glacial marine limit is given, is shown in C. Bann Valley ice was probably in contact with Scottish ice along the line of the Armoy moraine at the maximum of the Drumlin Readvance stage.

A tentative chronology is suggested for the various ice stages and their associated shorelines. With acknowledgment to G. F. Mitchell, F. M. Synge, and see also Stephens *et al.* 1975.

similar expansion of Irish Sea ice took place at this time into southeastern Co. Down, where drifts of Irish Sea provenance occur at the surface inland of Ballymartin. These deposits have been previously regarded as dating from the earlier Munsterian Cold Period (Stephens *et al.* 1975).

The arrival of relatively 'warm' marine waters in the northern Irish Sea contributed to the break-up of the extensive ice masses. But meanwhile the limited withdrawal of the inland ice was followed by a widespread readvance to the coast-lands, probably reflecting a major climatic change. The whole pattern of elevated late-glacial strandlines in eastern Ulster supports the concept of marine trans-gression while tongues of inland ice crossed the coast, and while ice masses occupying the northern Irish Sea Basin continued to contract. At the same time Scottish ice made a limited movement into northern Ulster to the line of the Armoy Moraine (Figs 1 and 2). The mountains of Donegal and the Mournes supported only restricted local ice-caps or valley glaciers, as did the Antrim Plateau, during the Midlandian Cold Period.

In Ulster the widespread readvance of the inland ice masses is known as the Drumlin Readvance, which was responsible for the deposition of a series of 'upper tills' involved in the construction of the drumlins (Vernon 1966) (Fig. 1). Outside the Drumlin Readvance moraines in northwestern Co. Donegal there are extensive areas of heavily cryoturbated tills, and thick head deposits, testifying to a long period of severe periglacial conditions. In this area such conditions

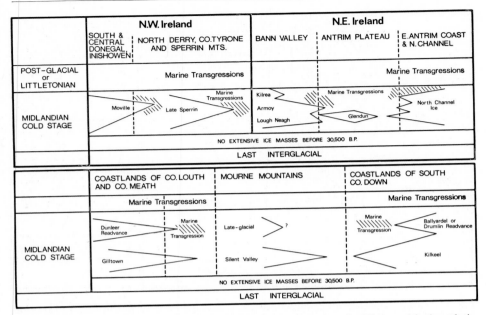

Fig. 2. Schematic summary charts of Midlandian ice movements in Ulster, with the relative expansion of the different ice-sheets and valley glaciers indicated. Overlaps of ice margins are supported by litho-stratigraphical evidence, and late-glacial marine transgressions are shown to be closely associated with certain ice limits, but it should not be assumed that all these limits are of precisely the same age.

Acknowledgment is made to Stephens *et al.* 1975.

probably lasted throughout the entire span of the Midlandian Cold Period. But from northwestern Co. Donegal to southeastern Co. Down the late-glacial shorelines are always closely related to the position of the ice-fronts at the Drumlin Readvance Stage. For example, at Moville, on Lough Foyle, and at Killough and Ardglass in Co. Down, red marine clay is interbedded with either morainic or outwash deposits from the Drumlin Readvance Moraines (Stephens and Synge 1965; Mitchell 1972; Stephens *et al.* 1975).

3.   The Late-Glacial shorelines and drumlin readvance

In Lough Swilly, Lough Foyle and Carlingford Lough, the early stage of the transgressive late-glacial sea was excluded by the presence of the Drumlin Readvance ice. On the north coast the Scottish ice advance to the Armoy Moraine excluded marine activity between Lough Foyle and Fair Head, and the North Channel coast was occupied by a combination of Irish and Scottish ice (Figs 1 and 4). Thus some late-glacial shorelines and their associated marine sediments are considerably restricted in area and extent, and their precise age is still very difficult to determine.

The heights at which these shorelines are found reflect the differing rates of isostatic uplift, and eustatic marine submergence and regression, while ice unloading continued at a time when world sea-level was still greatly depressed below that at the present day. Ice decay and withdrawal simply permitted marine transgression of the more deeply depressed parts of the coastal margins of Ulster, the Irish Sea Basin and the North Channel, at different times. It is not yet possible to calculate the total isostatic recovery of the coastal margins of Ulster since either the Main Midlandian Glaciation or the Drumlin Readvance Stage, but recovery could be in excess of 80 m since the end of Pollen Zone I, about 12000 years B.P. (Mitchell *et al.* 1973; Stephens *et al.* 1975).

The extent of the Drumlin Readvance ice has presented some difficulties of interpretation, although it is clear that topographical constraints such as the Donegal Mountains, the North Derry Hills, the Antrim Plateau, and the Mourne-Carlingford massif disrupted the geographical continuity of the ice-front. Until recently it was believed that the Drumlin Readvance (Irish) ice extended eastwards across northern Co. Down and Belfast Lough, probably joining with ice in the North Channel; the ice also moved south-eastwards to a well-marked morainic line extending from Ballyquintin Point at the southern tip of the Ards Peninsula, to Ardglass and Killough (Synge and Stephens 1966a; Hill and Prior 1968). The ice probably reached Dundrum Bay, but Slieve Croob and its adjacent hill masses may have constituted a barrier, as no drumlins occur above about 160 m, or between these hills and Dundrum Bay. The ice margin appears to have looped around the hills north and west of the Mourne Mountains, while major meltwater streams carried water away towards Newcastle and Dundrum Bay from the ice-front near Castlewellan.

The ice-edge lay west of the Mourne Mountains, but at a distinctly lower level than the maximum altitude of the Main Midlandian ice against the mountains. It extended as a large lobe to the entrance of Carlingford Lough at the Cranfield Point and Ballagan Point moraines. A further lobe occurred to the south of the Carlingford massif in Dundalk Bay, with limiting moraines at Johnstown, Templetown and Rathcor on the northern side of the bay and at Dunany Point on the southern side (Fig. 1). Outside these various morainic limits a single late-glacial shoreline was originally mapped (Stephens 1963, 1968) and only post-glacial shorelines were observed inside Carlingford Lough and inside the Dunany Point–Johnstown moraines in Dundalk Bay. However, recent mapping has indicated that more than one late-glacial shoreline is present, the highest 'shoreline' (A in Fig. 5) perhaps being associated with a more widespread ice cover along the Mourne coast (Figs 3 and 5). Therefore, if the age and significance of the Dunmore Head–Ballykeel ice limits are reconsidered this will necessarily alter the possible sequence of events during the withdrawal of Midlandian ice, and the subsequent Drumlin Readvance. While absolute dates are not yet available it is necessary to consider a possible alternative explanation of the moraine and shoreline sequences described above.

4.   The Late-glacial shorelines and a possible revised limit of the drumlin readvance in southeastern Ulster

The Drumlin Readvance ice may have extended southeastwards across Dundrum Bay to Dunmore Head with a giant lobe moving east from Carlingford Lough to Ballykeel and south to Clogher Head (Figs 1 and 3). At the southeastern end of the Carlingford Mountains the fluvio-glacial gravel complex in the Bush area of the Big River Valley may represent not the maximum extent of Midlandian ice but an inter-lobate accumulation between the ice lobes centred in Carlingford Lough and Dundalk Bay. The only ice-free coastal segments at the Drumlin Readvance stage would then have extended between Dunmore Head and Ballykeel (Co. Down) and south of Clogher Head (Co. Louth).

It is precisely in these areas that benches and notches in drift have been observed at about 26 m and 22 m (Figs 3 and 5). The outwash extending south from the Dunmore Head moraine to Mullartown and Annalong leaves no doubt that this segment of coast was ice-free at this stage. The benches and notches could be marine-eroded features, and therefore related to the high shorelines south of Clogher Head (see Synge, this volume), but equally they could represent merely the innermost edge of a very gently sloping outwash terrace against an irregular drift topography. There are insufficient sections to prove either case conclusively, although the fall in altitude of the highest notch ('Shoreline' A in Fig. 5) southwards suggests that it is more likely to be associated with outwash from the Dunmore Head Moraine together with some outwash from the Ballykeel Moraine.

The difference in height between the notches ('A' group in Fig. 5) north and south of Annalong (26–22 m) might be explained by a tilted shoreline, as indicated in Figure 5, but it has not proved possible to trace an equivalent shoreline between Ballagan Point and Cooley Point, Co. Louth. A shoreline developed to the south of Clogher Head, Co. Louth, cannot be related according to the tilt of 'shoreline' A (Fig. 5), and to the latest findings of F. M. Synge. Synge's Shoreline Diagram (this volume, Fig. 9) suggests that Shoreline B is the highest of the true marine notches and related to his LG2 shoreline (19 m).

Consequently it is reaffirmed that the Cranfield Point—Ballagan Point moraines at the entrance to Carlingford Lough and the Ballyquintin Point—Ardglass—Killough line to the northeast of the Mourne Mountains most likely represents the true Drumlin Readvance line (Stephens et al. 1975). The equivalent position of the ice lobe in Dundalk Bay would be the Johnstown—Templetown—Rathcor—Dunany Point Moraines. There is no doubt that in Co. Down this morainic line was related to a shoreline (B in Fig. 5) and washing limits within the height range 17–20 m, and this is the most prominent strandline between Port (south of Dunany Point, Co. Louth = Synge's LG2 shoreline) and Ballyquintin Point (southern tip of the Ards Peninsula, Co. Down). The strandline is marked by an extensive bench cut across outwash and morainic gravels between Cranfield Point and Derryogue and at Annalong and Mullartown, where notches occur between 17 m and 18 m. Similarly, benches, notches, and washing limits have been mapped on the coast at Minerstown, Killough, Ardglass (Fig. 6 A), Killard Point, Guns Island and Ballyquintin Point, within the height range 18–20 m. The upper terrace of the Kilkeel River seems to link to the 17 m shoreline at the coast. The inland Irish ice and North Channel ice had certainly withdrawn significantly, but still excluded the transgressing sea from Dundalk Bay, Carlingford Lough, and the bulk of the

Fig. 3. General directions of ice-movement and ice-limits are shown for the Midlandian Glaciation in southeastern Co. Down and the Carlingford Peninsula. Superimposed on this map is the alternative possibility of different ice-limits for the Drumlin Readvance—see text. The extent of the late-glacial strandlines A, B and C are indicated (see Shoreline Diagram—Fig. 5).
    The lower diagram shows the stratigraphical succession and certain ice limits between the Mourne Mountains and the coast of Ballymartin, where Shoreline B is recorded.
    Acknowledgment is made to Stephens et al. 1975.

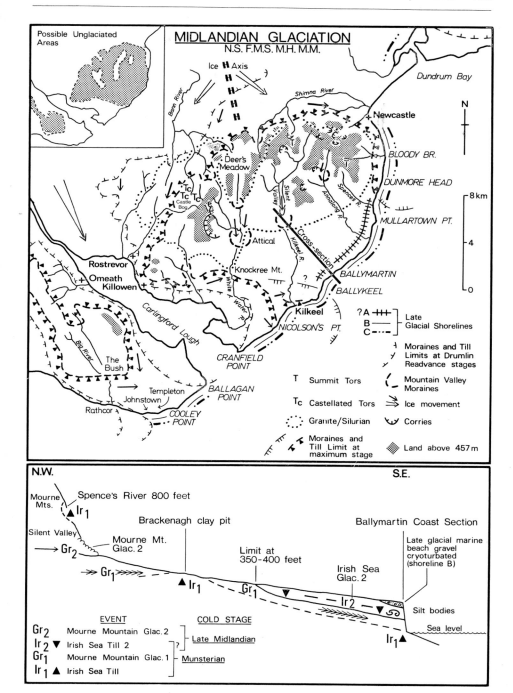

MIDLANDIAN GLACIATION
N.S. F.M.S. M.H. M.M.

Possible Unglaciated Areas

Ice H Axis

Dundrum Bay

Bann River

Shimna River

Newcastle

N

Deer's Meadow

BLOODY BR.

DUNMORE HEAD

Tc
Tc Tc
Castle Bog

Silent Valley

Spence's R.

MULLARTOWN PT.

8km

Attical

Kilkeel R.

Cross-section

4

Rostrevor

Knockree Mt.

Annalong R.

?

BALLYMARTIN

Omeath
Killowen

White

BALLYKEEL

0

Carlingford Lough

Water

Kilkeel

?A ++ } Late
B ——— } Glacial Shorelines
C –·–· }

Big River

NICOLSON'S PT.

The Bush

CRANFIELD POINT

⌐ Moraines and Till
⌐ Limits at Drumlin
⌐ Readvance stages

Templeton
Johnstown

BALLAGAN POINT

T   Summit Tors

⌐ Mountain Valley
  Moraines

Rathcor

COOLEY POINT

Tc   Castellated Tors

⟹ Ice movement

⋰⋱ Granite/Silurian

⌣ Corries

Moraines and
Till Limit at
maximum stage

Land above 457 m

N.W.   S.E.

Mourne Mts.   Spence's River 800 feet

▲Ir₁

Silent Valley   Ballymartin Coast Section

⟶ Gr₂   Mourne Mt.   Late glacial marine
                     Glac. 2   Brackenagh clay pit   beach gravel
                                                        cryoturbated
                                                        (shoreline B)

⟹ Gr₁⟹⟹   Limit at   Irish Sea
                     350-400 feet   Glac. 2

         ▲Ir₁ – Gr₁ ▽   ▽   Silt bodies

                   Ir₂ ▽   ⟹⟹

                                     Sea level

         EVENT   COLD STAGE

Gr₂ Mourne Mountain Glac. 2
                              ⌐ Late Midlandian   Ir₁▲
Ir₂ ▽ Irish Sea Till 2   ?┤
Gr₁ Mourne Mountain Glac. 1 └ Munsterian
Ir₁ ▲ Irish Sea Till

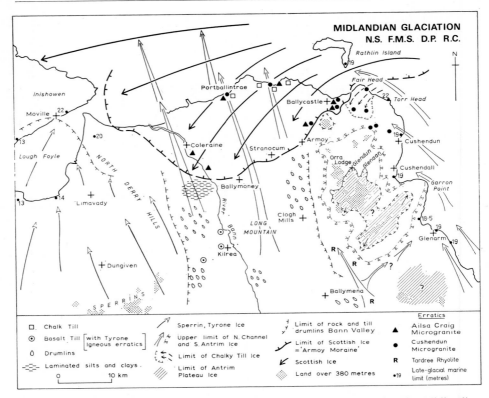

Fig. 4. General directions of ice-movement and ice-limits are shown for the Midlandian Glaciation in northern Co. Antrim, Co. Londonderry and part of the Inishowen Peninsula. The Armoy Moraine may represent a medial moranic complex between Scottish ice and Bann Valley ice at the Drumlin stage. The extent of Antrim Plateau ice is still conjectural.

The marine limit for the late-glacial shorelines is given for various localities, and it will be noted that differences occur on either side of certain ice limits.

Acknowledgment is made to Stephens *et al.* 1975.

coastal drumlin country north of Dundrum Bay, including Strangford Lough and the site of Roddans Port in the Ards Peninsula (Fig. 1).

Analysis of the shoreline pattern in Co. Down shows that one and possibly two further late-glacial shorelines (C and D) are present, although they tend to occur only as isolated fragments (Fig. 5) but may be closely related to shorelines LG3 and LG4 (Synge, this volume, fig 9). Shoreline C can be usually distinguished from the post-glacial shorelines at lower levels because of the absence of marine shells and the presence of severe cryoturbation of the beach sediments. Sometimes these two shorelines cut out the higher strandline (B), but neither has been found inside Carlingford Lough. A 12–13 m shoreline is seen at Cranfield Point as a

Fig. 5. The shoreline diagram shows the arrangement of the late- and post-glacial strandlines on the Mourne Coast between Carlingford Lough (Cranfield Point Moraine) and the Dunmore Head Moraine, together with possible extensions to Dundrum Bay, Killough, Ardglass, and Roddans Port. All heights have been corrected to British datum.

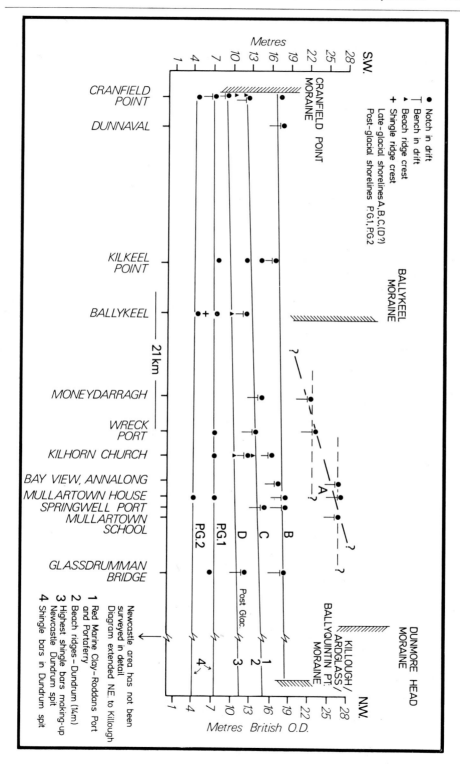

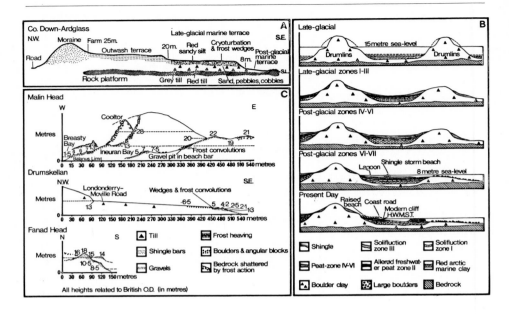

Fig. 6. Diagram A illustrates the relationship of the late- and post-glacial shorelines to the Drumlin Readvance morainic limit at Ardglass. Red sandy silt/clay is inter-bedded with the sands and gravels of the cryoturbated beach sediments.
   Diagram B indicates schematically the probable evolution of the Roddans Port site.
   Diagram C provides a number of coast profiles from the Inishowen Peninsula, Co. Donegal, where fine late-glacial strandlines are recorded.
   Acknowledgment is made to Stephens and Synge 1965; Morrison and Stephens 1965.

notch and a well-developed beach ridge, by notches at Kilkeel Point, Ballykeel, Kilhorn (14–15 m notch and beach ridge) and at Springwell Port (15 m notch). This shoreline appears to be increasing in elevation slightly to the north, and it may be represented by a massive shingle bar near Dundrum (crest at 14 m), and by washing limits and benches cut in drift north of the village. It seems likely that the ice front, at Killough, Ardglass and Ballyquintin Point would have been breached at this stage, thus allowing the sea to cut 13 m notches in some of the drumlins inside Strangford Lough, and to bring the red marine clay to inter-drumlin hollows as at Roddans Port (Fig. 6 B). The volume of ice remaining in the North Channel is not known at this stage.

   Very considerable coast erosion in post-glacial times has removed much of the evidence for the existence of shoreline D in Figure 5, and its validity as a separate strandline may be questioned. A notch at Cranfield Point (9 m), beach ridges at Ballykeel and Kilhorn (10 m), and a possible notch at Glasdrumman Bridge (11 m) constitute the rather restricted evidence. But the highest shingle bars making up the Newcastle-Dundrum spit reach 9 m, which is only a metre or so above the highest of the known shell-bearing post-glacial strandlines. There is therefore a possibility that on some exposed segments of the coast shoreline D may represent the very highest post-glacial level in spite of the lack of shell fauna in the beach deposits. Such a shoreline does not appear inside Carlingford Lough or Strangford Lough, presumably because of the protection afforded by the restricted entrances, and limited fetch inside the lough as compared with the Irish Sea.

5. The Late-Glacial shorelines and drumlin readvance limits in northeastern and northwestern Ulster

In northwestern Ulster the highest late-glacial strandlines occur to the northeast of Bloody Foreland, as well-defined elevated marine washing limits, notches and shingle bars occurring at Fanad Head, Malin Head and Ballyhillin (Fig. 6, C; Stephens and Synge 1965). The highest strandline (14–18 m at Fanad Head and 20–22 m at Malin Head and Moville) does not occur inside the massive morainic accumulations of the Drumlin Readvance limit in Sheephaven, Lough Swilly and Lough Foyle (Fig. 1). Lower strandlines (e.g. 13 m at Drumskellan, inside Lough Foyle) which occur inside the Drumlin Readvance moraine indicate that isostatic recovery was taking place before ice retreat occurred from this morainic line (Synge and Stephens 1966a; Colhoun *et al.* 1973), although it has not yet been established that these late-glacial shorelines are synchronous with those at a similar height in southeastern Ulster. However, it is tempting to regard the maximum of the Drumlin Readvance stage as a synchronous event throughout Ulster.

Similarly, strandlines at various heights have been mapped between Lough Foyle and Fair Head, and along the North Channel coast of Co. Antrim (Stephens 1963; Prior 1966), and attempts have been made to link these to similar features on the Scottish coast (Islay, Kintyre, Arran and Galloway), and with strandlines on the eastern coast of Co. Down, with varying degrees of success (Synge and Stephens 1966a; Stephens and Synge 1966b; Stephens 1968).

In northern Ulster (Fig. 4) the Main Stage of the Midlandian Glaciation carried northward moving ice beyond the present coastline. The Armoy Morainic complex probably represents the zone where Drumlin Readvance ice moving northwards in the Bann Valley met Scottish ice moving generally in a southwesterly direction. The net effect was to exclude the late-glacial sea from the coast between Lough Foyle and Fair Head. If the Armoy Moraine is contemporaneous with the Drumlin Readvance limits at Moville, in Lough Foyle, and in southeastern Ulster, then shorelines recorded inside the Armoy Moraine can be expected to differ in height and in age from those strandlines outside this ice limit. In fact, the interpretation of elevated shorelines on this segment of coastline is particularly difficult because of extensive cliffs, and the manner in which the basalt lavas weather in 'steps' and 'terraces' to simulate a series of marine-cut notches and platforms.

Prior (1968) has demonstrated that on the North Channel coast of Co. Antrim the Scottish ice advance to the Armoy Moraine was matched by a slight withdrawal southwards of Irish ice which had previously moved northwards along the coast (Hill and Prior 1968). If the Scottish ice remained at the Armoy Moraine for some considerable time then marine waters from the Irish Sea may have penetrated the North Channel, to cut shorelines somewhat earlier than was possible between Fair Head and Lough Foyle. Thus, once again, non-synchronous shorelines may occur at approximately the same height on the eastern and northern coasts of Ulster. Investigations currently in progress on an inter-tidal sequence of late- and post-glacial sediments at Carnlough (between Glenarm and Garron Point—Fig. 4) may provide useful information and comparison with events recorded at Roddans Port (Prior and Holland 1975).

6.   The age of the Late-Glacial shorelines

Shoreline correlations based upon altimetric considerations alone are fraught with many difficulties but attempts have been made to consider the shoreline sequence in relation to distinct ice limits such as the Drumlin Readvance moraine, and to suggest approximate dates for these events (Stephens *et al.* 1975; Gemmell 1973). F. M. Synge (this volume) has contributed another attempt to resolve some of the problems of shoreline correlation and the dating of the various strandlines to known ice limits. While there remain differences of opinion regarding the correlation in space and time across the Irish Sea and North Channel, this is probably the best 'approximation' that can be achieved on the basis of the evidence available.

The age of the earliest and highest late-glacial strandlines associated with the Drumlin Readvance on the coast of east-central Ireland is extremly difficult to determine. If marine transgression had followed withdrawal of the Midlandian ice too closely there may not have been time for the Giant Irish Deer (*Megaloceros giganteus*) to achieve widespread immigration and dispersal throughout the island. This herbivore is closely associated with late-glacial deposits which accumulated during Pollen Zone II (Allerød) throughout Ireland (Mitchell and Parkes 1949; Synge 1975). However, a marine transgression had to reach the Roddans Port site (as shoreline C in Fig. 5, and see Fig. 6 B) before the end of Pollen Zone I, some 12000 years ago, since the associated red marine clay underlies Pollen Zone I and Pollen Zone II sediments at this site (Morrison and Stephens 1965). The late-glacial sediments (Pollen Zones I–II) at Woodgrange, a low-lying inter-drumlin locality near Downpatrick, show no positive trace of marine transgression (Singh 1970), which implies exclusion of the sea by the Drumlin Readvance ice and sufficient isostatic recovery to elevate the site clear of later marine influence. Therefore, late-glacial shorelines B and C (Fig. 5) must be older, and so too the moraines of the Drumlin Readvance stage. Near Kirkmichael, on the West coast of the Isle of Man, sediments found in the base of a kettle hole have provided a ^{14}C date of 18990 ± 150 years B.P. (Birm. 270). Because the site was not subsequently over-ridden by ice this seems to establish effectively that the general glacial limit in the northern Irish Sea basin could be no further south than Kirkmichael about 18000–19000 years B.P. (Dickson 1970; Mitchell 1972). The first late-glacial shorelines (B and C) and the Drumlin Readvance moraine in southeastern Ulster must therefore post-date 18000–19000 years B.P. but probably by only a short time interval.

These shorelines, and the associated ice limits of the Drumlin Readvance stage, occurred long before ice wastage permitted the sea access to the Roddans Port site, prior to 12000 years B.P. Synge (1975) has suggested a date of 14000–15000 years for the Drumlin Readvance and associated shorelines, but it seems doubtful if a time interval of some 4000 years could have occurred between the shorelines outside the Kirkmichael limit (=Clogher Head limit on the Co. Louth coast?) and the first of the sequence of shorelines associated with the Drumlin Readvance in southeastern Ulster (Shorelines B and C on the Mourne coast).

If the Drumlin Readvance Stage was more or less a synchronous event throughout Ulster it could be argued that the highest shorelines at Fanad and Malin Head are the same general age as shoreline B in Co. Down (Fig. 5). However, it is worth recording that while red marine clay is known to be inter-bedded with sediments making up the Drumlin Readvance Moraine at Moville on Lough Foyle, and probably also at Killough and Ardglass in Co. Down, it has not been detected on the Mourne coast. If this distinctive red clay (Stephens 1963; Morrison and Stephens 1965) should prove to be a valuable 'marker' horizon, perhaps related to one major phase of marine transgression, then its absence from the highest marine sediments in Co. Down and Co. Louth could be significant. These shorelines could be older than the Malin Head shorelines because greater isostatic depression of the northern Irish Sea Basin and adjacent land areas, under the combined weight of Irish Sea and inland Irish ice masses allowed earlier flooding of the area. Alternatively, zones of structural weakness may have been re-activated during the phases of isostatic recovery, thus producing displacements of various shorelines between southeastern and northwestern Ulster, and making altimetric correlation impossible. Tectonic activity in the North Channel may also have occurred during isostatic recovery making it extremely difficult to correlate the shorelines in Ulster with those in Scotland (Stephens 1970; Peacock 1971).

There can be little doubt that the late-glacial upper marine limit on the Ulster coast is a metachronous feature, although often traceable as a synchronous feature over very short distances, as first tentatively suggested in 1966 (Stephens and Synge 1966b). As the ice downwasted and withdrew in the North Channel and in southwestern Scotland, the sea transgressed the isostatically depressed area, forming a series of strandlines which were closely related to local, phased ice-margin positions (Stephens *et al.* 1975; Synge 1975). It is difficult to prove that any of the sequences of shorelines in Ulster are synchronous because of their different geographic locations in relation to the various centres of ice dispersion, and because of possible variations in the rates of ice wastage.

7.   The Post-Glacial shorelines

The presence of buried or submerged peat beds at many places on the Ulster coast (Stephens 1970) indicates that following the late-glacial marine transgressions there was a period of relatively low sea-level. On both sides of the North Channel submerged peats are overlain by beach or estuarine sediments of post-glacial age. Isostatic uplift has elevated the strandlines to various levels above the modern shoreline (Bishop and Dickson 1970; Jardine 1971). Several elevated shorelines have been recorded on the Leinster coast (Synge, this volume) and on the Ulster coast, the beach sediments usually containing a shell fauna, with a species of *Littorina* often abundant (Praeger 1892, 1896). The raised beaches frequently contain man-made artifacts to which the general name Larnian has been given (Movius 1942). Mitchell (1971) has reviewed the importance of the Larnian culture and its relationship to the post-glacial beaches. Inside Lough Foyle, Belfast Lough and Strangford Lough considerable thicknesses of 'estuarine clay' have been recorded from borehole records, and from excavations for dock installations. The marine sediments generally rest directly upon glacial deposits, or upon peat layers overlying till or sands and gravels. Lenses of peat and wood fragments are frequently interbedded in 'estuarine clay'. In Belfast Harbour, and under the city, below 6 m O.D. the 'estuarine clay' consists of a soft, sandy grey unconsolidated mud, containing abundant marine shells (Praeger 1892), and is known locally as 'sleech'. A borehole at Castle Arcade, Belfast provided the following record, which is reproduced by courtesy of the Geological Survey of Northern Ireland (see also, Wilson 1972):

Ground level at approximately $+3$ *m* (*British*) *O.D.*

Thickness of strata given for each horizon

Made ground	0·00 to 1·20 m	1·20 m		
Grey silty sand	1·20 to 4·86 m	3·66 m		
'Sleech' or Estuarine clay	4·86 to 12·78 m	7·92 m	11·78 m	A piece of timber from this depth below the surface gave a ^{14}C date of 8715 ± 200 years B.P. (St. 3058).
Peat	12·78 to 15·83 m	3·05 m	14·94 m	Peat specimen from this depth below the surface gave a ^{14}C date of 9130 ± 120 years B.P. (St. 3057).
Red, stiff clay (glacial deposit)	15·83 to 17·96 m	2.13 m (base not determined)		

Too much emphasis should not be placed upon the heights of the various peat beds from site to site because this will vary considerably, and transgression of the peats will sometimes have depended upon the speed of the transgression or the ability of the waves to penetrate low barriers in a hummocky drift landscape. Furthermore, there is evidence to suggest that sometime between 6000 and 3500 years B.P. a eustatic transgression may have carried world sea-level to some 4 m above present (Mörner 1971). Clearly, if such a eustatic rise took place biogenic deposits much younger than the Belfast (Castle Arcade) peat would have been transgressed, but this is probably less apparent in areas where isostatic recovery was considerable during the post-glacial phase (Mitchell and Stephens 1974).

Several attempts have been made to produce shoreline diagrams for the post-glacial elevated beaches in Ulster (Synge and Stephens 1965, 1966a; Stephens 1968; Prior 1966), and in Leinster the most recent by F. M. Synge is shown on page 217. Each diagram indicates that there are several tilted strandlines, the greatest elevations occurring in the areas where the greatest isostatic recovery has taken place. Sissons (1976) has constructed a diagram of generalised isobases for the

Fig. 7. Post-glacial sites in Co. Down.
    A table of ^{14}C dated horizons is provided for some important sites: Woodgrange, Ringneill Quay, Dundrum and Ballyhalbert. Cross-sections at Ringneill Quay and Woodgrange are also shown in more detail.
    With acknowledgment to Dresser *et al.* 1973; Mitchell and Stephens 1974; Stephens and Collins 1960.

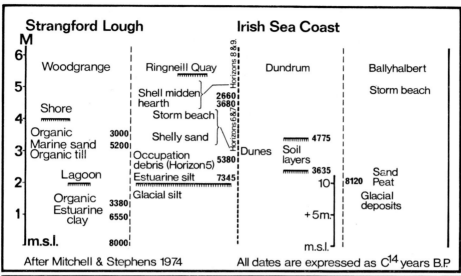

Strangford Lough   Irish Sea Coast

Woodgrange	Ringneill Quay	Dundrum	Ballyhalbert
	Shell midden hearth		Storm beach
Shore	Storm beach		
Organic	Shelly sand	Soil layers	
Marine sand	Occupation debris (Horizon 5)	Dunes	
Organic till	Estuarine silt		Sand
Lagoon	Glacial silt		Peat
Organic Estuarine clay			Glacial deposits

M
6
5
4
3
2
1
m.s.l.

Woodgrange dates: Organic 3000; Marine sand 5200; Organic 3380; Estuarine clay 6550; m.s.l. 8000

Ringneill Quay: Shell midden 2660; hearth 3680; Occupation debris 5380; Estuarine silt 7345

Horizons 8 & 9; Horizons 6 & 7

Dundrum: 4775; 3635; 10

Ballyhalbert: 8120

Irish Sea Coast: +5m.; m.s.l.

After Mitchell & Stephens 1974   All dates are expressed as C^{14} years B.P.

Ringneill Quay, Co. Down

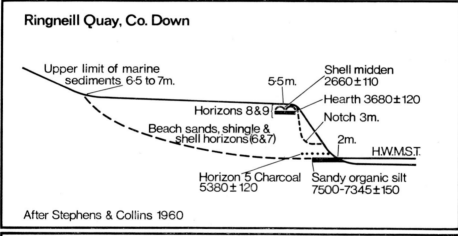

Upper limit of marine sediments 6·5 to 7m.

5·5m.

Shell midden 2660 ± 110

Hearth 3680 ± 120

Horizons 8 & 9

Notch 3m.

Beach sands, shingle & shell horizons (6 & 7)

2m.

H.W.M.S.T.

Horizon 5 Charcoal 5380 ± 120

Sandy organic silt 7500–7345 ± 150

After Stephens & Collins 1960

Woodgrange, Co. Down

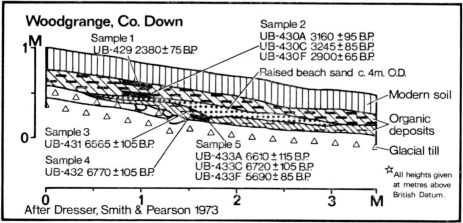

Sample 1 UB-429 2380 ± 75 B.P.

Sample 2 UB-430A 3160 ± 95 B.P. UB-430C 3245 ± 85 B.P. UB-430F 2900 ± 65 B.P.

Raised beach sand c. 4m. O.D.

Modern soil

Organic deposits

Glacial till

Sample 3 UB-431 6565 ± 105 B.P.

Sample 4 UB-432 6770 ± 105 B.P.

Sample 5 UB-433A 6610 ± 115 B.P. UB-433C 6720 ± 105 B.P. UB-433F 5690 ± 85 B.P.

☆ All heights given at metres above British Datum.

After Dresser, Smith & Pearson 1973

M
1
0
0   1   2   3   M

Main Post-glacial Raised Shoreline in Scotland, which suggests that this shoreline is downtilted from about 14 m near Glasgow to 8 m on the Antrim coast. However, heights of shingle beaches, notches and washing limits, are affected by a number of factors, including tidal range, exposure to the open sea and fetch. On the Ulster coast Spring tidal range varies from 5 m near the entrance to Carlingford Lough to 1 m at Fair Head, and 4 m in northwestern Co. Donegal.

The highest post-glacial strandline at Malin Head stands at 7·5 m, but declines in height southwestwards, and all trace of elevated shorelines disappears near Bloody Foreland. But it reappears again in Donegal Bay where the coastal drumlins are notched at a maximum height of 4 m. Inside Lough Foyle, at Drumskellan, the highest post-glacial shingle bar stands at 4 m O.D., and rests upon a thin forest bed containing an oak log which yielded a ^{14}C date of 6955±100 years B.P. (UB–206) (Colhoun *et al.* 1973). In spite of the close association of the shingle bar and forest bed, which implies that a transgression took place later than 7000 years B.P., the exact age of the beach bar is unknown. However, the low level of the beach, in comparison with the post-glacial raised beaches at Malin Head (7·5 m), Magilligan (6·8–7·6 m) (Carter 1973), Cushendun (8 m), Ballyhalbert (7·3 m), and Rough Island (6 m), is surprising for an area which is known to have experienced considerable isostatic recovery in the late-glacial period. It seems likely that there is an hiatus between the accumulation of the organic layer and the deposition of the beach, a situation also found at sites (Ringneill Quay and Woodgrange) inside Strangford Lough (Fig. 7). The marine transgression in Lough Foyle may have been delayed by the inability of the sea to penetrate the drift-choked lowland until much later than 7000 years B.P., and final penetration of marine waters may have resulted from an eustatic-induced surge about 5000 years B.P. It is interesting to find that the bulk of Magilligan Foreland, which all but closes the entrance to Lough Foyle, is probably of quite recent construction. A series of curving sand ridges are partially overlain by sand dunes, and the marine sands of the ridges are in places inter-bedded with peat and marl bands. These are exposed in section along the inner coast of the foreland, and one of the peat layers at c. 6 m O.D. gave a date of 1535±40 years B.P. (UB–547) (Pilcher 1973). The sequence of post-glacial marine sands, organic beds, fresh water marls and sand dunes rest upon a thick basement of glacial gravels, exceeding 50 m in thickness.

On the coastline of Co. Antrim the most prominent post-glacial shoreline (7–8 m) is marked by elevated notches in drift, and caves cut in the weaker basalt lavas (The Gobbins), Chalk (Garron Point and Larne), and Triassic conglomerate (Cushendall). Caves cut in Old Red Sandstone Conglomerate at Cushendun are probably also post-glacial in age. These features frequently reach an altitude of 8 m O.D. and occasionally notches, and more especially the crests of shingle bars may reach 10 m, for example, on Rathlin Island (8 m), at Cushendun (9 m), Cushendall (10 m), near Carnlough village (10 m), at Glenarn (9 m), and Larne (9 m) (Movius 1942; Prior 1968).

Much of the coastline of Co. Down north of Dundrum Bay consists of the drowned edge of an extensive drumlin 'field', although Palaeozoic bedrock is also seen at many places outcropping below till. 'Dead' cliffs are commonly found cut in the coastal drumlins, with notches marking the highest post-glacial shore-line occurring at 7–8 m O.D. Extensive shingle bars, with crests rising to 8 m, link some of the drumlins together, for example near Ballyhalbert and Kirkstown Castle. At Ballyhalbert a massive storm ridge rests upon the oldest known peat

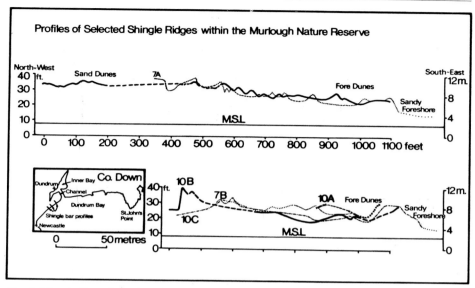

Fig. 8. The profiles of a series of shingle ridges forming the Newcastle-Dundrum spit are shown. The ridges decline in height towards the present shore, except that the ridges increase in height under the existing foredunes. This may indicate the action of a single storm or represent a phase of higher shingle ridge construction. Soil profiles containing charcoal are found throughout the dune sands resting upon the shingle bars, and archaeological sites are also present within Murlough Nature Reserve.

With acknowledgment to A. E. P. Collins, and G. F. Mitchell and N. Stephens (1974).

horizon outside Belfast Lough. The peat layer, at 2·5 m O.D., only just above present ordinary high water mark, has been dated at 8120±135 years B.P. (Q–214) (Morrison and Stephens 1960). A similar date, 8378±70 years B.P. (SRR 383), has been obtained for a lens of peat in a sand containing marine shells which stands at approximately 4 m O.D. at Killough (Wilson 1976). Thus the post-glacial transgression may well have taken place at a comparatively early date at points on the Irish Sea coast to seal these organic deposits, the dates of about 8000 years B.P. comparing closely with the age of some of the buried peats on the nearby Scottish coast (Jardine 1971).

However, within Strangford Lough, at sheltered sites among the drumlins the story is rather different. The highest post-glacial shoreline rarely exceeds 6 m as a notch in the drumlins or as the crest of a shingle beach, and it is often no more than 5 m above datum. At Woodgrange, near Downpatrick (Singh and Smith 1966, 1973; Dresser *et al.* 1973), estuarine clay at 1 m O.D. has been dated at 6550±300 years B.P. (LJ–903), and organic material underlying marine sand (ca. 4 m) gave a range of dates, from 6770±105 (UB–432) to 5690±85 years B.P. (UB–433F) (Fig. 7). But organic material resting upon the marine sand has given a range of dates, from 3160±95 (UB–430A) to 2900±65 years B.P. (UB–430F). There is little doubt that the beach sediments were laid down before about 3000 years B.P., and because the marine sand contains *Plantago lanceolata* the authors conclude that "the beach in all probability was laid down in post-Atlantic times, after about 5200 years B.P." (Dresser *et al.* 1973, caption to figure 1, p. 54; Fig. 7).

A similar hiatus has been shown to be present at Ringneill Quay, where estuarine silt at 2·5 m was dated at 7500±150 to 7345±150 years B.P. (Q–632) (Fig. 7). But resting upon the silt a Neolithic occupation level was dated at 5380±120 years B.P. (Q–770) and this is overlain by beach sediments (to 4·5 m O.D.) which are capped by a shell midden (2600±110 years B.P.) (Q–635) and hearth, the latter providing a date of 3680±120 years B.P. (Q–633) (Stephens and Collins 1960; Mitchell and Stephens 1974). This record is very similar to that at Woodgrange and is at variance with the Ballyhalbert evidence from the outer exposed coast. Thus a eustatic transgression between about 5500 and 3700 years B.P. may account for the bulk of the elevated strandlines in Co. Down, and it remains to be seen if there is any evidence to show that on the outer coast the transgression was at a significantly earlier date. It is of interest that the record provided by a sequence of relic shingle bars below sand dunes in Dundrum Bay (Fig. 8) indicates that 6–9 m shorelines may be preserved on a segment of coast well exposed to gales from the south-east. But the oldest C–14 dates so far obtained from buried soils in the dune sands overlying the shingle bars, which decline in height seawards, are 4775±140 (UB–412) and 3635±80 years B.P. (UB–352). Thus most of the shingle ridge complex may be no older than about 5000 years B.P. (Mitchell and Stephens 1974). The age of the maximum post-glacial transgression may thus vary from one part of the Ulster coast to another. At least two lower post-glacial strandlines have been recorded in Co. Donegal (Stephens and Synge 1965) and in Co. Antrim (Synge and Stephens 1966a; Prior 1966), while in Co. Down the general levels of beach ridges and notches occur at c. 4 m and c. 2 m. The precise age of these relic strandlines is unknown, but from the evidence within Strangford Lough, some of them are certainly younger than c. 3000 years B.P.

8.   Conclusion

The retreat and decay of the Late Midlandian ice in Ulster was interrupted by a major readvance, known as the Drumlin Readvance, sometime between 19000 and 12000 years B.P. It seems likely that the readvance may date from 16000–15000 years B.P., but no precise dates are available. While the exact limits of the Drumlin Readvance are still the subject of active research, it is clear that a series of late-glacial marine strandlines was closely related to the positions of the ice-fronts. The transgressing sea was excluded from segments of the coast by active ice lobes, while isostatic recovery took place.

Continuing isostatic recovery in Ulster brought about a period of relatively low sea-level, when organic layers accumulated, between about 11009 and 6000 years B.P. Subsequently, the post-glacial marine transgression (Littletonian in Ireland, or Flandrian in Great Britain) allowed beach and estuarine sediments to seal the coastal peats and forest beds, and continued isostatic recovery has elevated several strandlines to various levels above the modern shoreline.

On page 217 F. M. Synge has attempted to construct a shoreline diagram embodying all the available evidence for the eastern coast of Ireland as far north as Carlingford Lough. There remain some differences of opinion concerning the age of certain of the late-glacial shorelines, and their associated ice limits in the northern part of the Irish Sea. But the diagrams presented on pages 187 and 217 can be regarded as indicating the most up-to-date information available for the late-glacial and post-glacial elevated shorelines on the western side of the Irish Sea and North Channel.

Acknowledgments. The authors wish to express their thanks to Mr. F. M. Synge for his advice and critical comment concerning this paper. The Queen's University of Belfast and the Research Committee of the Northern Ireland Polytechnic provided financial support for the fieldwork, which is gratefully acknowledged. Our thanks are also extended to Mr. D. Houghton and Mr. R. Nelson (Northern Ireland Polytechnic) for drawing all the maps and diagrams. The authors also express their thanks to the Editor of *Irish Geography* (Professor G. L. Davies) for permission to reproduce in part or in whole material included in Figures 1, 2, 3 and 4.

References

BISHOP, W. W. and DICKSON, J. H. 1970. Radiocarbon dates related to the Scottish Late Glacial Sea in the Firth of Clyde, *Nature, Lond.* **227**, 480–482.

CARTER, W. 1973. Private communication.

CHARLESWORTH, J. K. 1939. Some observations on the glaciation of northeast Ireland. *Proc. R. Ir. Acad.* **45B**, 235–295.

—— 1963. Some observations on the Irish Pleistocene. *Proc. R. Ir. Acad.* **62B**, 205–322.

—— 1973. Stages in the dissolution of the last ice-sheet in Ireland and the Irish Sea region. *Proc. R. Ir. Acad.* **73B**, 79–86.

COLHOUN, E. A. 1970. On the nature of the glaciations and final deglaciation of the Sperrin Mountains and adjacent areas in the north of Ireland. *Ir. Geogr.* **6**, 162–185.

—— 1971. The glacial stratigraphy of the Sperrin Mountains and its relation to the glacial stratigraphy of northwest Ireland. *Proc. R. Ir. Acad.* **71B**, 37–52.

——, DICKSON, J. H., McCABE, A. M. and SHOTTON, F. W. 1972. A Midlandian freshwater series at Derryvree, Maguiresbridge, county Fermanagh, Northern Ireland. *Proc. R. Soc.* **180B**, 273–292.

—— and McCABE, A. M. 1973. Pleistocene glacial, glaciomarine and associated deposits of Mell and Tullyallen townlands, near Drogheda, eastern Ireland. *Proc. R. Ir. Acad.* **73B**, 165–206.

——, RHYDER, A. T. and STEPHENS, N. 1973. C–14 age of an Oak-Hazel forest bed at Drumskellan, Co. Donegal and its relation to Late Midlandian and Littletonian raised beaches. *Irish Nat. J.* **17**, 321–327.

DICKSON, C. A., DICKSON, J. H. and MITCHELL, G. F. 1970. The Late-Weichselian flora of the Isle of Man. *Phil. Trans. R. Soc.* **258**, 31–79.

DWERRYHOUSE, A. R. 1923. The glaciation of northeastern Ireland. *Q. J. geol. Soc. Lond.* **79**, 352–422.

DRESSER, P. Q., SMITH, A. G. and PEARSON, G. W. 1973. Radiocarbon Dating of the Raised Beach at Woodgrange, Co. Down. *Proc. R. Irish Acad.* **73B**, 53–56.

GEMMELL, A. G. M. 1973. The deglaciation of the Island of Arran, Scotland. *Trans. Inst. Br. Geogr.* **59**, 25–39.

HILL, A. R. and PRIOR, D. B. 1968. Directions of ice movement in northeast Ireland. *Proc. R. Ir. Acad.* **66B**, 71–84.

HUTCHINSON, J. N., PRIOR, D. B. and STEPHENS, N. 1974. Potentially Dangerous Surges in an Antrim Mudslide. *Q. J. Engng Geol.* **7**, 363–376.

JARDINE, W. G. 1964. Post-glacial sea-levels in southwest Scotland. *Scot. Geogr. Mag.* **80**, 5–11.

—— 1971. Form and age of late-Quaternary shorelines and coastal deposits of southwest Scotland: critical data. *Quaternaria* **14**, 103–114.

McCABE, A. M., MITCHELL, G. F. and SHOTTON, F. W. Alluvial freshwater deposit at Hollymount, Maguiresbridge, Co. Fermanagh, Northern Ireland. *In prep.*

—— 1972. Directions of Late-Pleistocene ice-flows in eastern countries Meath and Louth, Ireland. *Ir. Geogr.* **6**, 443–461.

—— 1973. The glacial stratigraphy of eastern counties Meath and Louth. *Proc. R. Ir. Acad.* **73B**, 355–382.

MITCHELL, G. F., PENNY, L. F., SHOTTON, F. W. and WEST, R. G. 1973. A correlation of Quaternary deposits in the British Isles. *Geol. Soc. Lond. Special Report* No. 4, 1–99.

—— and STEPHENS, N. 1974. Is there evidence for a Holocene Sea-Level higher than that of today on the Coasts of Ireland. *C.N.R.S. Les Methodes Quantitatives d'Etude des variations du Climat au cours du Pleistocene* **219**, 115–125.

—— and PARKES, H. M. 1949. The Giant Deer in Ireland. *Proc. R. Ir. Acad.* **52B**, 291–314.

—— 1971. The Larnian Culture: A Minimal View. *Proc. prehist. Soc.* **37**, 274–283.

—— 1972. The Pleistocene History of the Irish Sea: Second Approximation. *Sci. Proc. Roy. Dublin Soc.* A, **4**, 181–199.

MÖRNER, N.–A. 1971. The Holocene eustatic sea-level problem. *Geol. en Mijnb.* **50**, 699–702.

MORRISON, M. E. S. and STEPHENS, N. 1960. Stratigraphy and Pollen Analysis of the Raised Beach deposits at Ballyhalbert, Co. Down, N. Ireland. *New Phytol.* **59**, 153–162.
—— and STEPHENS, N. 1965. A submerged late Quaternary deposit at Roddans port on the north-east coast of Ireland. *Phil. Trans. R. Soc. Lond.* **B 249**, 221–255.
MOVIUS, H. L. 1942. *The Irish Stone Age.* Cambridge.
PEACOCK, J. D. 1971. Marine Shell Radiocarbon Dates and the Chronology of Deglaciation in Western Scotland. *Nature. Lond.* **230**, 43–45.
PRAEGER, R. L. 1892. Report on the Estuarine Clays of the north-east of Ireland. *Proc. R. Irish Acad.* **18**, 212–289.
—— 1896. Report on the Raised Beaches of the north-east of Ireland with special reference to their fauna. *Proc. R. Ir. Acad.* **4**, 30–54.
PRIOR, D. B. 1966. Late-glacial and post-glacial shorelines in northeast Antrim. *Ir. Geogr.* **5**, 173–187.
—— 1968. *The late-Pleistocene geomorphology of northeast Antrim* (Unpubl. Ph.D. Thesis, Queens University Belfast).
——, STEPHENS, N. and ARCHER, D. R. 1968. Composite mudflows on the Antrim Coast of northeast Ireland, *Georgr. Ann.* **50A**, 65–78.
—— and HOLLAND, S. 1975. A Late Quaternary Sediment Sequence at Carnlough, Co. Antrim. Abstract of a paper read at the *Irish Quaternary Research Meeting* 4–5.
SHOTTON, F. W. and WILLIAMS, R. E. G. 1973. Birmingham University Radiocarbon Dates VII. *Radiocarbon* **15**, 451–468.
SINGH, G. 1970. Late-glacial Vegetational History of Lecale, Co. Down. *Proc. R. Ir. Acad.* **69B**, 189–216.
—— and Smith, A. G. 1966. The Post-Glacial Marine Transgression in Northern Ireland—Conclusions from Estuarine and 'Raised Beach' Deposits: a Contrast. *Palaeobotanist* **15**, 230–234.
—— and SMITH, A. G. 1973. Post-glacial vegetational history and relative land and sea-level changes in Lecale, Co. Down. *Proc. R. Ir. Acad.* **73B**, 1–51.
SISSONS, J. B. 1976. *The Geomorphology of the British Isles: Scotland.* Methuen, London 150 pp.
STEPHENS, N. 1957. Some observations on the "Interglacial" platform and the early post-glacial raised beach on the east coast of Ireland. *Proc. R. Ir. Acad.* **58B**, 129–149.
—— 1958. The evolution of the coastline of northeast Ireland. *Advmt Sci.* **14**, 389–391.
—— 1963. Late-glacial sea-levels in northeast Ireland. *Ir. Geogr.* **4**, 345–359.
—— 1968. Late-glacial and post-glacial shorelines in Ireland and southwest Scotland. *Means of Correlation of Quaternary Successions* 437–456.
—— 1970. The coastline of Ireland *In* Stephens, N. and Glasscock, R. E. (Editors). *Irish Geographical Studies in honour of E. Estyn Evans*, Belfast. 125–145.
—— and COLLINS, A. E. P. 1960. The Quaternary deposits at Ringneill Quay and Ardmillan, Co. Down. *Proc. R. Ir. Acad.* **61C**, 41–77.
—— CREIGHTON, J. R. and HANNON, M.A. 1975. The late-Pleistocene period in northeastern Ireland: an assessment 1975. *Ir. Geogr.* **8**, 1–23.
—— and SYNGE, F. M. 1965. Late-Pleistocene shorelines and drift limits in north Donegal. *Proc. R. Ir. Acad.* **64B**, 131–153.
—— and —— 1966b. Pleistocene Shorelines. *In* Dury, G. H. (Editor) *Essays in Geomorphology.* Heinemann, London. 1–51.
SYNGE, F. M. and STEPHENS, N. 1966a. Late- and post-glacial shorelines, and ice limits in Argyll and north-east Ulster. *Trans. Inst. Br. Geogr.* **39**, 101–125.
—— 1970. The Irish Quaternary: current views 1969. *In*, Stephens, N. and Glasscock, R. E. (Editors), *Irish Geographical Studies in honour of E. Estyn Evans*, Belfast. 34–48.
—— 1975. Personal communication.
VERNON, P. 1966. Drumlins and Pleistocene ice flow over the Ards peninsula/Strangford Lough area, County Down, Ireland. *J. Glaciol.* **6**, 401–409.
WILSON, H. E. 1972. *Regional geology of Northern Ireland.* Geological Survey of Northern Ireland, H.M.S.O. Belfast. − 115.
—— 1976. Personal communication.

N. Stephens, Department of Geography, University of Aberdeen, St. Mary's, High Street, Aberdeen, AB9 2UF, Scotland.
A. M. McCabe, School of Biological and Earth Sciences, Ulster College, The Northern Ireland Polytechnic, Jordanstown, Newtonabbey, Co. Antrim. BT37 0QB.

The coasts of Leinster (Ireland)

F. M. Synge

During the maximum of the Last (Midlandian or Devensian) Glaciation a great glacier passed south down the Irish Sea basin, probably in the form of shelf ice. Subsequently sea-level fell below –100 m and a major readvance of ice from a source in the Irish midlands took place. As deglaciation progressed the rising sea flooded the isostatically downwarped marginal zone of the ice sheet while the drumlins were forming. With the disappearance of the Irish Ice Sheet isostatic uplift caused sea-level to drop to its present level by about 12000 B.P. in the north. This phase of uplift was terminated by a eustatic surge that culminated about 5000 B.P.

1. Introduction

In this contribution the 600 km coastline of the province of Leinster is described. The greater part of this coast forms the western margin of the Irish Sea and runs about north/south oblique to the strike of the main geological formations (Fig. 1). As the majority of these rock structures trend northeast-southwest, the lowlying coastline is diversified by headlands of harder rock interspersed by bays excavated from glacial drift or marine deposits. In spite of the effects of glaciation that prevailed both from the Irish Midlands or from the Firth of Clyde, the pre-existing coastal rock landforms produced by the prolonged erosion of differing rock types have survived more or less intact.

The correspondence between rock type and major coastal form would be much more apparent were it not for the thick deposits of drift. For example, Carboniferous Limestone, identified in the east of Ireland with low relief, is invariably

O

thickly covered by glacial drift. On the coast the limestone rock surface generally lies below present sea-level; only in North Co. Dublin and at Hook Head in Co. Wexford is the surface of the limestone high enough to form a cliffed coast.

Shales, slates and mudstones of the earlier or Lower Palaeozoic formations tend to form a low cliffed coastline such as seen in South Louth; in the counties of Wicklow and Wexford, where volcanic and igneous rocks of various types are interbedded with the above formations, a varied and picturesque coastline has evolved owing to differential marine erosion. The headlands, such as Wicklow Head, Mizen Head, Arklow Rock and Cahore Point owe their greater resistance to erosion to the presence of volcanic rocks. But it is another rock, quartzite, that forms the grandest and most striking cliffs of the coastline, viz., Ireland's Eye, Howth Head and Bray Head. Granite, though limited, forms a very characteristic coastline of jagged reefs and rough mammillated cliffs at the southern portal of Dublin Bay and the southeastern extremity of Ireland at Carnsore Point.

Glacial drift, as much as thirty or more metres thick almost completely masks the underlying rock surface round Dundalk Bay, in South Dublin at Shanganagh and also near Greystones as well as along most of the east coast of Co. Wexford. These thick deposits can be regarded as parts of morainic belts associated with the main glacial advances of the Last Glaciation. The present coastal form in such soft deposits is that of a straight or catenary cliffed shoreline.

In the major bays estuarine or littoral infill is dominant within the isostatically affected zone, that is to the north of Bray Head. Here sediments of lateglacial and Flandrian age predominate, and have been uplifted to a greater or lesser degree. Further south the progressive drowning of the coastlands has been more or less continuous since the disappearance of the ice sheet. In this case the rate of eustatic rise of sea-level has been greater than that of infill so that a distinctly 'drowned' coastline has evolved. One of the most striking effects of this drowning has been the advance of the tidal limit up the main rivers—about 19 km on the Barrow and 15 km on the Slaney. Although this phenomenon has been noted within the isostatic zone also, it is far less pronounced being only 4 km in the case of the Liffey and 8 km on the Boyne.

Evidence of high sea-levels before the maximum expansion of the Midlandian Ice Sheet has already been mentioned. This evidence is particularly clear along the south coast of Ireland. Beach gravels reaching 6–8 m mean sea-level (m.s.l.*) occupy gullies and embayments in the country rock. These are overlain by angular Head or local rock debris, capped by a type of glacial drift termed originally 'glacialoid' by Kinahan (1879) because the internal structure of this drift suggested rearrangement after deposition; this is one of the first references to periglacial activity on record.

Originally the beach was regarded as Pre-Glacial (Wright and Muff 1904) but later as an interglacial formation (Mitchell 1948). The position of the beach in the succession and the presence above of organic beds dated palynologically to the Gortian Interglacial (the equivalent of Hoxnian) emphasises the significance of this

---

*Mean sea-level based on records from tidal gauges at 22 stations round the coasts of Ireland lies at 8·218 feet (2·554 m) above Irish Ordnance Datum on the base of Poolbeg Lighthouse, Dublin Bay. This datum represents Low Water Spring Tide in Dublin Bay on 8 April 1837. Note that local mean sea-level in Dublin Bay is 8·53 feet (3·97 m) above Irish O.D., and local mean sea-level at Belfast (the new O.D. for Northern Ireland) is 8·9 feet (2·7 m) above Irish O.D.

beach as a very important stratigraphic marker in the Irish Quaternary. At present there is considerable doubt that this beach, now termed the Courtmacsherry Beach (Mitchell 1972), can be as old as the 'Great' Interglacial; the possibilities of a Last Interglacial or even younger age are now being considered.

Whatever the age of the Courtmacsherry Beach, an even older sloping beach platform extending from or even below present sea-level (Stephens 1957), up to a rock-cut cliff notch at 12–14 m m.s.l. is much more widespread. Regarded as an interglacial shore feature of pre-Gortian age, this rock-cut platform has a striated surface all along the east coast. Like the younger beach deposit that is associated with a small notch cut on its surface, this feature may also be much younger than previously thought; but it is clear that this platform cannot post-date the Last Glaciation.

Traditionally, the two main drift components of the east of Ireland—an earlier one derived from ice passing down the Irish Sea, and a later one from ice radiating east or southeast from the Irish Midlands were regarded as two glacial episodes separated by an interstadial (Farrington 1949) or interglacial (Farrington 1957). This break or interval in the succession has been confirmed by the more recent analysis of the drifts in Louth, Meath (McCabe 1973) and Dublin (Hoare 1975). However, doubts concerning the interglacial status of this interval were raised with the discovery that the uppermost shelly till in County Wexford overlies, at Shortalstown, an estuarine mud containing elm pollen. As such pollen is not typical of Gortian deposits, it has been ascribed to the Last Interglacial (Colhoun and Mitchell 1971). This event pre-dated the latest advance of Irish Sea ice and the interval mentioned above has been relegated to a minor break in the succession. Nonetheless the bulk of the Irish Sea drift is still regarded as belonging to the earlier or Munsterian Glaciation (Mitchell *et al.* 1973); but as shown later, this view might be an erroneous one.

Further difficulties in the interpretation of coastal sections may be encountered when the important role of slumping and mass movement, on both the ancient and modern coasts, is taken into account. The character and environment of such activity is, as yet, imperfectly understood. But there is little doubt that the natural succession can be much distorted and even overturned. In the light of these observations, a re-interpretation of certain critical coast sections gives a new chronological sequence that seems more plausible than the traditional one.

For reference, a summary of previously published interpretations of the Irish Quaternary is presented in Table 1.

There is general agreement on the upper part of the succession above the Shortalstown Estuarine Sand horizon, but the lower part is more questionable. The complexities of the drift sections can only be resolved by careful lithostratigraphical, morphological and structural techniques. Although the first of these has received considerable attention in recent years, the latter two have been neglected.

As yet the offshore deposits have not been fully investigated although such work has been in progress for a number of years. For example, the continuity of the drift belt of southeast Wexford across St. George's Channel to Pembrokeshire has been suggested (Garrard and Dobson 1974 and this volume). The origin of the linear zone of sandbanks that extend south parallel to the coast from Lambay and Howth is not known. These should not be confused with other banks composed of gravel that extend seawards from the Welsh coast. Such features, termed sarns, have been developed on lateral moraines by wave action (see Garrard, this volume).

Table 1. The Quaternary Succession in East and Southeast Ireland

Mitchell, Colhoun, Stephens and Synge 1973	McCabe 1973	Hoare 1975
Sutton Raised Beach		
Nahanagan Cirque Moraines		
Ballybetagh *Salix herbacea* clay/ Shortalstown Upper Silt		
Ballybetagh/Shortalstown Mud		
Dunany Moraine and Shortalstown Lower Silt	Dunany Formation	
Port Raised Beach		
Togher Moraine and associated earlier moraines	Boycetown F. Crockfean F.	
	(Gilltown F. Ardee, Collon, Gilltown and Balrath nembers)	Dublin F. (Cloghran, Jordanstown and Damastown members)
Athdown Moraine/Glenasmole granitic outwash		
Blessington/Killakee Moraine Killiney/Shortalstown Upper Shelly Till	Gordanstown F. (Laytown, Ben Head amd Gormanstown members)	Irish Sea F. (Balheary, Rush and Loughshinny members)
Derryvree mud		
Hollymount silt		
Shortalstown Estuarine Sand		
Tullyallen/Shelly Till and Killiney/ Shortalstown Lower Shelly Till	Tullyallen F.	Loughshinny F.
Brittas Moraine		
Mell Glacio-marine Series	Mell F.	
Drogheda Till/Bannow Till/ Kilmore Quay Upper Till	Drogheda F.	
Kilmore Quay Shelly Till		
Ennikerry Granitic Till		Slievethoul F.
Kilmore Quay Head		
Newtown Peat		
Kilmore Quay (Courtmacsherry) Raised Beach		
Ballykeerogemore Mud		

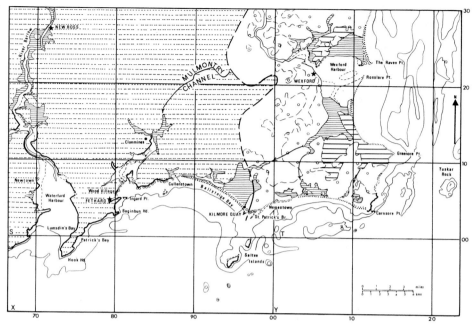

Fig. 2. Map of the south coast of Leinster.
Gravelly mixed drifts represented by the Crosstown, Screen, Killinick and Broadway soils
(Gardiner and Ryan 1964) indicate the terminal moraine belt of the Midland (Last) Glaciation.
The ground moraine is indicated by the Macamore Soil (Gardiner and Ryan 1964). Note that
many of the marine inlets have silted up or drained in post glacial times. Key to Symbols on
Figure 4.

2. The coastline beyond the limits of the Last Glaciation, from New Ross to Kilmore Quay

The limits of the Last Glaciation have long been considered to pass through
southeast Wexford. According to the most recent published work this limit has
been extended to Ballytrent near Carnsore Point (Colhoun and Mitchell 1971).
But even this limit should be extended further west to Kilmore Quay, where the
most westerly outcrops of Irish Sea drift coincide with the terminus of a boulder
fan from the Carne Granite outcrop. West of this point large boulders of this
rock are never found.

To the west this region is bounded by Waterford Harbour, an extensive ria
system of drowned valleys extending some forty kilometres north of the open
coast, guided by fault structures trending south across the northeast-southwest
strike of the country rock—lower Palaeozoic shales and slates of Ordovician age,
with inliers of Old Red Sandstone and interbedded volcanic rocks. The precipitous
sides of Waterford Harbour, up to sixty metres high, are buttressed by the downfaulted block of Carboniferous Limestone that forms Hook Head. To the east a
low coastline is indented first by Bannow Bay and then by Ballyteige Bay. Both
bays contain thick infills of marine sediment (Fig. 2).

Throughout south Wexford the Bannow Till was deposited by ice flowing south
from the Midlands. Originally this flow of ice was considered contemporaneous

with that moving down the Irish Sea (Synge 1964) and formerly termed the 'Eastern General' Glaciation (Farrington 1944). Because of the similarity between the soils derived from both tills, the Irish Sea ice was believed to have extended across south Wexford. The Rathangan soils, widespread in this area, suggested a mixing of the two ice streams (Gardiner and Ryan 1964). Although both types of soil developed from a calcareous subsoil, the primary source in the case of the Bannow Till was from the Midlands and not from the Irish Sea (Culleton 1976).

Beneath both the Bannow and Irish Sea drifts an even older deposit has been observed; an olive or ochre coloured silt containing rounded grains and pebbles of quartz. This silt was first studied in detail at Nemestown, where it was regarded as a gumbotil, weathered from an earlier till (Farrington 1939) or a periglacial deposit (Mitchell 1962). Similar deposits occur near Patrick's Bay and at Ballymadder.

Only at Cullenstown and Wood Village is thick drift being cliffed by the sea. Elsewhere rock cliffs or sandy infill form the present coast. At first sight the oldest drift deposit along this coast would appear to be the Courtmacsherry Beach (Wright and Muff 1904). Closer examination, however, shows that the cemented beach gravels and associated shore notch closely parallel the present coast. In many places recent cliff erosion has removed the old shore features, for example at Wood Village, Clonmines and between Clammers Point and Cullenstown. These sections show a fairly uniform thickness of 'glacialoid' drift (1–4 m), generally unsorted, although in places, especially east of Ballymadder Point, rude horizontal bedding occurred throughout. Invariably the ground surface at the top of all these cliff sections slopes gently seawards and no obvious signs of glacial erosion were ever found at the base of the 'glacialoid' drift.

At Cullenstown the drift is much thicker, about fifteen metres, but has a heavier texture and is full of large striated stones. The position of this cliff in close proximity to both the old cliff and beach suggests that the present coast now lies behind the old one, so that fresh till *in situ* is now being exposed. Furthermore, in the vicinity, the old beach here contains numerous erratics similar to those in the till (Fig. 3).

The inference to be drawn from these sections suggests that the drift overlying the head that seals the old beach and the covering layer of sandrock is not *in situ* but moved or slumped from a higher level *after* the head had been deposited. Only on Hook Head were striae observed along the tops of the low rock cliffs beneath *in situ* till. There were no indications that this surface was marine cut; no cliff notches or beach deposits were noted. On the other hand, the abrupt cut-off of the landforms between Baginbun Head and Lumsdin's Bay suggest the presence of a fault line scarp. Perhaps this limestone plateau subsided as a block after the formation of the raised beach.

Within Waterford Harbour, the 'bevel' or rounded profile of the old cliff top has survived in places where subsequent marine erosion has completely demolished all traces of the associated beach—viz., at Broomhill and Stonewall.

One site, Newtown, on the west side of Waterford Harbour, is absolutely crucial to the Quaternary stratigraphy of S.E. Ireland. At this locality, Ballyvoyle (equivalent to Bannow) Till was observed overlying peat containing pollen indicative of open pinewoods dated older than 38000 B.P. (Watts 1959; Mitchell 1970). Some of the organic material was incorporated in the later till as wisps and lenses. The peat was believed to overlie the cemented beach.

However, re-interpretation of the Newtown section in terms of a slumped succession suggests a different sequence (Fig. 3.) The presence, nearby, of a large striated erratic as well as several small ones completely enclosed within the Head suggests

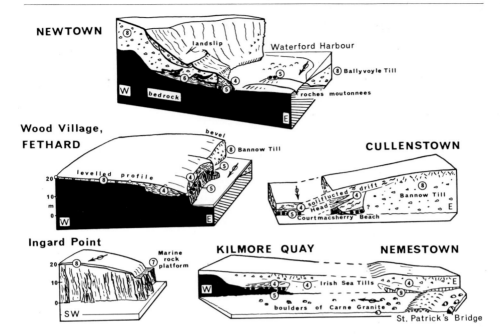

Fig. 3. Cliff sections in drift on the south Wexford coast, and beside Waterford Harbour.
The arrows indicate the direction the ice moved during the youngest glacial episode represented.
Except for the Kilmore Quay—Nemestown section, which was clearly visible, all other west-east sections are hypothetical reconstructions.
The succession represented is:

(1) Present beach.
(2) Drowned postglacial peat (formerly present at Newtown).
(4) 'Glacialoid drift'; Bannow till that has been frost heaved and soliflucted during the Last Glaciation.
(5) 6–8 m raised beach (at Newtown, Fethard, Cullenstown and Kilmore Quay), in places overlain by sand rock.
(6) Peat Bed at Newtown.
(7) 14 m cliff notch and platform, Ingard Point, near Fethard.
(8) Bannow Till (*in situ* at Cullenstown).
(9) The Nemestown Deposit—possibly derived from an older drift.

the formation and subsequent weathering of a rock cliff subsequent to the deposition of the till; otherwise an earlier glaciation would have to be invoked, for which there is little evidence elsewhere. Likewise the above peat could pre-date the old cemented beach, over which it slumped later. This peat is quite distinct from the very extensive spread that recently extended along the margins of the estuary below the cliff. The latter accumulated before the very recent rise of the sea to the present level and has only recently disappeared owing to diggings for fuel over the last few centuries.

At Ballykeerogemore (Map reference S 7319), a freshwater mud rich in wood contains remains of *Abies, Taxus* and other species indicative of the Gortian Interglacial. The relationship between this mud and the Bannow drift sheet is obscure (Mitchell, Colhoun, Stephens and Synge 1973).

In summary, the succession for southwest Wexford may be presented as follows:
(1)   Contemporary marine formation; coast being eroded for the first time since phase 5.
(2)   LITTLETONIAN (Flandrian). Sea-level below that of present.
(3)   MIDLANDIAN (Devensian). Glacial sea-levels below that of present.
(4)   Solifluction deposits.
(5)   6–8 m Raised Beach.
(6)   Interglacial 14 m cliff notch and Rock Platform.
(7)   MUNSTERIAN (Wolstonian), Bannow Till
(8)   GORTIAN interglacial. Newtown Peat.
(9)   Nemestown Silt

3. The coastline between the limits of the Last Glaciation and that of the first major readvance, from Kilmore Quay to Wicklow Head

This part of the coastline is characterised by great sweeps of coast developed in drift and interspersed by rocky sectors of low cliffs striking northeast/southwest. The latter are represented by the offshore gneissose Saltee Islands, by the granite of Carne and by the series of headlands of igneous and volcanic rock between Cahore Point and Wicklow Head. All these coastlands are dominated by extensive stretches of thick drift; only at Arklow Rock, Castletimon and Wicklow Head do rock hills over sixty metres high reach the actual coast, though isolated masses of volcanic rock such as Tara Hill and Ballymoyle Hill rise above the drift cover quite close to the coast.

Only two large rivers, the Slaney and Avoca, break the continuity of the coastline. The extensive drowned valley of the former, developed over the northern part of the depression of Carboniferous Limestone running northeast from Ballyteige Bay, effectively isolates the district of Carne from the rest of County Wexford (Fig. 4).

The oldest till observed on this sector of the coast occurs at Clogga, near Arklow (Fig. 6). This deposit plugs a small glen; unlike the overlying Irish Sea Till it is completely non-calcareous, containing an appreciable amount of Leinster Granite along with local stones (Farrington 1954). Named the Clogga Till, this deposit fails to cover the very clear rock platform scored by striae associated with the overlying Irish Sea tills. Instead, the phase of marine abrasion indicated by the rock platform apparently occurred after the deposition of the lower till as continuity of both surfaces could be proved by levelling. Hitherto the Clogga till was claimed to have occurred elsewhere on this coast directly overlying the Rock Platform (Synge 1964). This must now be considered doubtful as a rather similar till containing limestone, is now associated with the Irish Sea ice.

Some of the gravels intercalating the Clogga Till and the Irish Sea drifts originated from the underlying till and might be considered as beach gravels—the Cahore Beach of Synge (1964). In other cases, as at Seabank Point, the presence of flint, shell fragments and limestone in the gravels in this stratigraphic position should properly be associated with the upper drift sequence.

At Shortalstown (T 0214) a wedge or lens of estuarine sand appears to have been caught up in the terminal zone of the Irish Sea glacier. Push-moraine structures were observed near the terminus of this glacial event close to St. Patrick's Bridge,

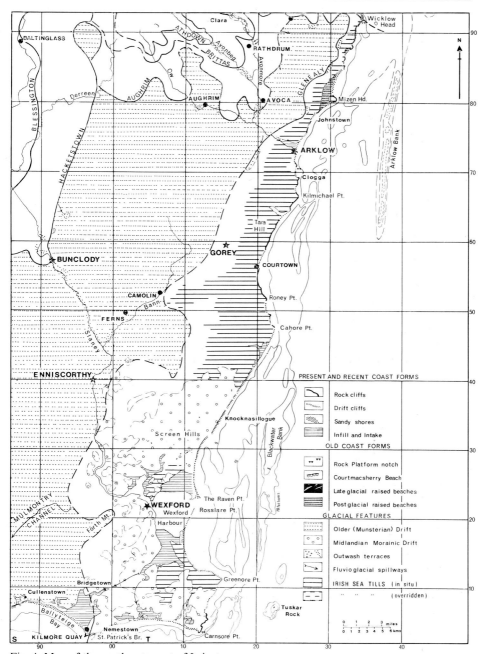

Fig. 4. Map of the southeast coast of Leinster.
The stippled area represents the older drift (Bannow and Clogga Tills) believed to be Munsterian in age.
The younger drift limits are represented by end-moraines at Glenealy (Irish Sea glacier), Hacketstown (Midland ice) and Aughrim (local Wicklow ice).
North of Arklow the postglacial marine limit is located above present sea-level—viz., in the vicinity of Mizen Head.

which is a sarn that connects Little Saltee to the mainland (Fig. 5). Associated with this same ice limit, a whole suite of marginal channels, overflow channels, deltas and kames suggest the damming of the Lower Slaney valley by ice and the diversion of the drainage to the south. The highest outlet from the Slaney catchment is indicated by the striking Mulmontry gorge that led drainage into Bannow Bay (Colhoun and Mitchell 1971).

One of the finest kame and kettle complexes in Ireland, the Screen Hills, was formed during the continued presence of the Irish Sea glacier after retreating from the line of maximum advance. The suggestion that this advance followed an interglacial is indicated by the large quantities of marine shells with temperate affinities in the glacial sands and gravels of this area.

The possibility that the Irish Sea ice advanced as a glacier of shelf ice is suggested by the evidence of very weak onshore movements and the complete absence of any lateral drop in level in a distance of eighty kilometres from Wicklow Head to Forth Mountain. All along this coastal plain a heavy boulder clay with few stones and derived from offshore marine muds gives rise to the characteristic sticky Macamore soils of east Wexford. The western edge of this till plain is associated with lateral moraine and other marginal deposits with a constant upper limit of about a hundred metres.

Beyond the above glacial limit a somewhat featureless mantle of older or Clogga drift is related to a glacial stream from the northwest, as testified by striae and boulder trails. On this relict drift surface pingos abound in certain areas, but never extend for any considerable distance within the younger ice limit (Mitchell 1973).

An earlier expansion of the local or Wicklow Mountain glaciers reached Aughrim and Avoca. Although no distinct drift landforms survive, the margins of this advance are clearly delimited by great concentrations of erratics of Leinster granite. The complete absence of such surface blocks from areas of older or Clogga drift places this local advance within the Last Glaciation (Synge 1973). At this time outwash seems to have escaped seawards unimpeded. During the later damming of these valleys by Irish Sea ice the local glaciers were less extensive.

Although the level drift surfaces of the coastal deposits north of Tara Hill simulate a raised beach (Davies 1960) there is no obvious evidence of marine activity. The highest fossil signs of such activity since the disappearance of the ice were noted as clear notches in rock undercutting the present rock cliffs between Wicklow Head and Harbour at about −1 m m.s.l. during low water spring tides. This level probably equates with that of the main postglacial beach, at a time when Mizen Head was an island and the lower part of the Avoca valley was a marine estuary.

In summary, the succession for southeast Leinster is as follows:

(1)   Contemporary marine formation; active erosion of drift cliffs.

(2)   LITTLETONIAN (Flandrian) sea-level at −1 m, Johnstown-Wicklow.

(4)   MIDLANDIAN (Devensian) Irish Sea tills at Shortalstown/Kilmore Quay.

(5)   6 m beach at ?Clogga and Kilmore Quay.

(6)   Interglacial 13 m cliff notch, and upper edge of marine rock platform.

(7)   MUNSTERIAN (Wolstonian). Clogga Till.

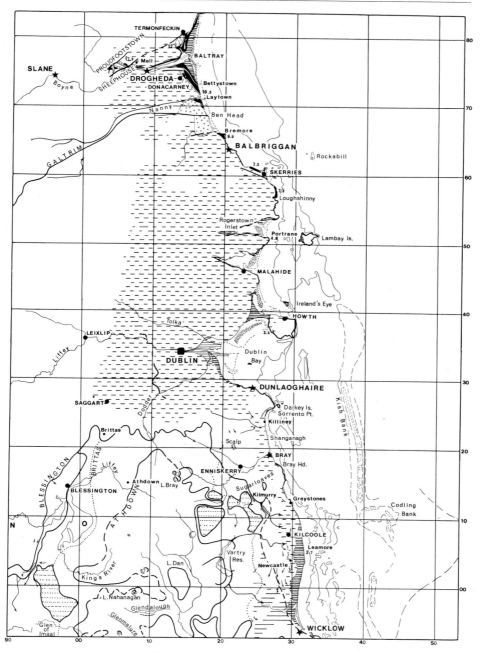

Fig. 5. Map of the east coast of Leinster.
Apart from nunataks at Church Mt, Djouce and Great Sugarloaf, the entire area was ice covered during the Last Glaciation. Within the limit of the Blessington Readvance, mixed drift derived from the till of the earlier advance has survived up to twenty kilometres inland.
The marine limit, 4 m m.s.l. in the south (at Leamore) is postglacial in age, but represents a much older shoreline at 17 m on the Boyne, 70 km further north. This latter beach passes off the present coast at Loughshinny (8–10 m); the southern continuation of this marine limit seems to coincide with the line of the Kish, Codling and Arklow Banks. Key to symbols on Figure 4.

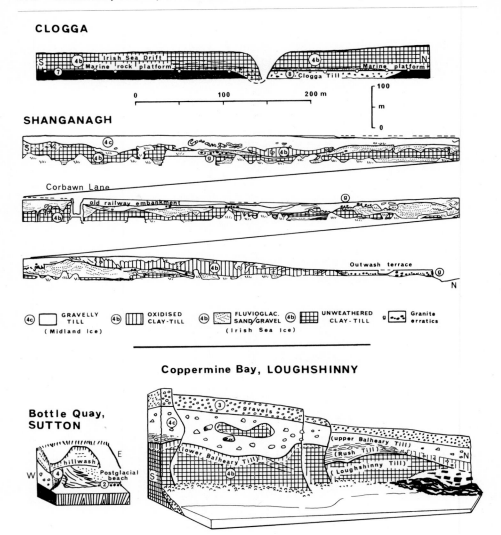

Fig. 6. Cliff sections in drift on the coasts of Wicklow and Dublin.
At Clogga the marine shore platform has been developed across shale, slate and sandstone bedrock, and across an old non-calcareous till. A bed of gravel between this surface and the overlying Irish Sea tills may or may not be a remnant of beach gravel.
At Shanganagh and Loughshinny an upper till related to midland ice from the west overlies a drift sequence associated with the earlier movement of the ice down the Irish Sea. The letter names Balheary Rush and Loughshinny refer to the drift member classification of Hoare (1975). At Sutton the Irish Sea drift was cliffed by the postglacial transgression at 5 m m.s.l.

4. The coastline between the first major readvance limit and that of the Dunany ('Drumlin') readvance from Wicklow Head to Dunany Point

Between the dominant headlands of Wicklow Head (84 m), a promontory of shale and slate, and Dunany Point (32 m), composed entirely of moraine, the first section of the coast extends almost straight for 32 km to the granite reefs of Sorrento Point, apart from the dolerite dykes of Greystones and the massive quartzite cliffs of Bray Head (Fig. 5). Further north a number of large bays indent the coast— Dublin Bay, Malahide inlet and Rogerstown inlet all of which are drowned valleys in limestone. From Rush low limestone cliffs extend to Skerries; shales, slates and sandstones to Balbriggan and volcanics to Gormanstown. Between the latter and Clogher Head, a wide 18 km sweep of sandy foreshore is broken only by the estuary of the Boyne. North of Clogher Head, a promontory of volcanic rock, another sweep of both sand and shingle extends for 6 km to Dunany Point (Fig. 6).

On the coast, the first major or Blessington Readvance limit does not form any prominent feature. Instead shingle presumably derived from the morainic deposits between Wicklow and Kilcoole, impounds a large former lagoon. Morphologically this limit is defined by a network of small marginal drainage channels and clear end-moraines at Kilmurry and Enniskerry (Farrington 1944).

Traditionally, in the drift cliffs between Bray and Killiney, a threefold sequence has been recognised—an upper stony drift full of limestone derived from the north-west (Farrington's Midland General or Upper Drift at Enniskerry), a middle unit of outwash gravels and sands, and a lower unit of very clay-rich tills from the east or northeast (Farrington's Eastern General drift) intercalated with sand and gravel horizons (Lamplugh *et al.* 1903). But according to a more recent view of this Shanganagh sequence (Hoare 1977) based mainly on till colour and matrix charac-teristics only one main glacial episode from the east was recognised.

In the writer's view, a reassessment of Hoare's results and an appreciation of the structural geology of this section confirm Lamplugh's original interpretation. Above the main structural break, Hoare's upper tills could be regarded as the ground moraine of ice from the northwest that incorporated lenses and rafts of the weathered surface of the lower till, 4b. Both the stone counts and fabrics support this interpretation (Fig. 7). The swarms of striae running NW-SE on Hill of Howth, Killiney Hill and Bray Head indicate very clearly that strong glaciation from the Midlands was the latest ice movement in this area. The same arguments apply to the type site of Hoare's Loughshinny Till Member at Copper Mine Bay (Fig. 7).

If the lower till from the Irish Sea in the Dublin area is regarded as a single unit, then the presence of this deposit as far west as Leixlip (Hoare 1975) suggests that the Blessington Readvance was considerable, amounting to 34 km at least.

Covering a wider area, there is clear evidence to show that each expansion of the ice sheet corresponded with an advance of the local glaciers in the Wicklow Hills (Farrington 1949; Synge 1973; Hoare 1975), although not quite in phase (Fig. 10). The youngest phase of glaciation, the Nahanagan advance, radiocarbon dated as younger than 11600–11500 B.P. from organic material caught up in the outer-most moraine, occurred as a small cirque glaciation (Colhoun *et al.* in press).

No obvious moraines can be associated with the retreat of the Last Ice Sheet across county Dublin. Minor features, however, suggest that the ice front ran east-west at that time, suggesting a re-appearance of the Irish Sea glacier. By the time the ice front lay between Skerries and Clogher Head the interplay between ice from the Midlands and ice from the Irish Sea had become more obvious; this is apparent from several multi-till sequences (McCabe 1973). In this area, the oldest moraine runs from Skerries to Gormanstown; the next, occurs at Ben Head and is followed by one along the line of the Nanny river. Each of these phases is associated with an outwash terrace, and all belong to the Balrath Member of the Gilltown Formation, identified with Midland ice. Further north, at the mouth of the Boyne the latest ice was of Irish Sea origin. Older deposits of similar origin occur near Drogheda as the Tullyallen Formation, composed of till members similar to those of the lower tills round Dublin.

In the valley of the Lower Boyne the association of moraines and outwash terraces influenced by a rising sea-level is particularly clear (Fig. 8). While the ice stood at the Sheephouse Moraine, the Drogheda outwash terrace built into the sea at 7–8 m. Withdrawal of the ice front by 4 km to Proudfootstown corresponded with the deposition of an ice-marginal delta at 17 km in an arm of the sea as the highest shoreline was forming at Baltray on the open coast. While the ice front still lay at Proudfootstown sea-level fell rapidly according to the gradient of the lower outwash terraces (Synge 1976).

The marine transgression mentioned above, subjected the drift coastline to a considerable amount of erosion. A wide shore platform was developed between Laytown and Clogher Head, and on the north side of the latter, the drift was entirely removed below the level of the marine limit. This highest shoreline, LG 1, (Fig. 9) has been traced from Reynoldstown (O 1486) at 21 m to Loughshinny (O 2757) at 7 m, a distance of 31 km. At this time the ice sheet had already disappeared from the coast south of Dunany, but still blocked the valley of the Lower Boyne, causing marginal drainage to pass down the Nanny to form a delta in the sea at Laytown. Lower shorelines, LG 2 (8 m at Bettystown), and LG 4 (8–9 m at Port) lie at two and seven metres respectively below LG 1.

This clear evidence of a high glacial sea-level would seem related logically to the glacio-marine deposit discovered at Mell (O 7707) at 35 m. The possibility that this deposit was transported as a large erratic during a late readvance of the ice has been suggested (Synge 1976). Yet the uniform east-west pattern of the retreat moraines (McCabe 1973) does not fit with the pronounced lobation implied by the fabric of the overlying Tullyallen Till. The original contention that the glacio-marine beds are *in situ* and are related to a much earlier event could be correct. According to the faunal assemblage, deposition took place in very cold water of normal salinity at depths of 90–110 m (Colhoun and McCabe 1973); the fauna

Fig. 7. Map of the coast of Louth, northeast Leinster.
The whole of Louth was ice covered during the Last Glaciation. During the early or Glenealy phase Irish Sea ice pushed into the Big River valley, on Cooley Peninsula, from the south. During the Blessington Readvance, Midland ice entirely surrounded the Carlingford Mountains as a nunatak.
Shortly before the Dunany Readvance the sea reached its highest levels, c. 18 m m.s.l. at Clogher Head and Templetown.
Retreat moraines at Ballaghan Point and Greenore, at the mouth of Carlingford Lough are associated with beaches at 14 m and 11 m respectively. No high beaches are recorded within the innermost moraine girdle.
Lostglacial beaches are well developed up to 7 m m.s.l. in Dundalk Bay and around Carlingford Pough. Key to symbols on Figure 4.

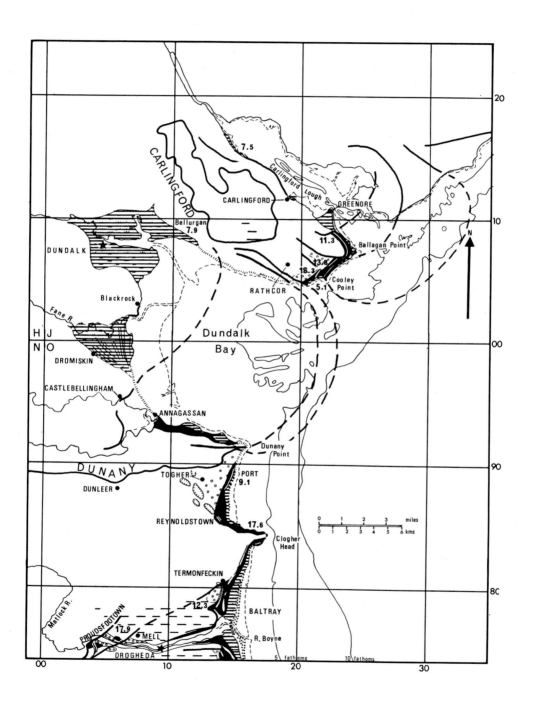

closely resembles that of the Bridlington Crag (Catt and Penny 1966) and that of silts found near present sea-level beneath the Screen Hills morainic complex at Knocknasilloge in Co. Wexford (D. Huddart, *personal communication.*) Correlation of the latter deposit with the Mell Formation suggests that a marine incursion occurred before the Irish Sea glacier had reached its maximum. According to the altitude of the glacio-marine beds, however, isostatic uplift had not yet occurred to any significant degree since the deposition of the underlying Drogheda Till.

The latest marine transgression cycle is represented by a very clear postglacial raised beach, 6 m at Port, 4 m at Bettystown, 4–5 m near Skerries, and at Portrane, and 2–3 m at Sutton (Fig. 6). In section these younger beaches are readily disting- uished from the older series; the latter are frost-heaved. At Sutton the highest postglacial raised beach rests upon an old marine cut rock platform (Stephens and Synge 1958) and also overlies charcoal in a kitchen midden dated 5250 ± 110 (I-5067) B.P. (O 2639) on the west side of the former island of Howth now connected to the mainland by a tombola. To the south an alternating series of estuarine and lagoonal deposits have been noted in Dublin Bay from borings (Naylor 1965). At Leamore (O 3105), even further south, there is evidence for a marine incursion across coastal marshes at about 3–4 m m.s.l. by 5000 years B.P. (Mitchell and Stephens 1974). This transgression post-dated the growth and development of mixed oak forest, a fine example of which is exposed at low water spring tides off Bray in a drowned remnant of the Dargle valley.

In summary, the Quaternary succession for east Leinster is as follows:

(1) Contemporary sea-level actively eroding drift cliffs.
(2) LITTLETONIAN (Flandrian) Sea-level curve rises from 2–3 m at Sutton to 6 m m.s.l. at Port.
(3) MIDLANDIAN (Devensian) Lateglacial marine limit rises from 7 m at Loughshinny to 21 m m.s.l. near Port.
(4) Retreat moraines at Proudfootstown, Sheephouse, Donacarney, Laytown, Ben Head and Gormanstown overlying a regional drift sheet (Gilltown Formation) that terminates at the Blessington end-moraine. This forma- tion overlies Irish Sea drift (Tullyallen Formation).
(5) Sea-level ca. 140 m at Drogheda (Mell Formation). Early Midland drift (Drogheda Formation).

5. The coastline within the limits of the Dunany ('Drumlin') readvance from Dunany Point to Omeath

This part of the coastline extends in an arc round Dundalk Bay and continues north round the end of the Cooley Peninsula, a highland area of Tertiary volcanics bounded by downfaulted troughs of Carboniferous Limestone. Around Dundalk Bay, apart from low reefs of slate and sandstone at Blackrock, the coast cuts across either moranic drift (at Annagassan and between Rockmarshall and Rathcor), or postglacial beach and estuarine deposits at the head of the bay, and on the end of the peninsula. Along the northeast side of the latter a rocky foreshore, at first in limestone (to Carlingford), and then in Lower Palaeozoic shales, slates and sand- stones, extends to Omeath (Fig. 7).

McCabe noted that the tills on opposite sides of Dundalk Bay can be matched. Outside the end-moraine that encircles the Carlingford nunatak, Irish Sea till similar to that of the Tullyallen Formation outcrops on the Big River valleys

RATHCOR

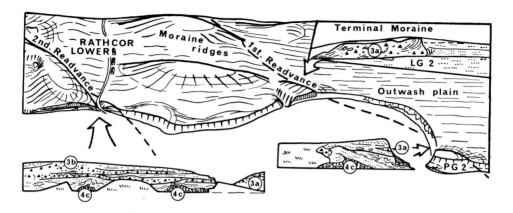

LOWER BOYNE VALLEY

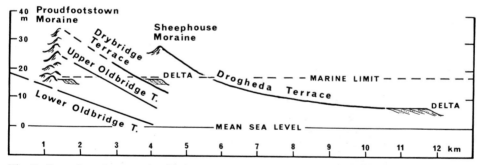

Fig. 8. Coastal sections on the Cooley Peninsula and profile of fluvioglacial outwash terraces along the lower Boyne.
At Rathcor the lowest till (4c) was deposited by the Carlingford Lough glacier after the Blessington Readvance. Later, during the Dunany Readvance, the Dundalk Bay glacier advanced into a lateglacial sea where it deposited outwash. When the readvance reached its maximum (Outer Rathcor Moraine. 3a) sea-level had fallen somewhat and outwash formed. A second readvance deposited the Inner Rathcor Moraine, 3b. The drift cliffs have been notched by the younger postglacial transgression, P.G.2 at 5 m m.s.l.
In the Lower Boyne a sequence of outwash terraces bracket a late glacial marine transgression indicated by delta fragments at 19 m m.s.l. Rising sea-level transgressed deltaic outwash associated with a stand of the ice sheet at the Sheephouse Moraine by about 10 m.

(Stephens *et al*. 1975). On the coast, the basal till at Rathcor is a grey boulder clay with erratics of Newry granite; this can be correlated with the Boycetown Till lying on the distal side of the Dunany moraines. A second grey till associated with the outer readvance moraine at Rathcor postdates the highest raised beach on the peninsula at 18 m m.s.l. (Stephens 1963, 1968). The till characteristics of this unit agree with those of the Dunany member which is also associated with the outermost moraine at Dunany Point (Figs 7 and 8).

P

The inner readvance moraine at Rathcor, identified with a brown stony till, relates to a lower beach at 11 m m.s.l. and a moraine of the Carlingford Glacier at Greenore Point and Greencastle The younger readvance till member (Dunleer) of the Dunany suite of moraine corresponds with this phase (McCabe 1973). A third girdle of moraine crosses the narrows at the mouth of Carlingford Lough between Greenore and Greencastle in association with a beach at 11 m m.s.l. Within this glacial limit no beaches higher than the postglacial series could be found. Identification of this limit as a broad belt of dead-ice landforms is apparent between Greenore and Carlingford; between Giles Quay and Rockmarshall and between Annagassan and Dromiskin (the Kilsaran Member in McCabe 1973). Within this morainic zone drumlins are first encountered at Omeath and Dundalk.

The marine limit of the postglacial transgression is identified at most places by a clear feature rising northwards from 6 m at Port to 7 m at Omeath. Another distinct beach some two metres lower was observed near Greenore. The beach ridges of this transgressional cycle attain their finest development at Greenore (at the mouth of Carlingford Lough), and beside Dromiskin, at the mouth of the Fane river. At the latter locality no less than four separate spit formations were identified migrating seawards. The oldest stands at 5 m, the next at 7 m, the next at 5 m and the last one also at 5 m. These magnificent features have not been analysed or dated.

In summary the succession for northeast Leinster is as follows:

(1)   Contemporary sea-level actively eroding at most places. Accretion of muds and silts occur at the head of Dundalk Bay.

(2)   LITTLETONIAN (Flandrian). Four raised beaches recognized between 5 m and 7 m m.s.l.

(3)   MIDLANDIAN (Devensian). Three moraine phases identified with particular raised beaches:
    11 m beach. Inner Rathcor and Greenore Moraine (=Dunleer Member).
    14 m beach and Outer Rathcor/Ballagan Moraine (=Dunleer Member).
    18 m beach following on the Cooley Point Moraine (=Dunany Member).

(4)   Basal till at Rathcor (=Boycetown Member).
    Carlingford Nunatak Moraine (=Blessington Moraine).
    Irish Sea Till by Big River (=Tullyallen Formation).

6. Conclusion

The account of the Quaternary history of the Leinster Coast outlined above suggests some quite radical changes to the story that has been usually told. One of these, an idea that the Courtmacsherry Beach is very much younger than previously thought is not completely new (Bowen 1973). But the concept that this is related to a glacio-marine sequence many metres above present sea-level in the Middle Devensian may not be so acceptable! Yet there is a strong suggestion that much of the Irish succession is younger than was previously realized (Fig. 10).

Notably, the Courtmacsherry Beach survives intact at or beyond the limits of the Irish Sea drifts (Farrington 1966) but hardly ever inside those limits. On the south coast the survival of this beach has been ascribed to its position protected in the lee of cliffs from the glacial streams. A closer examination of some of these sites shows that this explanation is not wholly adequate. For example, such sites as Wood Village, near Fethard, and Newtown, both in an area where the last

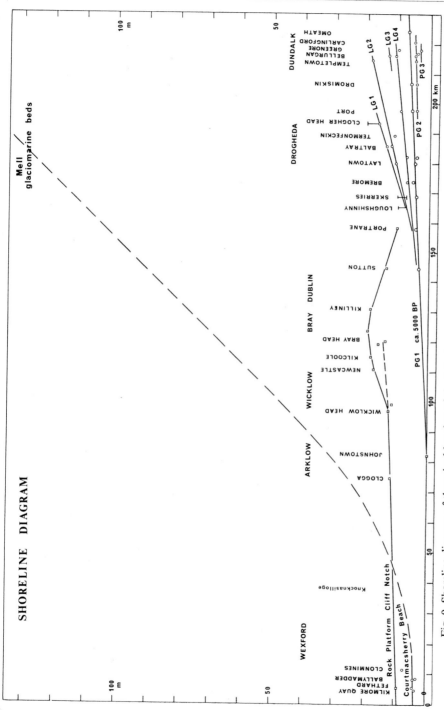

Fig. 9. Shoreline diagram of the raised beaches of the Leinster coasts.
Four series of raised beaches can be identified. The oldest, associated the Courtmacsherry Rock Platform occurs throughout the entire coastline and could belong to the Last Interglacial. The next, about Mid-Midlandian in age is equated with the Courtmacsherry Beach on the south coast and with the isostatically uplifted Mell glaciomarine deposits near Drogheda. The third series, consisting of four tilted shorelines, LG 1–4, were associated with the recession of the ice front from the Boyne to Dundlak Bay. And finally three postglacial beaches PG 1–3 tilt at a lesser angle from south Wicklow to Carlingford Lough.

glaciation moved almost due south are reasonably unprotected and open to the north! That glacial erosion *after* deposition of the beach and overlying Head is clear on the coasts round Cork Harbour; a striated and glacially polished rock platform associated with a wholesale removal of the beach at unprotected situations is quite evident in that area.

This beach is clearly widespread in the unglaciated south coast of England and north coast of France at about six metres above present mean sea-level (Wright 1937). This should not be confused with the higher beach (12 m at Brighton) associated with a Last Interglacial fauna (West 1968) and with the main marine rock platform. Presence of erratics in the former has been cited as evidence of icefloes (George 1932). But the association of a high eustatic sea-level with a glacial episode has been deemed impossible. Recently, however, global evidence of such a level of the sea during the 35000–22000 B.P. time range is growing. Evidence occurs in Greece (Kraft *et al.* 1975), in Florida (Osmund *et al.* 1970) and Georgia (MacNeil 1950) both on the Atlantic seaboard of the United States, and finally in the Gower, South Wales (John and Ellis-Gruyffydd 1970). Could this mean that in Scotland and Ireland an ice sheet built up quickly *before* the great continental ice caps grew to a large size, or was the growth of the ice sheets so rapid that large quantities of water were displaced before isostatic sinking took effect?

Along most coasts absence of the Last Interglacial beach would be expected because of marine erosion at the time the Courtmacsherry Beach was being formed. Only parts of the old cliff and abrasion platform would tend to survive.

The presence of a high sea-level at c. 140 m in the Lower Boyne valley during the Middle Devensian, immediately before the advance of an Irish Sea glacier, can be explained by isostatic depression caused by a previous expansion of the ice sheet. This early advance of the ice is represented by an Irish Midland till (Drogheda Formation) in N.E. Leinster. Beneath the drumlins of Co. Fermanagh interstadial deposits dated $30500 \pm^{1170}_{1030}$ (Birm. 166) from Derryvree (Colhoun *et al.* 1972) and others dated older than 41500 B.P. (Mitchell 1976) from Hollymount separate the upper tills from a lower one. Therefore if the Drogheda Formation is related to a high sea-level of 30000 years it must be younger than the lowest Fermanagh till. On this basis the latter should be regarded as Munsterian in age. According to this correlation the Mell glacio-marine series near Drogheda correspond to the Derryvree Interstadial.

This evidence suggests that the subsequent advance of the Irish Sea glacier from the Firth of Clyde occurred in the form of shelf ice. Such a glacier would tend to move onshore, normal to the coastline, a movement testified by the striae. Also a floating glacier of this type could have filled the entire sea area between Ireland and Cornwall. In fact, the Irish Sea Till as an apparent single unit can be identified on the Plains of Glamorgan in South Wales, at Fremington in North Devon, on the north edge of the Scilly Islands and east of Cork Harbour (at Ballycroneen) on the south coast of Ireland. This limit would seem to indicate the maximum extent of the ice during the Last Cold Period about 22000 years B.P.

The succeeding readvance, so well displayed round Dublin, probably occurred about 19000–18000 B.P. according to a ^{14}C date from the basal organic layers of a kettle hole in one of the moraines of this glacial phase in the Isle of Man (Shotton and Williams 1971). In Leinster the limit of this readvance is represented by the Blessington Moraine in Co. Wicklow, and by the Carlingford Nunatak Moraine in North Louth.

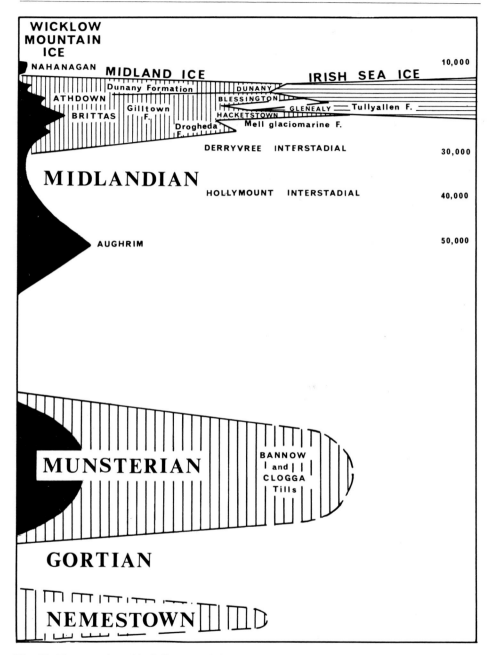

Fig. 10. Time-stratigraphical diagram of the Pleistocene succession in Leinster.
The Pleistocene of Leinster is dominated by glaciation from the Irish Midlands, from the Firth of Clyde in Scotland (Irish Sea Ice) and to a lesser degree from local ice in the Wicklow Hills.

The next major readvance occurred shortly after the formation of the highest lateglacial raised beach, and is represented by the Dunany Moraine in East Leinster. A similar event, also associated with a rapid rise in sea-level, has been recorded in northeast United States and in Denmark at c. 15000 years B.P. (Mörner 1969). Once the collapse of the ice sheet has gained momentum, rapid uplift caused a falling sea-level as milder conditions persisted until the next cold phase at c. 11000–10300 years B.P. Increased precipitation of snow before this date may have caused the reappearance of glaciers in certain cirques as early as 11500 years B.P. (Nahanagan advance).

By about 5000 years B.P. the rising eustatic sea-level reached a transgressional peak that is expressed as a raised beach north of Bray Head and as a drowned feature to the south. Therefore isostatic rebound continued after this date. But at the present day the extraordinary rapid erosion of all the major drift cliffs of the Leinster coast suggests that a eustatic surge is now active.

Acknowledgments. The author wishes to acknowledge permission from the Director of the Geological Survey of Ireland to publish this chapter and from Dr. P. G. Hoare to comment on his doctoral thesis. He also wishes to acknowledge assistance from W. Warren of the Geological Survey of Ireland in mapping the cirques of Co. Wicklow and also the considerable help of the students employed at the base camp during the Geological Survey's drift survey of Co. Wicklow. Finally, the author thanks Mrs. Diana Large for completing Figures 2, 4, 5 and 7.

References

BOWEN, D. G. 1973. The Pleistocene Succession of the Irish Sea. *Proc. Geol. Ass., Lond.* **84**, 249–272.

CATT, J. A. and PENNY, L. F. 1966. The Pleistocene deposits of Holderness, East Yorkshire. *Proc. Yorks. geol. Soc.* **35**, 375–420.

COLHOUN, E. A., DICKSON, J. H., McCABE, A. M. and SHOTTON, F. W. 1972. A Middle Midlandian freshwater series at Derryvree, Maguiresbridge, County Fermanagh, Northern Ireland. *Proc. roy. Soc.* B **180**, 273–292.

—— and McCABE, A. M. 1973. Pleistocene glacial, glaciomarine and associated deposits of Mell and Tullyallen townlands, near Drogheda, Eastern Ireland. *Proc. R. Irish Acad.* B **73**, 165–206.

—— and MITCHELL, G. F. 1971. Interglacial marine formation and lateglacial freshwater formation in Shortalstown townland, Co. Wexford. *Proc. R. Irish Acad.* B **71**, 211–245.

—— SYNGE, F. M. and WATTS, W. A. (*in press*). The cirque moraines at Lough Nahanagan, Co. Wicklow, Ireland. *Proc. R. Irish Acad.* B.

CULLETON, E. 1976. *Pleistocene deposits in South Wexford and their classification as soil parent materials.* Unpublished Ph.D Thesis, Dublin University.

DAVIES, G. L. 1960 Platforms developed in the boulder clay of the coastal margins of counties Wicklow and Wexford. *Irish Geogr.* **4**, 107–116.

FARRINGTON, A. 1939. Glacial geology of south-eastern Ireland (337–344) *In*, L. B. SMYTH *et al.* The geology of south-eastern Ireland, together with parts of Limerick, Clare and Galway. *Proc. Geol. Ass., Lond.* **50**, 287–351.

—— 1944. The Glacial Drifts of the district around Enniskerry, Co. Wicklow. *Proc. R. Irish Acad.* B **50**, 133–157.

—— 1949. The Glacial drifts of the Leinster Mountains. *J. Glaciol.* **1**, 220–225.

—— 1954. A note on the correlation of the Kerry-Cork glaciations with those of the rest of Ireland. *Irish Geogr.* **3**, 47–53.

—— 1957. The ice age in the Dublin District. *J. Inst. Chem. Ire.* **5** 23–27.

—— 1966. The Early-Glacial Raised Beach in County Cork. *Sci. Proc. R. Dublin Soc.* A **2**, 197–219.

GARDINER, M. J. and RYAN, P. 1964. Soils of Co. Wexford. *Soil Survey Bulletin No. 1*, An Foras Taluntais. 171 pp.

GARRARD, R. A. and DOBSON, M. R. 1974. The nature and maximum extent of glacial sediments off the west coast of Wales. *Marine Geology* **16**, 31–44.

GEORGE, T. N. 1932. The Quaternary beaches of Gower. *Proc. Geol. Ass., Lond.* **43**, 291–342.

HOARE, P. G. 1975. The pattern of glaciation of County Dublin. *Proc. R. Irish Acad.* B **75**, 207–224.

—— 1977. The glacial stratigraphy in Shanganagh and adjoining townlands, south-east, County Dublin. *Proc. R. Irish Acad.* B (in press).

JOHN, B. S. and ELLIS-GRUFFYDD, I. D. 1970. Weichselian stratigraphy and radiocarbon dating in South Wales. *Geologie Mijnb.* **49**, 285–296.

KINAHAN, G. H. 1897. The County of Wexford *Mem. geol. Surv. Ireland.*

KRAFT, J. C., RAPP, G. and ASCHENBRENNER, S. E. 1975. Late Holocene Paleogeography of the Coastal Plain of the Gulf of Messenia, Greece, and Its Relationships to Archaeological Settings and Coastal Changes. *Bull. geol. Soc. Am.* **86**, 1191–1208.

LAMPLUGH, G. W., KILROE, J. R., McHENRY, A., SEYMOUR, H. J. and WRIGHT, W. B. 1903. The geology of the country around Dublin. *Mem. geol. Surv. Ireland.*

MACNEIL, F. S. 1950. Pleistocene shore lines in Florida and Georgia. *Prof. Pap. U.S. geol. Surv.* **221-F**, 95–107.

McCABE, A. M. 1972. Directions of Pleistocene ice-flow in eastern Counties Meath and Louth, Ireland. *Irish Geogr.* **6**, 443–461.

—— 1973. The glacial stratigraphy of eastern counties Meath and Louth. *Proc. R. Irish Acad.* B **73**, 355–382.

MITCHELL, G. F. 1948. Two interglacial Deposits in South-East Ireland. *Proc. R. Irish Acad.* B **52**, 1–14.

—— 1962. Summer Field Meeting in Wales and Ireland. *Proc. Geol. Ass. Lond.* **73**, 197–213.

—— 1970. The Quaternary deposits between Fenit and Spa on the north shore of Tralee Bay, Co. Kerry. *Proc. R. Irish Acad.* B **70**, 141–162.

—— 1972. The Pleistocene history of the Irish Sea: second approximation. *Scient. Proc. R. Dubl. Soc.* **10**, 187–199.

—— 1973. Fossil pingos in Camaross Townland, Co. Wexford. *Proc. R. Irish Acad.* B **73**, 269–282.

—— 1976. *The Irish Landscape.* Collins, London. 240 pp.

——, COLHOUN, E. A., STEPHENS, N., and SYNGE, F. M. 1973, Ireland *in* MITCHELL *et al.* A correlation of Quaternary deposits in the British Isles. *Geol. Soc. Lond.*, Special Report No. 4, 99 pp.

—— and STEPHENS, N. 1974. Is there evidence for a Holocene sea-level higher than that of to-day on the coasts of Ireland?, 115–125 *In*, Les Methodes quantitatives d'etude des variations du climat au cours du Pleistocene. *Colloques Internationaux C.N.R.S.* 219.

MÖRNER, N.-A. 1969. The Late Quaternary history of the Kattegat Sea and the Swedish West Coast. *Sver. geol. Unders. Afh.* C **640**, 487 pp.

NAYLOR, N. 1965. Pleistocene and Post-Pleistocene Sediments in Dublin Bay. *Sci. Proc. R. Dublin Soc.* A **2**, 175–188.

OSMUND, J. K., MAY, J. P. and TANNER, W. F. 1970. Age of the Cape Kennedy barrier and lagoon complex. *J. Geophys. Res.* **75**, 469–479.

SHOTTON, F. W. and WILLIAMS, R. E. G. 1971. Birmingham University Radiocarbon Dates V. *Radiocarbon* **13**, 141–156.

STEPHENS, N. 1957. Some observations on the 'Interglacial' Platform and the Early Post-Glacial Raised Beach on the east coast of Ireland. *Proc. R. Irish. Acad.* B **58**, 129–149.

—— 1963. Late-Glacial sea-levels in North-East Ireland. *Irish Geogr.* **4**, 345–359.

—— 1968. Late-Glacial and Post-Glacial Shorelines in Ireland and Southwest Scotland. *Means of correlation of Quaternary Successions*, 8, *Proc. VII Congress INQUA*, 437–456.

——, CREIGHTON, J. R. and HANNON, M. A. 1975. The late-Pleistocene Period in North-Eastern Ireland: An assessment 1975. *Irish Geogr.* **8**, 1–23.

STEPHENS, N. and SYNGE, F. M. 1958. A Quaternary succession at Sutton, Co. Dublin. *Proc. R. Irish Acad.* B **59,** 19–27.

SYNGE, F. M. 1964. Some problems concerned with the glacial succession on South-East Ireland. *Irish Geogr.* **5,** 73-82

——— 1973, The glaciation of south Wicklow and the adjoining parts of the neighbouring counties. *Irish Geogr.* **3,** 216–222.

——— 1976. Records of sea-levels during the Late Devensian, *Proc. Roy. Soc. Lond.* (in press).

WATTS, W. A. 1959. Deposits at Kilbeg and Newtown, Co. Waterford. *Proc. R. Irish Acad.* B **60,** 79–134.

WEST, R. G. 1968. *Pleistocene Geology and Biology.* Longmans, London, 377 pp.

WRIGHT, W. B. 1937. *The Quaternary Ice Age.* Macmillan, London, 478 pp.

——— and MUFF, H. B. The Pre-Glacial Raised Beach of the South Coast of Ireland. *Sci. Proc. R. Dublin Soc.* **10,** 250–324.

F. M. Synge, Geological Survey of Ireland, 14 Hume Street, Dublin 2, Ireland.

The coast of Wales

D. Q. Bowen

Lithostratigraphic, biostratigraphic and chronostratigraphic relations of coastal Pleistocene deposits in Gower are outlined. Other successions in the Bristol Channel, Celtic Sea, Cardigan Bay and North Wales are then related to Gower where a number of potential boundary stratotypes exist. The event sequence inferred from the coastal Pleistocene succession is: (1) complete glaciation, (2) the high sea-level event of the raised beach episode (Ipswichian), and (3) incomplete glaciation during the Late Devensian which was antedated and postdated by periglacial conditions. Initiation of the rock coastline is probably older than the Ipswichian but this cannot be shown on stratigraphic evidence in Wales. No more than the broad outline of the Holocene sea-level rise can be indicated at present.

1. Introduction

From largely coastal deposits, the Pleistocene succession of Wales has been exhaustively examined in recent years (Bowen 1973a, 1974) and related to a wider area (Bowen 1973b). It has emerged that the succession exposed along the Gower coast, where rock-stratigraphic, biostratigraphic and chronostratigraphic relations are determinable with greater clarity than elsewhere, possesses all the attributes for establishment of boundary stratotypes applicable to the entire principality. Precise proposals are pending, and the following contribution attempts to relate the Pleistocene deposits elsewhere on the Welsh coastline (Fig. 1) to those of Gower, by considering in turn the coasts of the Bristol Channel, Celtic Sea, Cardigan Bay and North Wales. Because a comparatively limited and geographi-

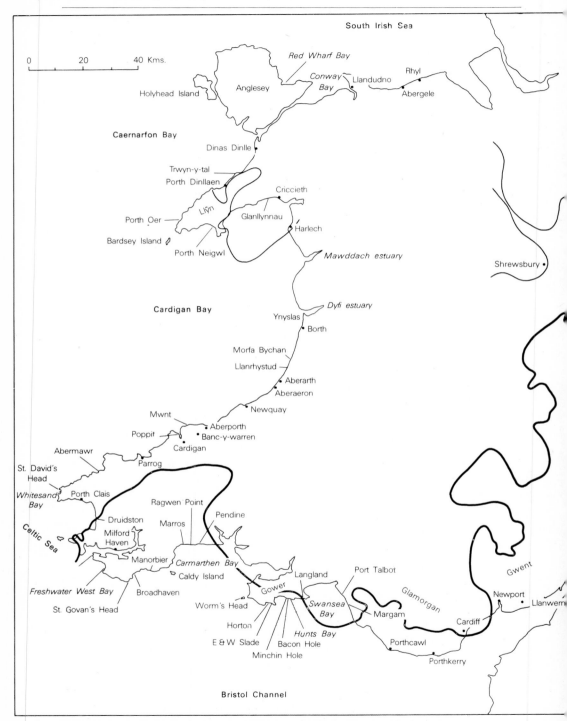

Fig. 1. Map of localities mentioned in the text together with the maximum extent of Late Devensian Glaciation (heavy line) and Late Devensian Readvance (thin line).

cally uneven amount of work has been carried out on coastal Holocene deposits
discussion of them is deferred until Section 7.

2. Gower

The general rock-stratigraphic succession and its classification are as follows:

Table 1

South Gower (west of Langland Bay)	Southeast Gower (East of Langland Bay)		Chronostratigraphy
'upper loam'	'upper loam'	⎫	
Head, including several litho-facies.	Head	⎬ (Late Devensian)	Devensian
	Glacial drift	⎭	
	Head		
Colluvial silts	Colluvial silts		
Raised beach (*Patella*)	Raised beach (*Patella*)		Ipswichian

2a. The *Patella* beach

The *Patella* beach of George (1932) forms a marker horizon along the south
coast of Gower. Characteristically it consists of a resistant conglomerate, cemented
by carbonate of lime, which represents a former storm beach. Other lithofacies
occur below this conglomerate but the majority of exposures are as described
above. Pebbles for the most part consist of local Carboniferous Limestone with a
variable proportion of lithologies foreign to the immediate area. Many of these
are of South Wales provenance. Others may be matched with parent outcrops in
west Dyfed. Most exposures contain a locally prolific molluscan fauna including:
Littorina littorea (Linné), *L. rudis* Maton, *L. neritoides* Linné, *Patella vulgata*
Linné, *P. athletica* Bean and *Purpura lapillus* Linné which are common; more
rarely *Buccinum undatum* Linné, *Ostrea edulis* Linné, *Chlamys varia* (Linné),
Trochus sp. (George 1932); and *Cardium edule* Linné, *Cyprina sp., Tellina
(Macoma) baltica* (Linné) (Prestwich 1892).

The *Patella* beach lies on shore platforms ranging in height from just over
50 ft (15·24 m) OD to approximately present day mean high-water mark 12 ft
(3·6 m) but most commonly on a platform at 10 m OD, as for example east of
Langland Bay or at Horton. This high platform is often separated by a distinct
cliff from the present inter-tidal one, but at some localities, e.g. Hunts Bay or
Overton Mere, is co-extensive with the contemporary inter-tidal one and may be
partly submerged at high tide. It is possible that a similar conglomeratic deposit
at Heatherslade Bay and in the bay immediately to the west, both lying on the

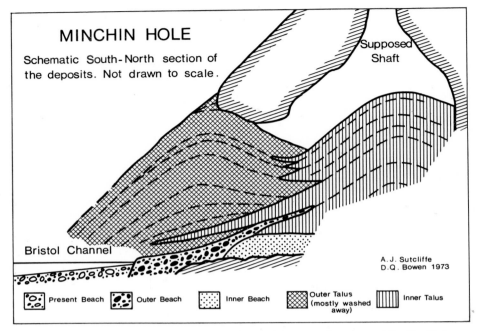

Fig. 2. Schematic longitudinal section through the Pleistocene deposits at Minchin Hole cave (after Sutcliffe and Bowen 1973).

present intertidal platform and submerged daily, are part of the *Patella* beach formation, though George (1932) believed they were younger and of Holocene age.

Potentially more complete information on the raised beach unit is available from the deposits preserved in the caves which open out on to the 'raised' plat- forms. Gower contains nearly 50 such caves of various sizes. Most have been visited by collectors of mammalian remains, but unfortunately collections were acquired with scant reference to deposits. Two notable exceptions are Bacon Hole and Minchin Hole caves where excavations have been conducted since 1972.

Minchin Hole Cave is a fissure cavern extending some 37 m from the entrance, where its mouth is 22 m high. The sequence of deposits is shown diagrammatically as Figure 2 and on Table 2. Broadly three sets of deposits can be distinguished: a basal beach unit, an inner talus and an outer talus, but in more detail seven units occur:

(1) an inner beach consisting of grey sand with fine shingle layers and occasional pebbles and containing: *Patella vulgata* Linné, *Littorina littorea* (Linné), *L. littoralis* (Linné), and *Nucella lapillus* (Linné), all characteristic of the middle shore intertidal area; and a foraminiferal assemblage found in temperate waters around the British coast today. It lies on the marine abraded limestone floor of the cave at 9·9 m and extends to 11·8 m OD.

(2) 'Lower Red Cave Earth', a tongue of the inner talus with some mammalian remains. Previous discoveries of *Dicerorhinus hemitoechus* (Falconer) (Murchison 1868) came from this unit (Plate 1).

(3) Outer beach: a coarse, shelly sand with large cobbles and rolled bone frag- ments. It contains *Patella vulgata* Linné, *Littorina littorea* (Linné), *L. saxatilis*

Table 2

Minchin Hole	Bacon Hole
Inner (c. 17) and Outer (c. 20) talus cones	Cemented breccias ($>$4)
Sandrock (0·7)	
Brown cave earth (0·9)	Upper Brown cave earth (1·0)
Pink sandy clay (0·6)	
Outer beach (1·2)	Upper Brown sands (0·25)
	Grey clays and sands (0·75)
Red cave earth (0·9)	Lower Brown cave earth (0·9)
Inner beach (1·8)	Yellow brown sands (beach) (0·7)
? Breccia (pre-outer beach)	

(Olivi), *L. littoralis* (Linné) and *Nucella lapillus* (Linné), representative of the upper middle and middle shore which is submerged daily. It is coextensive with George's (1932) *Patella* beach exposed at the cave entrance, who with Bowen (1970a) thought it was the oldest deposit in the cave and it ranges in height between 9·40 and 11·89 m OD. Stratigraphically it lies unconformably on units (1) and (2) into which its associated low cliff was cut (Fig. 2).

(4) On the west wall of the cave the outer beach is overlain by a pink sandy-clay (probably equivalent to George's (1932) 'Neritoides beach'), which contains *Littorina saxatilis* (Olivi) and *L. littoralis* (Linné), in its lower layers, with the addition of *L. neritoides* Linné in the upper ones, with the land mollusca *Clausilia bidentata* (Ström), *Cochlicopa* spp. and *Helicella* spp.

(5) Brown cave earth with angular limestone fragments (Plate 2). It contains a large number of *L. saxatilis* (Olivi) but decreasing upwards; some *L. littoralis* (Linné) (mostly in the lower part) and a few *L. neritoides* Linné. More significantly land snails occur throughout the unit, dominated by *Clausilia bidentata* (Ström). Also present are *Pyramidula rupesrtris* (Draparnaud), *Helicella* sp., *Pupilla muscorum* (Linné), *Vallonia* sp. and *Vertigo pygmaea* (Draparnaud.)

(6) Sandrock.

(7) Breccia (head) the base of the outer talus cone proper.

Some units correlate with the succession on the east wall of the cave: units (7) and (6) are clearly continued. Units (5) and (4) correlate with George's 'ossiferous breccia' from which he obtained teeth of *D. hemitoechus* (Falconer). A relationship of considerable interest is that, on the east side, the outer (*Patella*) beach is clearly emplaced in a notch which undercuts an isolated pillar of breccia with blown sand cemented on to the wall of the cave. It probably represents the oldest exposed deposits in the cave.

The upper part of the cave sequence which comprises the inner talus interfingers with the outer one (Fig. 2). At one time the cave mouth was blocked.

At Bacon Hole the succession is not dissimilar (Stringer 1975) (Table 2):

(1)   Basal sand (1 m) containing mollusca, *Equus* sp. and *Microtus oeconomus* (Pallas).

(2)   Lower Brown cave earth (0·9 m) with marine and littoral mollusca, various rodents (including water vole, *Avricola terrestris* Linné), *Palaeoloxodon antiquus* Falconer and Cautley, bison, red deer and fallow deer.

(3)   Grey clays and sands (0·75 m) with relatively few mollusca, northern vole, water vole, *Palaeoloxodon antiquus* Falconer and Cautley, *D. hemitoechus* (Falconer), hyaena, roe deer.

(4)   Upper Brown sands (0·25 m) with foraminifera and water vole.

(5)   Upper Brown Cave Earth (1 m) with terrestrial mollusca, water vole and northern vole, *P. antiquus* Falconer and Cautley, *D. hemitoechus* (Falconer), red deer, bison.

(6)   Breccias (head) (4 m).

At the entrance to the cave, separate from the above succession, is (i) marine sand, (ii) head or scree, with (iii) shingle beach lying unconformably across (i) and (ii) in a manner reminiscent of the stratigraphic relations at Minchin Hole (see Fig. 2).

Unit (6) of Minchin Hole, the sandrock, is represented widely along the coast, lying on the *Patella* beach as at Caswell Bay. At Shirecombe the cemented bedding planes of this aeolian deposit are aligned from east and southeast to northwest (George 1932). A variable proportion of blown sand occurs in the succeeding head deposits, as well as in colluvial beds, which often overlie the *Patella* beach.

2b. Colluvial silts and Head

Red beds identified as 'colluvial silts' by Bowen (1970a) overlie the beach (Plate 4), as well as occurring as discrete beds or matrix in the head deposits. Several litho-facies occur. Characteristically they are bedded and thickest in the centre of bays; while towards bay sides disappear or thin rapidly. Ball (1960) examined thin sections of the material from Worm's Head and pronounced them *terra fusca;* but elsewhere field and stratigraphic relations show them to be colluvial. At Port Eynon Bay their mineralogy is similar to that of local glacial deposits and an outlier of Keuper Marl. Locally their constituents reflect a provenance determined by immediate catchment area. In general, the redder the colour, the higher the clay content.

As well as being diversified by variable amounts of colluvial deposits (Plate 3), three principal lithofacies of head occur; coarse blocky head with boulders up to 0·5 m in diameter often mixed with colluvium (litho-facies IIa); finer calibre head consisting of limestone chips (litho-facies IIb); recycled old glacial deposits (litho-facies I). Litho-facies (IIa) and (IIb) are contaminated to a greater or lesser extent by scattered erratic pebbles. The blocky material (IIa) lies at the foot of the abandoned limestone cliff and is invariably related to the close proximity of

Plate 1
Stratified intertidal sand, with rounded pebbles, overlain by red cave earth at Minchin Hole cave (scale in feet).

Plate 2
Brown cave earth overlying pink *Neritoides* clay (lighter tone), and overlain by base of sandrock (transition at upper marker). (Scale in cm and inches.)

Plate 1

Plate 2

Plate 4

Plate 3

bedrock. Finer calibre material (IIb) overlies this and replaces it with increasing distance from the fossil cliff base.

Lithofacies (I), formerly termed the 'Older Drift of South Wales' by George (1932) and Zeuner (1945), occurs in a highly characteristic mode (Figs 3 and 4). In any quantity the deposit lies infilling small coastal valleys and at Eastern and Western Slades expands in the drift terrace in a manner reminiscent of an alluvial fan. It is bedded and its clasts lie at low angles parallel to this, as well as displaying a strong preferred orientation. Away from the valley entrance the recycled drift merges into, and is replaced by, head limestone lithofacies (IIb).

All the aforementioned characteristics are consistent with an alluvial-solifluctional origin, and the deposit is clearly contemporaneous with local limestone head.

2c. Post-beach glacial deposits and 'upper head'

These occur at both eastern and western ends of the peninsula and enable the extent of the post-*Patella* beach glaciation to be outlined (Bowen 1970a). At Langland, on the east side of the Bay (Rotherslade) and extending towards Mumbles up to 14 m of glacial drift lies on head and *Patella* beach. It is of south Wales provenance and its erratics characterise it as a typical Brecknockshire glacial drift (Bowen 1971a). At Bryn House, Derwen Fawr, Trottman (1964) described a kettle hole in this formation which contained Devensian Late-glacial sediments.

In Rhossili Bay outwash from an ice-margin running across the northern part of Carmarthen Bay (Bowen 1970a) is interbedded between lower and upper heads of Old Red Sandstone lithofacies.

Wherever site factors were appropriate, that is where a source rock was exposed above the glacial deposits, head formed after deglaciation, being time-equivalent to the upper layers of the solifluction terraces in the unglaciated part of the penin-sula (Fig. 1). At Langland Bay some glacial drift was re-sorted. But in Rhossili Bay, at the base of a west-facing 190 m high coastal slope of well-jointed Devonian Beds, some 12 m of head accumulated.

2d. Discussion

The coastal Pleistocene deposits of Gower are of cardinal importance for the following reasons: (1) they occur outcropping more or less continuously across the boundary which delimits areas last glaciated before the raised beach event, and those ice-covered subsequently; (2) biostratigraphic characterisation of the raised beach formation is possible in a manner not possible elsewhere around the shores of the Irish Sea; (3) the glacial formation which postdates the raised beach

Plate 3
Coastal Pleistocene succession at Hunts Bay (western side) showing: *Patella* beach conglomerate, red colluvial silts interbedded with head, and head (containing some erratics). Note bedding of the deposits off adjacent slope. Height of drift terrace 7·5m.

Plate 4
Coastal Pleistocene succession at Hunts Bay East showing: extensive outcrop of *Patella* beach conglomerate on wide shore-platform which is co-extensive with the present inter-tidal one, thick colluvial silts at base of drift terrace, overlain by head. Height of drift terrace 9m.

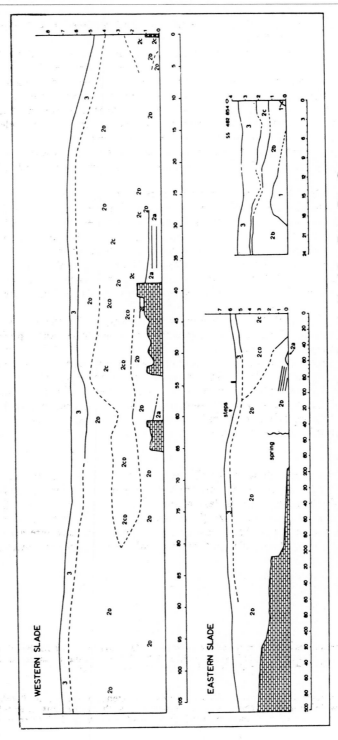

Fig. 3. The Quaternary succession exposed on the north side of Port Eynon Bay, Gower (after Bowen 1971a). Key: 1 = raised beach gravel, 2a = colluvial silts, 2b = head (lithofacies IIa), 2c = head (lithofacies IIb), 2d = redistributed glacial drift (head lithofacies I), 2cD = admixture of 2D and 2b, 3 = 'upper loam' (loess, cover sand and colluvium).

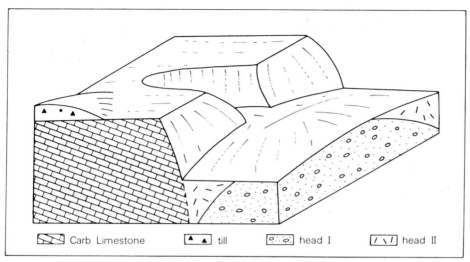

Carb Limestone   till   head I   head II

Fig. 4. Model of lithostratigraphic relationships at Eastern Slade, type site of redistributed glacial drift lithofacies (head I); head II = limestone scree.

event may be dated with respect to its upper limit by basin deposits near Swansea; (4) the recycled nature of the older drift of south Wales may be established at its type locality in Eastern Slade; (5) the origin of the red beds, variously interpreted in the past, can be established as colluvial.

That the raised beach represents an interglacial 'high' of ocean level is confirmed by the faunal remains from Minchin and Bacon Hole caves. Fauna of interglacial aspect occur at Minchin Hole in the inner beach, red cave earth, outer beach and overlying cave earths. While the work there is not yet complete, it seems reasonable for the present to assign all of these beds to the same interglacial, namely the last or Ipswichian (Eemian) temperate stage (Sutcliffe and Bowen 1973). Sutcliffe (1960) has argued that the presence of *D. hemitoechus* Falconer and the absence of *D. kirkbergensis* Jäger is indicative of such an age; but what is more significant at Minchin Hole is the record of the water vole, *Avricola terrestris* Linné, from the Brown cave earth. This species first appeared in the British Pleistocene in the Wolstonian cold stage (Stuart 1974): hence a Hoxnian age for the raised beach beds is excluded. Similarly at Bacon Hole units (1)–(5) are regarded as Ipswichian and unit (6) as Devensian (Stringer 1975).

Following from the dating of the raised beach as Ipswichian is the corollary that the overlying Pleistocene deposits are of Devensian age. Such a proposition is directly enforced by the dating of the Langland glacial deposits as Late Devensian: and indirectly by an absence of beds which demonstrate temperate climatic conditions in the deposits younger than the beach. Thus the head deposits of Gower outside the Langland area represent all of Devensian time. Basal colluvial silts indicate the onset of cold conditions when sheet washing and soil erosion occurred on a more or less treeless land surface. Following the removal of fine grained regolith in this way, subsequent frost action detached larger particles which accumulated at the cliff base as a 'blocky' litho-facies. This material had been subject to preparatory chemical weathering during the previous temperate episode. Exposure of fresh unweathered bedrock to frost action produced the finer head litho-facies. At the same time previous glacial deposits were reworked and deposited by alluvial and solifluctional processes as part of the head formation.

Attempts have been made to recognise greater chronostratigraphic complexity by Wirtz (1953) and Mitchell (1972). At Horton the former author and at Hunts Bay (Table 7) the latter sought to confer separate stage chronostratigraphic status on the lower (blocky) and upper (fine calibre) head facies. It has been shown, however, that head sequences fining upwards are a consequence of a cyclical denudation of adjacent rock slopes, and in particular no trace of unconformity or disconformity (wherein Mitchell would lose a glacial and interglacial event) are apparent (Bowen 1973a, 1973b). In any event, a recognition of a lower Wolstonian and an upper Devensian head is equating on a one to one basis the lithostratigraphy with a chronostratigraphy (Bowen 1973a, 1973b). Similarly any attempt to recognise separate events on the basis of presence or absence of derived erratic content either in the head of *Patella* beach may be criticised on the same basis (George 1932; Bowen 1973a).

3. The Bristol Channel

3a. Glamorgan and Gwent (excluding Gower)

With the exception of an exposure of sandrock cemented on to the Lias cliff at Porthkerry (Strahan and Cantrill 1902), and the gravel described by Knight at the foot of Newton Down, Porthcawl (Knight 1853) no outcrops of the raised beach are known between Gower and Chepstow. Between Newport and Cardiff a buried rock platform, now covered by Holocene deposits has been interpreted as wave-cut by Williams (1968) and its landward termination as a former cliff-line by Steers (1964). Resting on this surface immediately below Holocene deposits is a gravel layer interpreted by Locke (1974) as an ancient beach. However the beach gravels may be Holocene lying on a re-trimmed, older surface.

On the Glamorgan coast, which is formed largely of comparatively unresistant Lower Jurassic rocks, appreciable rates of Holocene cliff retreat have been established. Iron Age hill-top forts, originally circular, square or rectangular in plan, have been bisected by cliff-retreat (North 1929); while historical data also demonstrate recession, though at somewhat slower rates (Trenhaile 1972). On the more resistant Carboniferous Limestone and Trias cliffs, however, rock ledges have been described, ranging in height from MHWS to 12 m OD, and up to 20 m wide at Sker, Porthcawl, Ogmore Black Rocks, Sully Hospital and Sully Island (Trenhaile 1971). Trenhaile has argued that these may represent the remains of interglacial shore-platforms, and indeed that much of the 'modern' platform may be an inherited feature (Trenhaile 1972).

3b. Carmarthen Bay

Carboniferous Limestone outcrops along the coast west of Pendine between Dolwen and Gilman Points. In the lee of Dolwen Point, sand rock is cemented against the cliff and at the cliff foot is overlain by head; both deposits pass below contemporary beach level. At Gilman Point a considerable thickness of limestone head mantles the former cliff. At the base it consists of a colluvium with boulders, overlain by blocky head which is prominently cemented by stalagmite and which 'fines' upwards to beds which dip off the limestone outcrop.

Just above HWM at Ragwen Point, raised beach gravel, consisting of local quartzite pebbles, is overlain by sandrock, colluvium and head (Bowen 1970a). A short distance to the west a subangular beach gravel lies on a raised platform which consists of a stripped bedding plane of Millstone Grit Quartzite. The beach is overlain by dune sand with silt lenses containing pollen, which include grains of *Pinus, Salix, Betula nana, Hippophaë, Alnus, Carpinus, Picea, Taxus, Tilia* and *Tsuga* (Mitchell, in Bowen 1970a). The sand is overlain by head, which is magnificently exposed at Marros Sands. No trace of erratic material occurs in these exposures. Yet the plateaux adjacent to the coastline bear unmistakable boulder trains and outliers of till and fluvioglacial deposits which demonstrate former glaciation by an Irish Sea ice sheet from north-northwest to south-southeast (map in Bowen 1970a).

West of Marros exposures of raised beach have only been recorded at the southern end of the Esplanade, Tenby, as consolidated pebbles overlain by sand, in the roof of Merlin's Cave; the southern end of the Burrows, Penally, where the cliffs commence (Dixon 1921); and at Giltar Point (Leach 1909). Leach (1933) commented, however, that the wide mouthed cave at the southern end of Trevayne Sands demonstrates the virtual identity of present and raised beach coastlines.

3c. Caldy Island and Giltar Point to St. Govan's Head

Several exposures of raised beach occur on Caldy Island where they lie on a platform 3·6 to 5 m above HWM (Dixon 1921) (c. 8–10·5 ft OD), the general succession being (1) raised beach with *Littorina littorea* Linné, *L. obtusata* Linné and *Patella vulgata* Linné; (2) blown sand; (3) head. Dixon recorded 1·8 m of raised beach between Sandy Bay and Little Sound which he sub-divided into (i) a lower 'chiefly subangular rubble'; (ii) a middle unit with large rounded boulders of limestone; (iii) an upper subangular rubble: no head was recorded (Dixon 1921). Between Sandy Bay and Rat Island, Leach recorded raised beach. The shingle contained *L. littoralis* Linné, *Mytilus edulis* Linné and *Patella* sp., flints and red sandstone pebbles (Leach 1934). Of some significance, however, is his record of beach shingle overlying a thin unit of 'angular stones in red clay earth', thus inviting comparison with the Gower cave succession. Overlying the beach is head, but at present there is no adjacent high ground from which the sandstone fragments could have been derived. The same situation occurs along the southern shore of St. Margret's Island where a stack, projecting some 6 m above the shore, bears beach material and large limestone blocks which must have been derived from upslope along a connection to the island now removed by marine erosion (Dixon 1921) presumably during the Holocene.

Pleistocene deposits are only infrequently preserved at the base of the high cliffed coast between Giltar Point and St. Govan's Head, though platform remnants [for example northwest of Whitesheet Rock, Lydstep, on the neck of Black Mixen Peninsula overlain by head, at Swanlake Bay, and perhaps concealed by head at Freshwater East (Dixon 1921)] demonstrate once again that the raised beach coastline is similar to the present one. The beach deposits that survive confirm this.

At Manorbier Bay a more or less continuous raised platform at about 5·5 m above HWM bears raised beach on the eastern side of the Bay, overlain by a metre of head (red sandstone). In an adjacent exposure the head is underlain by colluvium. Despite the well preserved platform at Swanlake Bay, beach is exposed at one site only where a metre of shingle is overlain by 9 m of head.

According to Dixon the raised platform at Broadhaven is somewhat lower, being only some 2·5 to 4 m about HWM. Compared with the above named bays, however, the fossiliferous conglomeratic beach is extensively exposed and is overlain by 2 m of sandrock, 1 m of head and 2 m of aeolian material.

Finally, between Broadhaven and St. Govan's Head raised beach shingle, with angular material on its surface, lies on steep and precipitous stacks rising sharply from the present inter-tidal platform. As on Caldy Island, they serve to demonstrate the locally extensive amount of post-raised beach marine erosion.

3d. Discussion

Any comparison between the Gower coast and that east of Swansea Bay is based almost entirely on morphology, thus any event sequence for the latter is only tentative. On the other hand, outcrops of Pleistocene deposits west of Gower allow ready correlation. The fundamental succession of raised beach overlain by sand or sandrock, colluvium and head (several lithofacies), invites clear correlation with the rock stratigraphic succession of Gower west of Langland. As in Gower, these show quite unequivocally that the last local glaciation of the region (by Irish Sea ice) antedated the raised beach event (Bowen 1973a, 1973b).

Chronostratigraphic correlation with Gower (Table 7) shows: (1) a pre-raised beach glaciation, of unknown age; (2) the raised beach interglacial event of Ipswichian age; (3) the Devensian cold stage when the area lay exclusively in the periglacial zone. Ice of central Wales provenance penetrated no farther southwest than the eastern side of Carmarthen Bay, and the area around St. Clears (Bowen 1970a).

4. The Celtic Sea

4a. St. Govan's Head to West Angle Bay

Between St. Govan's and Linney Head the Carboniferous Limestone cliffs present some of the finest coastal scenery in Britain. Despite a virtual absence of raised beach outcrops in this high energy storm wave environment, there can be no doubt of their Pleistocene age for raised beach has been fortuitously preserved near the Devil's Barn at Saddle Head (Dixon 1921) and discontinuous fragments of a raised platform occur at their base, as for example near Bullslaughter Bay.

At Freshwater West Bay, however, raised beach is extensively preserved along some 0·9 km of the bay. It lies on an extensive platform rising to 5 m above HWM, which is coextensive with the present inter-tidal platform some 180–275 m wide. Platform and beach are covered by head up to 6·6 m thick. The beach shingle contains some blocks up to 0·6 m in diameter, igneous and flint pebbles, which Dixon speculated had been derived from an offshore submarine outcrop. Immediately overlying the beach is a partly cemented blown sand with snail shells (Dixon 1921). A short distance inland, northwest of Castlemartin, Dixon (1921) recorded raised beach below head at 9 m OD.

4b. West Angle Bay

Dixon and Leach examined the Pleistocene deposits of the bay together and the succession which they recorded is shown in Table 3 (Dixon 1921). The black clay contained silicified shells and may be lower limestone shales rotted *in situ*; but Dixon suggested it could be estuarine mud due to the presence of nests of pyrites and

Table 3

Dixon (1921)	John (1968, 1970a)	Ribbon and Bowen et al. (in preparation)
Limestone pebble shingle (0·3)	Thin head	6. Limestone single (0·2)
Gravel and sand (1·8)	Red gravel (glacial)	5. Red gravel (3·9)
Buff fine loam (0·6) Grey fine loam (with remains) (1·5)	Dark grey clay Bright grey silts White sands and silt (1·8—2·4) ▲	Orange loam (2·0) Grey loam (0·8) 4. Peat (0·04) Purple loam (0·8)
Gravel and shingle (1·2)	Sands with pebbles and head ▼	3. Gravel (0·8)
Sand (0·9)	Cemented raised beach (0·5)	2. Sands (1·28)
Boulder clay (1·8)		1. Till (1·5)
Black clay (1·5)		

selenite. The till contains rounded and subrounded striated pebbles including igneous rocks in a stiff purple clay, and has been exposed since Dixon (1921) described it (Bowen 1973a, 1974), though John (1974) ignored Dixon's account (*op. cit.* p. 190). The sand, gravel and shingle overlying the boulder clay was considered to be a possible representative of the raised beach, though 'no agreement was reached (with Leach), as to which, if any of the beds should be taken' as such (Dixon *op. cit.* p. 197). A short distance west of the exposure Dixon note 0·3 m of coarse gravel overlain by loamy "head" and thus suggested that the sand (2) and gravel (3) of his succession (Table 3) "may represent the Raised Beach" (*op. cit.* p. 98).

Subsequently the deposits were investigated by John (1968, 1970a) (Table 3), who did not recognise Dixon's boulder clay, but pronounced the sand and shingle to be raised beach, and the overlying loams ('sands, silts and clays') to be estuarine, the entire sequence having formed during an interglacial transgression. Head (in the sand unit) was still forming during beach formation. Pollen analysis of the 'loam' units revealed a temperate flora with *Quercus, Alnus* and *Corylus* (John 1968). The red gravel he interpreted as glacial. His interpretation was cited by others (Bowen 1973a, 1973b, 1974; Peake *et al.* 1973), but work still in progress shows it requires radical revision (Ribbon and Bowen *et al.*). Dixon's till has been re-exposed and a succession directly comparable with his re-established (Table 3). Unit 2 is cross-bedded and consists of 5 sub-units with angular and subangular sand grains, generally well-sorted, but less so upwards. This characteristic is paralleled by an increase in the angularity of the grains; wood fragments were also recorded. Unit 3 is poorly sorted and includes angular fragments. Unit 4 contains no shell fragments and no foraminifera though numerous wood fragments and a discrete peat horizon occur. The succession above the till is interpreted as terrestrial: the sands and gravels as fluvial deposits, probably having accumulated in a cold climate, and the 'loam' sequence as lacustrine—there are no data to support a marine origin.

Fabric analysis confirms the red gravel as a periglacial slope deposit. The upper shingle is a modern deposit as concluded by Dixon (1921). In essence, therefore, the interglacial beds lie between two periglacial units, the lower of which is under-lain by till.

4c. Milford Haven

Within Milford Haven the raised platform lies about 2·7 m above HWM rising locally to a maximum of 3·6 m above that datum (Dixon 1921). It may occur as a ledge at the foot of the coastal slope, or as an extensive feature (up to 23 m wide in Angle Bay) bounded landward by an abandoned cliff, or as basal to a prominent drift terrace, for example, south of Burton Mountain. Beach shingle or estuarine clay, which compares well with offshore tidal mud (Dixon 1921), occurs on the platform at several localities (Dixon 1921) and is overlain exclusively by colluvium and head of various litho-facies.

4d. Milford Haven to St. David's Head

Very few instances of the raised shore platform occur on this coast. At Porth Clais, Leach estimated that it lay between 4·5–6 m above HWM (Leach 1911), but despite a recent paper on that site (John 1970b) the precise height has not yet been surveyed. Unfortunately the same is true for all the other platform remnants though in geomorphological mode they compare with others farther south and in Gower. The principal remnants are found on Grassholm Island (Dixon in Cantrill *et al.* 1916), at Porth Clais (Prestwich 1892) where Leach described striations across the platform (Leach 1911), Whitesands Bay (Prestwich 1892), Pendaladeryn, Porth-cadno, Ogof Henllys, Ogof Lle sugn (John 1970b), Trwynhwrddyn and on the mainland opposite Carreg-fran (D. P. Davies, unpublished).

Exposures of raised beach occur at Druidston Haven, the southernmost ins-tance west of Milford Haven, near Porth Clais (two exposures), Carreg-fran, and Whitesands Bay. All consist of local rocks with a subordinate number of erratics, which Leach believed may have been introduced by floating ice (Leach 1911). A feature of particular interest is the occurrence of large and smoothed boulders near the base of the beach at Whitesands and Porth Clais, where Leach argued they were certainly part of the beach deposit. Towards the top of the beach an increasing proportion of angular fragments derived from local cliffs indicates the close of the high sea-level event. At Whitesands, beach and overlying head are intermixed (D. P. Davies, unpublished).

Jehu (1904) studied the Pleistocene deposits of the northern part of this area and recognised the following succession:

(3) Upper boulder clay and rubbly drift

(2) Sands and gravels

(1) Lower boulder clay

though subsequently the occurrence of head below the boulder clay was noted by Leach (1911), while John (1970a) showed that the upper boulder clay and rubbly drift were slope deposits (Head). The complete succession is thus:

(5)   Sands and loams

(4)   Upper head

(3)   Till and outwash

(2)   Lower head

(1)   Raised beach

At many sites the glacial beds have been recycled by solifluction and now constitute 'head'. For example, at Pendaladeryn and Trwynhwrddyn the head includes local clasts and erratics in a matrix of silt and clay (D. P. Davies, unpublished). At Porth Clais, Prestwich (1892) simply recorded raised beach overlain by head though Leach recognised till above the head, and John (1970) till between an upper and lower head. Bowen (1966), however, while also recognising the 'unstratified gravelly' deposit of Leach had argued that such glacial material was not *in situ*, having been reworked by slope processes, a conclusion endorsed by D. P. Davies who also recognised the same facies along the cliffs at Caerfai, Porth-y-rhaw and Cwm Mawr Bays. She thought it unlikely that the distinctive Irish Sea till, present at Dowrog Common some two miles to the north (Jehu 1904) would have acquired a 'land (litho) facies' over such a short distance, as John (1970) had argued. Jehu (1904) had anticipated this conclusion for he believed the rubble drift and yellow brown clay to have been rearranged and modified by 'sub-aerial agencies'. At many exposures, therefore, Unit 3, the glacial deposits, are represented only by fragmentary re-worked materials. It is significant that these localities are found at the base of steep slopes where, even if any appreciable quantity of glacial drift had been deposited, it would hardly have survived for long in place, for example at West Dale Bay. On the other hand, in deep, steep-sided coastal valleys, not scoured by Holocene marine action, Irish Sea till has been extensively preserved as for example at Druidston and Portmelgan.

4e. Discussion

The event sequence for this area has a clear parallel in Gower (Table 1) where the chronostratigraphic stages recognised are represented by similar sediments. The succession of till, fluvial deposits, freshwater loams and head, of West Angle is correlated directly with the basically three-fold stage sequence recognised (Table 7). This means that the West Angle till of Dixon antedates the raised beach event and is thus correlated with the plateau drift deposits of south Pembrokeshire and east Carmarthenshire. Unfortunately, the pollen evidence from West Angle has not allowed comparison with data from elsewhere. While it may seem surprising that the West Angle deposits (the top of the till lies at 3·6 m OD) have survived erosion during the high sea-level of the Ipswichian (shingle beaches up to 10 m OD on the outer coast), it is worth noting that they have similarly survived Holocene denudation, during which time considerable marine erosion occurred in the lee of Long Mathew Point on Caldy and Worm's Head.

In the south of this area Devensian time is represented by the head deposits of varying facies which overlie the raised beaches. Occasionally, as at West Angle, these local deposits contain erratics derived from the redistribution of glacial deposits originally emplaced during the pre-beach glaciation. Farther north, however, the coastal succession is correlated directly with that of east Gower, that is, of beach, head, glacial drift, and head, thereby demonstrating that post-beach glaciation occurred in this area. The limit of such glaciation by Irish Sea ice, determined stratigraphically, lies at Druidston Haven. South of that locality no exposures of raised beach allow continued control until they are overlain by head only, as in Milford Haven. A reconstructed ice margin was inserted along the coast south of Druidston, however, to take into account extensive till and outwash outcrops on and just above the coastal slope (Bowen 1967, 1970b, 1973a). Inland the limit lies more or less against the Roch-Trefgarne anticlinal upland north of which glacial deposits are extensive and from which limit the outwash trains of the Western Cleddau and Camrose Brook commence whereas south of the limit glacial deposits are limited (map, in Bowen 1970b). In alternative non-stratigraphic speculative reconstruction, John has successively suggested that, (i) the post-beach glaciation covered Pembrokeshire completely (John 1970a); or (ii) terminated between Milford Haven and Bosherston-Linney Head coast (John 1971), or (iii) terminated a short distance north of Milford Haven (John 1974). All of these views are expressly discounted by the available stratigraphic evidence and distribution of glacial drift.

At present there are no means of dating the glaciation (as in Gower) but to the north kettle holes lie on fluvioglacial deposits at Mathry and their basin deposits are to be subject to analysis. The numerous radiocarbon determinations on shell debris from glacial drifts, such as those collected by John (1970a), are not considered to be reliable by British workers (Shotton 1967; Boulton 1968; Mitchell 1972; Bowen 1971b, 1974).

5. Cardigan Bay

As a marker horizon the raised beach is only exposed in the southern part of this area. Fortunately, however, the general succession may be readily correlated, using the raised beach sites as stratigraphic control points for this region. The general succession (for its correlation see Table 7) is:

(4) Periglacial deposits (several lithofacies)
(3) Glacial deposits of both Welsh and Irish Sea provenance
(2) Periglacial deposits (several lithofacies)
(1) Raised beach

Whereas the beach unit is limited in distribution, other members of the succession are well exposed in sea cliffs cut into terraces of drift which mantle a cliffed coast-line in rock. Much controversy obtains on the age of the rock-cut coastline as well as the chronostratigraphic classification of the superficial deposits which bury it.

5a. St. David's Head to the Teifi Estuary

In general this rocky coastline consists of high cliffs separated by bays of com-paratively limited extent which preserve what little remains of the Pleistocene succession. The raised rock platform is found at only five localities: Pendeudraeth (Abermawr), Parrog and Newport Sands, Poppit and Gwbert on each side of the Teifi estuary. At Parrog and Poppit it bears raised beach deposits.

The most important exposure is at Poppit where a raised shore platform and raised beach deposits (Jones 1965) are exposed for about a kilometre along the southern shore of the Teifi estuary. The platform is perfectly preserved and bears all the signs of recent exhumation from beneath a drift cover. It lies 1·7–3 m above HWM according to John (1970a). A feature of the platform is the planation of both the shale and grit beds quite unlike the present shore platform where the shale beds are being differentially eroded. The principal beach exposures are between Cei Bach and Trwyn Careg Ddu. Like elsewhere in north Pembrokeshire the beach is characterised by large boulders near its base, though most exposures show well marked stratification of both shingle and sand units. Hitherto it has been found to be unfossiliferous.

Overlying the beach is a succession of blocky head, sometimes exceeding 5 m in thickness derived from the high backslope which in places reaches 183 m OD. This is overlain by 1 m or so of Irish Sea till mostly concealed by vegetation on the degraded surface of the drift terrace. On the opposite side of the estuary at Gwbert the platform described by Jones (1965) is covered by a thin basal head, Irish Sea till and an upper head unit.

In Newport Bay, at Parrog, Pleistocene deposits are discontinuously exposed, but reveal a succession of (i) raised beach, (ii) head, (iii) red Irish Sea till, (iv) a thin upper head unit.

With the exception of the raised beach the most complete and extensively exposed succession is at Abermawr. On the north side of the bay a shore platform is preserved at Pendeudraeth, but raised beach is missing. It may, however, lie at the base of the exposure at Abermawr for the bottom of the Pleistocene succession has not been proved. The succession exposed is:

(8) sandy loam (0·6 m).
(7) 'rubble drift'—a head deposit which incorporates re-worked elements of (6) (2 m).
(6) fluvioglacial sands and gravels (4·6 m).
(5) Irish Sea till (2 m).
(4) blocky head of local bedrock with some scattered erratics (3·6 m).
(3) small water worn gravels (3·6 m) (largest fragments less than 25 cm).
(2) blocky head with scattered erratics (1·5 m).
(1) shale head (2·7 m).

Units (1)—(4) are periglacial slope deposits, though the precise significance of the gravels is not clear; they may represent climatic amelioration. Synge (1970) thought that unit (3) represented glacial materials redistributed through solifluction and units (5) and (6) were separate glacial deposits. On this basis he postulated three glaciations: Elster (Unit 3), Saale (Unit 5) and Weichsel (Unit 6). Others, however, interpreted the succession as representing the Devensian stage only with scattered erratics in the lower periglacial deposits having been derived from a former glaciation; unit (6) being the outwash material related to deglaciation of the Irish Sea glaciation (5) (John 1970a; Bowen 1971b, 1974). A feature of the succession is that it is repeated on the south side of the bay with the exception of unit (3). This indicates the gravels are simply a local facies variation of the head units (1–4) as a whole. Furthermore, the Irish Sea till (5), which is calcareous in the northern exposure, is fully decalcified in the south. This feature reflects again different site characteristics.

5b. The Teifi estuary to Llanrhystud

With relation to exposures containing the raised beach unit described hitherto, it is clear that an Irish Sea till maintains a consistent stratigraphical relationship to the beach and lower and upper heads. On this basis it continues to function as a stratigraphical marker along the coastline at least as far north as Llanon. The raised beach is unfortunately not exposed, though it may lie concealed at the base of the rock coastline now buried by the coastal drift terrace. Williams (1927) thought that the raised platform might be exposed in the stream bed at Aberarth where the Arth breaks through the drift terrace, but this is not conclusive. Williams also erected a tripartite succession of glacial deposits overlying a basal head:—

(4)   Upper boulder clay

(3)   Middle sands and gravels

(2)   Lower boulder clay

(1)   Head

Subsequent work by Mitchell (1962) and Watson and Watson (1967) modified this succession.

Irish Sea till occurs preserved in small bays, such as Mwnt, Aberporth and Gilfach-yr-halen, as far north as Aberaeron, though at Newquay it is extensively exposed in a wide terrace. Table 4 shows some of the interpretations of the Newquay succession:

Table 4

Williams	Mitchell	This paper
Upper boulder clay	(5) Pencoed boulder clay	Head†
Middle sands and gravels*		
Lower boulder clay	(4) Fremington boulder clay	Irish Sea till
	(3) Newquay boulder clay	Solifluction deposit?
	(2) Head	Scree (head)
	(1) Raised beach	

* Locally present only east of afon Llethi.
† Watson and Watson, 1967.

Mitchell suggested the presence of raised beach on admittedly meagre evidence and it is more likely that the scattered pebbles referred to are derived from unit (3), his Newquay boulder clay. Watson has shown by stone orientation analysis that the upper unit (5) is a slope deposit composed of local bedrock fragments. But considerable uncertainty still attaches to unit (3): is it a solifluction deposit or a Welsh till of local origin? If the former, it occurs some distance seawards of the abandoned rock cliff and is somewhat thicker than the unequivocable periglacial deposits seen beneath Irish Sea till farther south.

At Aberaeron the general sequence of a lower and upper periglacial formation separated by calcareous Irish Sea till is exposed south of the harbour though northwards to Aberarth (Fig. 5) the succession differs in that any upper head unit of slope origin is replaced by the alluvial gravels of Afon (River) Arth. The succession north of the mouth of Afon Arth, however, has been subject to changing interpretation over the years as Table 5 shows (see Fig. 5).

Table 5

Watson 1969	Mitchell 1962	Williams 1927
Upper head (6·1 m)	Head (3 m)	Upper boulder clay (1·8–4·6 m)
Irish Sea till (9·1 m)	Grey boulder clay (7·6 m) (Pencoed)	
Sands, silts and gravels (<6 m) ⎱ (7·6 m) Coarse gravels ⎰	Silts, sands and gravels (2·4 m)	Sands and gravels (0·9–1·2 m)
Solifluction deposit (>15 m)	Boulder clay (6·1 m) (Newquay?)	Welsh boulder clay (7·6 m)
Soliflucted till		
Rubble head (6 m)	Head (3·6 m)	

The most recent work by Watson, however, is based on careful survey, using ladders to inspect hitherto inaccessible cliff exposures (Fig. 5). Overall the succession consists of (i) a basal periglacial 'complex', which includes redistributed calcareous till, scree, soliflucted beds and various sized fluvial slope deposits, (ii) an Irish Sea till, and (iii) an upper periglacial unit, this continuing the fundamental lithostratigraphic succession established farther south. The presence of Irish Sea till incorporated into the lower periglacial deposits, however, points to glaciation which probably antedates the raised beach event.

At the northern end of the Aberarth exposure the buried cliff emerges and with a pronounced coastal bevel (Wood 1959), the result of periglacial erosion, continues to Morfa where it swings inland and is buried by the Llanon-Llanrhystud drift terrace. The succession exposed in the terrace consists of:

(2)   Alluvial gravels of the Wyre, Peris and Clydan and Morfa-mawr streams.

(1)   Till of Welsh origin but with discrete lenses of Irish Sea material.

Mitchell (1962) recognised a zone of weathering between the two units which he named the Llantsantffraid Soil, but subsequent examination has not confirmed its pedological origin.

A feature of the alluvial gravels is the degree to which they have been subject to cryoturbation in the form of large scale involutions and vertical realignment of their stones (Watson 1965; Watson and Watson 1971). It has been suggested that drifts subject to such extensive and deep modification are older than the last (Devensian) cold stage (Watson 1965; Watson and Watson 1967; Mitchell 1962). On the other hand, it has been pointed out that similar features characterise some Devensian deposits elsewhere and, in any event, any such features at best simply indicate the former possible presence of permafrost (Bowen 1973a, 1974).

5c. Llanrhystud to Portmadoc

Perhaps the most enigmatic exposures around Cardigan Bay are found at Morfa Bychan south of Aberystwyth where a drift cliff, some 50 m high, mantles an old coastline backing on to an immediately adjacent coastal hinterland rising to some 185 m OD in less than 0·55 km. Wood (1959) described the relationship between the buried coastline and the superficial deposit which he believed was boulder clay, some of which had been re-arranged 'by solifluction at the end of the last glacial period'. Subsequently, in a detailed study of the Morfa bychan deposits, Watson and Watson (1967) recognised the following succession (1) Yellow head, (2) Blue head, (3) Lower brown head, (4) Loess, (5) Upper brown head (Fig. 5). Some uncertainty attaches to the blue head for its clasts are frequently striated, and its preferred orientation often resembles that found in till. Thus the basal layers, at least, of the blue head may well include glacial drift, originally deposited upslope of their present position. The site factor of a steep coastal slope may thus account largely for the absence of unequivocal glacial deposits as seen elsewhere along the coast. Technically the deposits consist of 'head', but in the case of the blue head unit, it seems more than likely that it represents Welsh till recycled shortly after its original deposition. From Morfa Bychan northwards, scattered exposures of periglacial deposits are seen at Clarach, glacial drift at Wallog but the wide drift terrace is not seen until north of the Dyfi estuary where it forms a wide, flat area between Aberdyfi and Tonfannau. Unpublished work on this area by D. Morris shows that the general succession is (1) a basal head complex, including re-cycled glacial materials, (2) Welsh till and outwash.

North of the Mawddach estuary the outstanding features are those of Morfa Dyffryn and Morfa Harlech, two extensive sandy forelands, the latter having grown north from Harlech and the former from Llanaber. An abandoned rock coastline is buried in turn by glacial deposits and Holocene sands (Steers 1964). At Mochras, near Harlech, the onshore continuation of Sarn Badrig, till is exposed and a borehole proved a total thickness of some 75 m of glacial beds. Subdivision and correlation of the succession has not proved possible (Woodland and Wood 1971).

5d. The southern Llŷn peninsula

Whittow and Ball's classification (1970) of coastal and inland deposits recognised a periglacial phase followed by the main glacial advance, then after deglaciation readvance of lesser extent, followed by further periglaciation (Table 6). This

Table 6

West		East
Periglaciation		
		Arvon Readvance
Deglaciation, erosion, weathering and permafrost		
Irish Sea advance		Criccieth advance
		Periglaciation

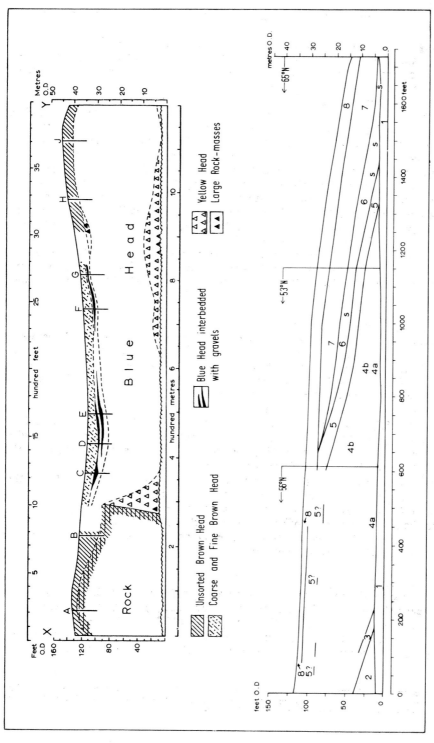

Fig. 5. *Upper*: The Pleistocene succession at Morfa Bychan, Aberystwyth (after Watson and Watson 1967). *Lower*: The Pleistocene succession due north of Afon (River) Arth, Aberarth (after Watson 1969). Key: 1 = modern beach, 2 = bedrock, 3 = rubble head, 4a = soliflucted calcareous till, 4b = local ⌠solifluction deposits, 5 = muddy sandy coarse gravel, 6 = sands, silts and washed gravel, 7 = Irish Sea till, 8 = upper head and gravels, s = slump structures in 6.

recognition of a major readvance follows to a large extent the earlier work by Jehu (1909).

Through correlation with the northern side of the peninsula (Table 7) all these events postdate that of the raised beach episode. In southern Llŷn the raised beach is not exposed and the raised shore-platform is only to be seen on Bardsey Island at about 6 m OD (Matley 1913).

Evidence of deglaciation and subsequent readvance rests very largely on the Criccieth and Glanllynnau exposures (Table 7). At both localities the surface of the Criccieth till is weathered and penetrated by ice-wedge pseudomorphs infilled by the finely divided weathered material. The periglacial horizon is overlain by glacigenic sediments which include laminated beds, gravels, and the Llanystumdwy till (Arvon Readvance) (Whittow and Ball 1970; Simpkins 1968; Saunders 1968). As an alternative explanation Synge (1964, 1970) suggested that the horizon of weathering was an iron-pan effect and that the wedge structures were load cast features: in other words, all the deposits are classified in terms of one advance only. Similarly, Boulton (1972, 1977) argued that the entire sequence had been deposited from one glacier advance. On the other hand the regional distribution of two quite discrete till units (cited by the above authors) and Morris's conclusion that the Llangelynin till farther south is somewhat older than the Mochras (=Llanystumdwy) till makes the readvance hypothesis plausible (Bowen 1974).

5e. The floor of Cardigan Bay

Geophysical survey supplemented by borehole data has enabled the Pleistocene geology of the bay floor to be established. Two glacial provinces occur: (1) Irish Sea till in the western part of the bay up to 40 m thick, and exceptionally 70 m thick, (2) Welsh till, which in comparison is poorly sorted and contains angular and subangular clasts (Garrard and Dobson 1974). Welsh till described as 'immature' by Garrard and Dobson lies 13 km off the Dyfi estuary, 6 km off Llanrhystud, 5 km off Aberarth, and 3 km off Newquay (Fig. 1). It is therefore pertinent to note that such textural characteristics compare favourably with deposits called till by Mitchell (1962) and Williams (1927) at Aberarth, but solifluction drift by Watson and Watson (1967).

Linear ridges (sarns) of glacial deposits lie more or less perpendicular to the present coastline. Much speculation on their origin has occurred in the past (Steers 1964) but they are now regarded as moraine ridges. With the exception of Sarn Badrig, which runs offshore from Harlech, and forms the limit of the 'upper boulder clay' (Llanystumdwy till), readvance of Llŷn, (Fig. 1), the precise mode of origin of the others is unknown.

5f. Discussion

On a rock-stratigraphic basis the general succession of Cardigan Bay of (1) raised beach, (2) periglacial beds, (3) glacial deposits, (4) periglacial beds, may be correlated directly with Gower (Bowen 1973a, 1973b, 1974). Certain contrary views, however, would seemingly imply that such a correlation is subject to homo-taxial error. These arise due to criteria principally of a non-stratigraphic kind:

Fig. 6. Reconstructed limits of Late Devensian Glaciation after various authors. Note that Mitchell delimits an Irish Sea Ice sheet and a Welsh one leaving an unglaciated area between. The limit of Bowen was proposed in 1967 and 1970b.

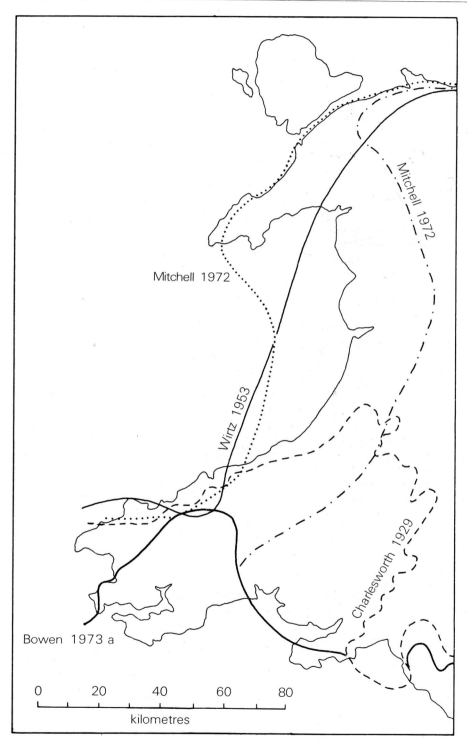

Mitchell 1972

Mitchell 1972

Wirtz 1953

Charlesworth 1929

Bowen 1973 a

0   20   40   60   80

kilometres

R

that is, by suggesting a more restricted advance of ice during the last glaciation, and by attributing to certain rock-stratigraphic units a chronostratigraphic status based largely on matters of opinion.

With the exception of a small area near the Teifi estuary, Mitchell (1972), Wirtz (1953) and Watson and Watson (1967) have considered most of Cardigan Bay to have been ice-free during the last glaciation (Fig. 6), thereby implying that the glacial deposits which mantle the interglacial coastline are Wolstonian or older, and that no record of the Ipswichian sea occurs on this coastline. Their arguments are three-fold. Firstly, because glacial deposits outside their limit of last glaciation are disturbed by periglacial structures (including pingo scars), and inland, solifluc-tional effects are widespread particularly in mid-Wales. Secondly, because two Irish Sea tills are recognised; one only in the mouth of the Teifi estuary, and the older one to the south (e.g. Abermawr) and to the north (e.g. Newquay). This distinction seems to have arisen due to the morphologically 'youthful' landforms at Banc-y-warren, near Cardigan, which first motivated Wirtz (1953) to insert a lobe of last glaciation ice across Cardigan Bay. (On this basis the oldest till that has been 'soliflued' at Aberarth is implied to be Elsterian in age (Watson 1969), just as the slope gravels with erratics at Abermawr (Synge 1970) are also.) Thirdly, because the solifluction deposits of mid-Wales (Watson 1968), and the periglacial deposits at Morfa Bychan (Watson and Watson 1967) are deemed to represent all of Devensian time.

Periglacial structures such as ice-wedge casts or pingo scars, however, simply record the past existence of permafrost and no more; they cannot be used as chronostratigraphic artefacts. In any event such phenomena exist inside the last glaciation limit elsewhere (Bowen 1973a, 1973b). Equally, a distinction recognising two Irish Sea tills is not based on lithostratigraphic evidence and till outcrops north and south of the Teifi estuary occur in the same position with respect to underlying and overlying deposits. Chronostratigraphically, they cannot be regarded as homotaxial without clear proof of separate age: none such is forth-coming.

Attribution of the periglacial deposits at Morfa Bychan deposits and those of mid-Wales to the entire span of Devensian time (Watson and Watson 1967, Watson 1968) has been challenged respectively by Bowen (1973a, 1973b, 1974) and Potts (1971). The latter maintained that Late Devensian glaciation occurred and that subsequent to deglaciation till was redistributed by solifluction along the predominantly steep slopes of the region. At Morfa Bychan, Bowen (1973a) followed Wood's (1959) view in suggesting that the Blue Head was similarly redistributed glacial material. In both these reassessments the site factor is funda-mental in significance. The controversial deposits invariably occur at the base of slopes in circumstances of locally high relief. It seems inconceivable that downslope movement would not have occurred, leading in some cases to the elimination of the glacial horizon, as for example at Trwynhwrddyn and Pendeudraeth (close by Abermawr), or substantial re-sorting as at Porth Clais and Westdale valley— both localities glaciated during the post-raised beach advance. Local site factors are further illustrated between the exposures at Abermawr north and Abermawr south, some 0·4 km apart; at the latter the unit of slope gravels does occur and in contrast to the former the Irish Sea till is completely decalcified.

The status of proposed readvance in Llŷn and Tremadoc Bay while based on lithostratigraphy (e.g. Whittow and Ball 1970) is under strong challenge from Boulton (1972, 1977). But what is established is that deglaciation occurred shortly before 14,468±300 (Birm. 212), a radiocarbon date from moss preserved in solifluction clays from a kettle hole at Glanllynnau (Coope *et al.* 1971). Offshore, Garrard and Dobson (1974) suggested the submarine glacial deposits were Late Devensian in age. Thus, after due allowance for site factors has been made, the lithostratigraphic succession of Cardigan Bay, assisted minimally by biostratigraphic data from Glanllynnau, allows ready correlation with the Pleistocene succession of the Bristol Channel area (Table 7).

6. North Wales

6a. Caernarfon Bay

Shore platforms at 6·6 m and 3 m OD are found on both sides of the bay. The 6·6 m platform (complete with notch at Porth Dinllaen) was described by Jehu (1909) and Matley (1936), in Llŷn by Greenly (1919) and Hopley (1963) in Anglesey and in both areas by Whittow (1960, 1965) who distinguished a lower 3 m platform. He mapped the 6·6 m platform at 19 localities, and the 3 m platform at 8; four of them directly below the 6·6 m platform. The lower platform is at present washed by the tide, but near Holyhead it passes below the till which forms drumlins at Gorsedd-y-Penrhyn and Porth Dylusg. Traced laterally from those sites it is notched into the higher platform (Whittow 1965). Only at Porth Oer, near the tip of the Llŷn peninsula, does the higher platform bear raised beach deposits where partly cemented sands and shingle are overlain by glacial drift (Whittow 1960).

The succession of coastal Pleistocene deposits has been correlated with those of southern Llŷn (Saunders 1968; Whittow and Ball 1970; Peake *et al.* 1973) (See Table 7). The basal till is a typical calcareous Irish Sea shelly clay, sometimes associated with calcareous sands and gravels. As in the south of the peninsula indications of ice withdrawal and readvance occur in the form of a 'solifluction layer and boulder pavement' at Trwyn-y-tal (Whittow and Ball 1970) together with a widespread horizon of weathering (Saunders 1968). The upper till consists of two lithofacies: in the south west at Port Nevin by a till with erratics from Anglesey and Caernarfon Bay, and farther north east by the Clynnog till with erratics from Snowdonia. At Dinas Dinlle all the beds are associated with overthrusting and overfolding, which may have been caused by readvance.

Unlike the coastline of eastern Anglesey, Greenly (1919) was unable to sustain his general tripartite classification of a Lower boulder clay, Middle sands and Upper boulder clay for the Caernarfon Bay coast of the island. The existing till consists of preponderantly local rocks, with subordinate amounts of far-travelled material. It is associated with a passage of ice from northeast to southwest across the island. Whittow and Ball (1970) classified this till as part of the Main Anglesey Advance (i.e. the 'upper boulder clay' glaciation of northern Llŷn). There is no trace of the Irish Sea advance which deposited the 'lower boulder clay' (calcareous till) of northern Llŷn on the island.

6b. The southern Irish Sea (Anglesey, Conway Bay and Clwyd)

On the east coast of Anglesey Greenly (1919) established a tripartite succession of lower boulder clay, middle sands and gravels and upper boulder clay on which he erected his classification for the entire island. He suggested that the advance responsible for the upper boulder clay removed any earlier deposits, a suggestion later endorsed by Hopley (1963) who attributed the moulding of drumlin fields in the north of the island to that advance. At Red Wharf Bay on a platform some 3 m above HWM is a raised beach of Carboniferous Limestone pebbles lying partly on the platform, but mostly emplaced in a notch within cemented limestone head which is overlain by a red-brown till (Whittow and Ball 1970). They suggested that the lowest head antedates the raised beach, but it is possible that beach and head are penecontemporaneous (Bowen 1973b).

Along the mainland of north Wales between Aber Ogwen and the coastal plain east of Llandulas, is a complex zone where Irish Sea ice and Snowdonian ice were in contact. The available exposures were discussed by Whittow and Ball (1970), who showed Irish Sea till overlying Welsh till at Llandudno, Welsh till on Irish Sea till in the Aber valley, or lenses of Irish Sea till within Welsh till at Deganwy and Conway. Only east of Llandulas is Irish Sea till found exclusively where it is coextensive with that of the Vale of Clwyd, and beyond that the Cheshire-Shropshire lowland. Head was described beneath Irish Sea till on the Little Orme, Llandudno by Hall (1870), but any raised beach deposits antedating glacial deposits are unknown.

6c. Discussion

The regional generalised succession consists of (1) raised beach (penecontemporaneous with head formation), (2) head, (3) glacial deposits, (4) zone of weathering, erosion, head accumulation and permafrost, (5) glacial deposits, (6) head. As in southern Llŷn, the status of readvance (Unit 5) is debatable, but a penetration of Irish Sea ice into the Vale of Clwyd at a topographic level considerably lower than previous occupation by Welsh ice, which in turn seems contemporaneous with high-level Irish Sea glacial deposits on the eastern side of the Clwyd uplands, and south to Oswestry (Bowen 1973a, 1974), may indirectly support a readvance concept. In turn this is correlated with the Welsh readvance in the Shrewsbury district (Wills 1938), where Welsh till is extensive overlying Irish Sea fluvioglacial deposits. Other readvances postulated, such as the Liverpool Bay phase (Whittow and Ball 1970) or Llay readvance in Clwyd (Peake 1961) are not as securely based (Bowen 1974).

7. The Holocene Transgression

A comparatively limited amount of information is available on the Holocene sea-level rise in Wales (Bowen 1974). This results not from an absence of data, but lack of systematic sampling and dating.

In Swansea Bay, Godwin (1940) was able to incorporate the work of earlier workers such as von Post (1933), George (1936) and Strahan (1896) into the results of his pollen analytical investigation of alternating peats and clays largely recovered from bore-hole samples. He showed that the transgression occurred above 16 m OD during the later Boreal, with peat beds indicating stages of halt or regression. Later radiocarbon dates on borehole samples from the same area confirmed that fresh water organic muds and peats accumulated on a surface of glacial sediments at -18 m OD from 11980 ± 180 (Q–664) to 8970 ± 160 (Q–663) (2)[*] when they were submerged by the transgression (Godwin and Willis 1961).

[*] Numbers refer to points on Figure 7.

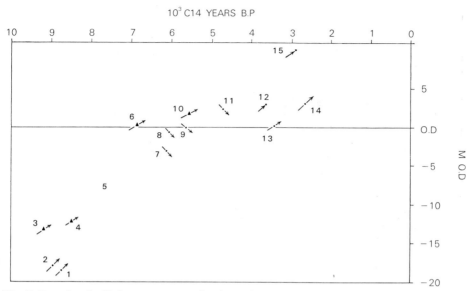

10^3 C14 YEARS B.P

Fig. 7. Data for the Holocene transgression in Wales. Arrows indicate transgression or regression contacts. All points are based on radiocarbon dates with the exception of 3, 4, 5, 6 and 10, which refer to the correlation of transgressions on the Clwyd coast with those in adjacent Lancashire (Tooley 1974). Numbers refer to the text, excluding: 1, Cardigan Bay 8,740±100 Birm. 400 (Garrard and Dobson 1974); 12, Moreton, Cheshire, 3,695±110 Q.620 (Godwin and Willis 1964); and 15, Borth Bog 2,900 ± 110 Q.712 (Godwin and Willis 1964).

At Margam the main sea-level rise ended about 6184±143 (Q–275) (7) at −3 m OD: but further transgression occurred after 3402±108 (Q–265) (13) (Godwin and Willis 1964), and probably indicated by Gibson's record of *Scrobicularia* clay overlying peat at Port Talbot between 0·3 and 0·61 m OD (Strahan 1907). Comparable estuarine/marine beds lie at this height throughout south Wales. At Llanwern marine clay extends to 4·4 m OD and is underlain by a peat at 3 m OD dated as 2660±110 (Q–691) (14) (Godwin and Willis 1964; Locke 1974). On the Clwyd coast of northeast Wales, M. J. Tooley has recognised five marine transgressions based largely on Strahan's log of the Foryd borehole, near Rhyl, (Strahan 1885), and his own work there and at Abergele (Tooley 1969, 1974 and unpublished). Marine conditions affected the coastal zone between 8500 and 4700 BP. The transgressions are correlated with those which he recognised in Lancashire (Tooley 1974), as follows:

North Wales I: (=Lytham I). At Foryd −13 to −12·7 m OD, 9200 to 8600 BP (3)

North Wales II: (=Lytham II). At Foryd −12·2 to −9·1 m OD, 8500 to 7800 BP (4)

North Wales III: (=Lytham III). At Foryd −8·9 OD, and at Abergele −2·8 m OD, 7600 to 7200 BP (5)

North Wales IV: (=Lytham IV). At Abergele 0·24 to 0·77, 6885 to 6025 BP (6)

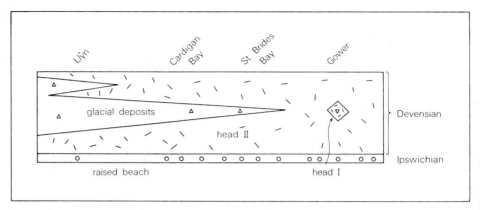

Fig. 8. Diagrammatic and reconstructed section from north to south illustrating principal lithostratigraphic relationships and chronostratigraphic classification (after Bowen 1975). Length of section, circa 200 km. Head I=recycled glacial drift. Head II=local rock debris.

North Wales V: (=Lytham VI). At Abergele 1·76 to 4·10 m OD, 5580 to 4725 BP (10)
ending at Rhyl beach at 2·42 m OD, 4725±65 BP (HV. 4348) (11).

He found no stratigraphic evidence of a marine incursion above 4·10 m OD, thus called into question the so-called 'Bryn-carrog' coastline of Rowlands (Rowlands 1955; see also Whittow 1965).

Other radiocarbon dates from scattered localities enable only the broad trend of sea-level change to be established for the principality (Fig. 7). At Freshwater West (5960±120 Q–530 (9)) and Ynys Las (6026±135 (Q–380) (8)) the end of the main transgression is indicated (Godwin and Willis 1964). At Borth two bore-holes at 19·8 and 28·9 m below the surface near the foreshore showed silty sands below *Scrobicularia* clays and a forest layer. The sands contained a warm microfauna which, taken with the Ynys Las data, suggested accumulation during the period c. 8000 to 6000 BP (Haynes and Dobson 1969). The same authors compared the sequence with that of Godwin's at Swansea and attributed the absence of peat beds to a site always below HWM at the edge of the estuary.

On Anglesey Holocene raised beach gravel recorded by Greenly (1919) and Hopley (1963) were re-examined by Whittow (1965). Occasionally the beach re-occupies the 'pre-glacial' shore platform, and may lie as high as 5·5 m OD (mean elevation of 18 sites is 4·24 m OD). It contains a fauna including *Patella vulgata* (Linné) and *Cardium edule* (Linné). On Bardsey Island beach gravel lies at 6 m (Matley 1913); while near Barmouth George (1933b) suggested that a pebble ridge could be a Holocene raised beach relic.

Thus, data at present available on Holocene coastal deposits allows no detailed pattern of either sea-level change or isostatic movements in Wales (Fig. 7). Considerably more work is required before such a goal is attained. In the meanwhile, claims of precise secular tectonic movement, such as those by Churchill (1965), must be regarded as speculative.

Similarly, the nature of the Holocene transgression must have varied from locality to locality along the coast due to the operation of local factors. But there is substantial disagreement between those who interpret each peat unit and marine clay bed as indicating respectively regression and transgression, and those who regard them as local phenomena which reflect variations in the relative rate of sea-level rise and sedimentation.

8. Conclusions

The Pleistocene history of the coastline is based exclusively on a repetitive and widespread sequence of deposits. Two distinct provinces occur: (1) where the last local glaciation antedated the raised beach event, as in most of Gower, the Bristol Channel and the southern Celtic Sea coastlands, and (2) where the last local glaciation postdated the raised beach event, as in the northern Celtic Sea coastlands, Cardigan Bay and north Wales (Table 7). In both these provinces direct and indirect evidence for glaciation antedating the raised beach event exists.

Figure 8 is a diagrammatic and reconstructed section from north to south. It shows a beach formation which on the faunal evidence from Gower is Ipswichian (last interglacial) in age. There is no known evidence to support a postulated existence of two separate interglacial beaches (Mitchell 1972). The two sea-level events at Minchin Hole, Bacon Hole, on Caldy and to be inferred elsewhere, have been assigned to the same interglacial (Sutcliffe and Bowen 1973), though the later of the two may be related to a rapid and short lived rise of ocean level (Bowen and Hollin, in preparation).

Head deposits in the south of the area represent all of Devensian time. These include several lithofacies, notably the redistributed glacial material of west Gower formerly termed 'older drift'. In the north, after initial head accumulation (Early and Middle Devensian) glaciation obtained during the Late Devensian (Bowen 1973a); the main pulse was followed by deglaciation and subsequent readvance. After deglaciation further periglacial deposits and structures formed particularly when coastal sites were topographically appropriate. Much periglacial modification took place during the cold episode prior to 10000 years B.P. (Late-Devensian chronozone III) though the cold phase probably lasted longer than the radiocarbon determination of that chronozone (see Coope *et al.* 1971).

Pre-Ipswichian glaciation is indicated by the dissected glacial drifts of south Wales, overlain by freshwater interglacial deposits at West Angle Bay, and Welsh (Late Devensian) till at Pencoed (Bowen 1973a), by erratics in raised beach and head deposits, and by recycled glacial materials which at present constitute head lithofacies. Only the broad outline of this glaciation may be reconstructed (see maps of ice movement in Bowen 1970a, 1973a).

The age of rock-platform(s) is indeterminate. Raised beach is separated from subjacent shore-platform by head at one site only, Red Wharf Bay; but even there both deposits may be penecontemporaneous. While there is evidence elsewhere in the Irish Sea which shows that platform and beach are not the same age, any confirmation of this from Wales must await further discoveries.

Data at present available for the Holocene sea-level rise allow no detailed pattern of either sea-level change or isostatic movements in Wales (Fig. 7).

Acknowledgments. The author acknowledges palaeontological information for Minchin Hole from J. Haynes, A. R. Lord, J. G. Evans and C. French.

References

BALL, D. F. 1960. Relic-soil on Limestone in South Wales. *Nature Lond.* **187**, 497–8.

BOWEN, D. Q. 1966. Dating Pleistocene events in south-west Wales. *Nature Lond.* **211**, 475–476.

—— 1967. On the supposed ice-dammed lakes of South Wales. *Trans. Cardiff nat. Soc.* **93**, 4–17.

—— 1970a. South-east and central South Wales. *In,* Lewis, C. A. (Editor). *The Glaciations of Wales and adjoining regions.* London. 197–227.

——1970b. The Palaeoenvironment of the 'Red Lady' of Paviland. *Antiquity* **44**, 134–136.

—— 1971a. The Quaternary succession of South Gower. *In,* Bassett, D. A. and M. G. (Editors). *Geological Excursions in South Wales and the Forest of Dean.* 135–142.

—— 1971b. The Pleistocene succession and related landforms in north Pembrokeshire and south Cardiganshire. *ibid.* 260–266.

—— 1973a. The Pleistocene history of Wales and the borderland. *Geol. J.* **8**, 207–224.

—— 1973b. The Pleistocene succession of the Irish Sea. *Proc. Geol. Ass.* **84**, 249–272.

—— 1973c. The excavation at Minchin Hole 1973. *Gower.* **24**, 12.

—— 1974. The Quaternary of Wales. *In,* Owen, T. R. (Editor). *The upper Palaeozoic and post-Palaeozoic rocks of Wales.* Cardiff 373–426.

—— 1975. Chronological and spatial aspects of Welsh Pleistocene deposits as a basis for classifying soil parent material. *In,* Adams, W. A. (Editor). *Soils in Wales.* Welsh Soils Discussion Group. 1–8.

BOULTON, G. S. 1968. A Middle Würm Interstadial in South-West Wales. *Geol. Mag.* **105**, 190–191.

—— 1972. Modern Arctic Glaciers as Depositional Models for Former Ice Sheets. *J. geol. Soc. Lond.* **128**, 361–393.

—— 1977 *in litt. In,* Bowen, D. Q. (Editor). Studies in the Welsh Quaternary. *Cambria.*

CANTRILL, T. C., DIXON, E. E. L., THOMAS, H. H. and JONES, O. T. 1916. The Geology of the South Wales Coalfield, Part XII. The country around Milford. *Mem. geol. Surv. U.K.*

CHURCHILL, D. M. 1965. The displacement of deposits formed at sea-level 6,500 years ago in Southern Britain. *Quaternaria* **9**, 239–249.

COOPE, G. R., MORGAN, ANNE and OSBORNE, P. J. 1971. Fossil Coleoptera as indicators of climatic fluctuations during the Last Glaciation in Britain. *Palaeogeog. Palaeoclim. Palaeoecol.* **10**, 87–101.

DIXON, E. E. L. 1921. The Geology of the South Wales Coalfield. Pt. XIII. The Country around Pembroke and Tenby. *Mem. geol. Surv. U.K.*

GARRARD, R. A. and DOBSON, M. R. 1974. The nature and maximum extent of glacial sediments off the west coast of Wales. *Marine Geol.* **16**, 31–44.

GEORGE, T. N. 1932. The Quaternary Beaches of Gower. *Proc. Geol. Ass.* **43**, 291–324.

—— 1933a. The glacial deposits of Gower. *Geol. Mag.* **70**, 208–32.

—— 1933b. The submerged forest near Barmouth. *Proc. Swansea Sci. Fld nat. Soc.* **1**, 187–191.

—— 1936. The geology of the Swansea main drainage excavations. *Proc. Swansea Sci. Fld nat. Soc.* **2**, 23–48.

GODWIN, H. 1940. A Boreal transgression of the sea in Swansea Bay; data for the study of post-glacial history. VI *New Phytol.* **39**, 308–321.

—— and WILLIS, E. H. 1961. Cambridge University Natural Radiocarbon Measurements III. *Radiocarbon* **3**, 60–76.

—— and —— 1964. Cambridge University Natural Radiocarbon Measurements VI. *Radiocarbon* **6**, 116–137.

GREENLY, E. 1919. The Geology of Anglesey. *Mem. geol. Surv. U.K.*

HALL, H. F. 1870. On the glacial and post-glacial deposits in the neighbourhood of Llandudno. *Geol. Mag.* **7**, 509–513.

HAYNES, J. and DOBSON, M. R. 1969. Physiography, Foraminifera and Sedimentation in the Dovey estuary (Wales). *Geol. J.* **6**, 217–256.

HOPLEY, D. 1963. *The coastal geomorphology of Anglesey.* Unpublished M.Sc. thesis, University of Manchester.

JEHU, T. J. 1904. The glacial deposits of northern Pembrokeshire. *Trans. R. Soc. Edinb.* **41**, 53–87.

—— 1909. The glacial deposits of western Caernarvonshire. *Trans. R. Soc. Edinb.* **47**, 17–56.

JOHN, B. S. 1968. Age of Raised Beach Deposits of South-western Britain. *Nature Lond.* **207**, 622–623.

—— 1970a. Pembrokeshire. *In,* Lewis, C. A. (Editor). *The Glaciations of Wales and adjoining regions.* London. 229–265.

—— 1970b. The Pleistocene drift succession at Porth Clais, Pembrokeshire. *Geol. Mag.* **107**, 439–457.

—— 1971. Glaciation and the West Wales landscape. *Nature in Wales* **12**, 1–18.

—— 1974. Ice Age events in South Pembrokeshire. *Nature in Wales.* **15**, 66–68.

JONES, O. T. 1965. The glacial and post-glacial history of the lower Teifi Valley. *Q. J. geol. Soc. Lond.* **121**, 247–281.

KNIGHT, H. H. 1853. Account of Newton Nottage. *Archaeol. Cambrensis.* **4**, 97–99.
LEACH, A. L. 1909. Excursions to Tenby, Easter 1909. *Proc. Geol. Ass.* **21**, 177.
—— 1911. On the relation of the glacial drift to the raised beach near Porth Clais, St. David's. *Geol. Mag.* **8**, 462.
—— 1933. The geology and scenery of Tenby and the south Pembrokeshire coast. *Proc. Geol. Ass.* **44**, 187.
—— 1934. The geology and archaeology of the Isle of Caldey. *Proc. Geol. Ass.* **45**, 189.
LOCKE, S. 1974. The post-glacial deposits of the Caldicot Level and some associated archaeological discoveries. *The Monmouthshire Antiquary* **3**, 1–16.
MATLEY, C. A. 1913. The geology of Bardsey Island. *Q. J. geol. Soc. Lond.* **69**, 514–533.
—— 1936. A 50-foot coastal terrace and other Late-glacial phenomena in the Lleyn peninsula. *Proc. Geol. Ass.* **43**, 222–233.
MITCHELL, G. F. 1962. Summer Field Meeting in Wales and Ireland. *Proc. Geol. Ass.* **73**, 197.
—— 1972. The Pleistocene history of the Irish Sea: Second Approximation. *Scient. Proc. R. Dubl. Soc. A.* **4**, 181–199.
MURCHISON, C. 1868. *Palaeontological Memoirs and Notes of the late Hugh Falconer.* London.
NORTH, F. J. 1929. *The evolution of the Bristol Channel.* Cardiff (2nd Ed. 1955).
PEAKE, D. S. 1961. Glacial changes in the Alyn river system and their significance in the glaciology of the North Welsh border. *Q. J. geol. Soc. Lond.* **117**, 335–366.
——, BOWEN, D. Q., HAINS, B. A. and SEDDON, B. 1973. Wales. *In*, Mitchell, G. F. *et al.* A Correlation of Quaternary Deposits in the British Isles. *Geol. Soc. Lond., Special Report No. 4* 59–67.
POST, L. von 1933. A Gothiglacial transgression of the Sea in south Sweden. *Geogr. Annlr.* **15**, 21–36.
POTTS, A. S. 1971. Fossil cryonival features in Central Wales. *Geogr. Annlr.* **53**, A. 39–51.
PRESTWICH, J. 1892. The raised beaches and "head" or rubble-drift of the south of England: their relation to the valley drifts and to the Glacial Period; and on a late post-Glacial submergence. *Q. J. geol. Soc. Lond.* **48**, 263.
ROWLANDS, B. M. 1955. *The glacial and post-glacial evolution of the landforms of the Vale of Clwyd.* Unpublished M.A. thesis, University of Liverpool.
SAUNDERS, G. E. 1968. A Fabric analysis of the ground moraine deposits of the Lleyn Peninsula of South West Caernarvonshire. *Geol. J.* **6**, 105–118.
SHOTTON, F. W. 1967. The problems and contributions of methods of absolute dating within the Pleistocene period. *Q. J. geol. Soc. Lond.* **122**, 357–383.
SIMPKINS, K. 1968. *Aspects of the Quaternary history of central Caernarvonshire.* Unpublished Ph.D. Thesis. University of Reading.
STEERS, J. A. 1964. *The coastline of England and Wales.* Cambridge at the University Press. (2nd ed.).
STRAHAN, A. 1885. Geology of the coasts adjoining Rhyl, Abergele and Colwyn. *Mem. geol. Surv. U.K.*
—— 1896. On submerged land-surfaces at Barry, Glamorgan, with notes on the fauna and flora by Clement Reid. *Q. J. geol. Soc. Lond.* **52**, 474–489.
—— 1907. The geology of the South Wales Coalfield. Part VIII. The Country around Swansea. *Mem. geol. Surv. U.K.*
—— and CANTRILL, T. C. 1902. The Geology of the South Wales Coalfield. Part III. The country around Cardiff. *Mem. geol. Surv. U.K.*
STRINGER, C. 1975. A preliminary report on new excavations at Bacon Hole Cave. *Gower.* **26**, 32–37.
STUART, A. J. 1974. Pleistocene history of the British Vertebrate Fauna. *Biol. Rev.* **49**, 225–266.
SUTCLIFFE, A. J. 1960. Joint Mitnor Cave, Buckfastleigh. *Trans. Torquay nat. Hist. Soc.* **13**, 1–26.
—— and BOWEN, D. Q. 1973. Preliminary report on excavations in Minchin Hole, April–May 1973. *Newsletter. William Pengelly Cave Studies Trust* **21**, 12–25.
SYNGE, F. M. 1964. The glacial succession in west Caernarvonshire. *Proc. Geol. Ass.* **75**, 431–444.
—— 1970. The Pleistocene Period in Wales. *In*, Lewis, C. A. (Editor). *The Glaciations of Wales and adjoining regions.* London, Longmans. 315–350.
TOOLEY, M. J. 1969. *Sea-level changes and the development of coastal plant communities during the Flandrian in Lancashire and adjacent areas.* Unpublished Ph.D. thesis, University of Lancaster.
—— 1974. Sea-level changes during the last 9,000 years in North-west England. *Geog. J.* **140**, 18–42.
TRENHAILE, A. S. 1971. Lithological control of high-water ledges in the Vale of Glamorgan, Wales. *Geog. Annlr.* **53A**, 59–69.
—— 1972. The shore-platforms of the Vale of Glamorgan. *Trans. Inst. Br. Geogr.* **56**, 127–144.
TROTTMAN, D. 1964. *Data for Late-Glacial and Post-Glacial history in South Wales.* Unpublished Ph.D. thesis, University of Wales.
WATSON, E. 1965. Periglacial structures in the Aberystwyth region. *Proc. Geol. Ass.* **76**, 443–62.
—— 1968. The periglacial landscape in the Aberystwyth region. *In* E. G. Bowen *et al.* (Editors). *Geography of Aberystwyth.* Cardiff.

WATSON, E. 1969. Fig. 17 in Bowen, D. Q. (Editor). *Coastal Pleistocene Deposits in Wales.* Quaternary Research Association.
—— and WATSON, S. 1967. The periglacial origin of the drifts at Morfabychan, near Aberystwyth. *Geol. J.* **5,** 419–440.
—— and —— 1971. Vertical stones and analogous structures. *Geogr. Ann.* **53A,** 107–114.
WHITTOW, J. B. 1960. Some comments on the raised beach platform of South-west Caernarvonshire and on an unrecorded raised beach at Porth Neigwl, North Wales. *Proc. Geol. Ass.* **71,** 31–39.
—— 1965. The Interglacial and Post-Glacial Strandlines of North Wales. *In,* Whittow, J. B. and Wood, P. D. (Editors). *Essays in Geography For Austin Miller.* 94–117. Reading.
—— and D. F. BALL. 1970. North-west Wales. *In,* Lewis, C. A. (Editor). *The Glaciations of Wales and adjoining regions.* London, Longmans. 21–58.
WILLIAMS, G. J. 1968. The Buried Channel and Superficial Deposits of the Lower Usk, and their correlation with similar features in the Lower Severn *Proc. Geol. Ass.* **79,** 325–348.
WILLIAMS, K. E. 1927. The glacial drifts of western Cardiganshire. *Geol. Mag.* **64,** 205–227.
WILLS, L. J. 1938. The Pleistocene development of the Severn. *Q. J. geol. Soc. Lond.* **94,** 161–242.
WIRTZ, D. 1953. Zur Stratigraphie des Pleistocäns in Westen der Britischen Inseln. *Neues Jb. geol. Paläont. Abh.* **96,** 267.
WOOD, A. 1959. The erosional history of the cliffs around Aberystwyth. *Lpool and Manchr geol. J.* **2,** 271–279.
WOODLAND, A. W. and WOOD, A. 1971. The Llanbedr (Mochras Farm) Borehole. *Rep. Inst. Geol. Sci.* No.71/18.
ZEUNER, F. E. 1945. *The Pleistocene Period.* London.

D. Q. Bowen, The University College of Wales, Department of Geography, Llandinam Building, Penglais, Aberystwyth, Dyfed, Wales SY23 3DB.

The coast of South West England

C. Kidson

The coasts of South West England bear witness to the presence of 'Irish Sea' ice during the Wolstonian glaciation possibly as far south as the Scilly Isles. Till from this glaciation has been described in a number of locations. In addition erratics including giant erratics from the same glaciation encircle the peninsula though some of this erratic material may have survived from earlier glacial episodes. The Devensian was a period without ice presence in the area but solifluction processes have produced a thick mantle of head and thin discontinuous loess deposits. There is evidence of two high Ipswichian sea-levels but some authors would place the main raised beach, regarded here as Ipswichian, in the Hoxnian Interglacial. Evidence of Devensian, and perhaps earlier glacial, low sea-levels takes the form of submerged cliffs at -43 to -44 metres and deep buried rock channels in all the rivers. High Devensian sea-levels and sea-level fluctuations throughout the Quaternary are also discussed. Varying views on Quaternary chronology and stratigraphy are considered. The evidence of early Pleistocene events is reviewed but many questions remain unanswered for lack of sedimentary evidence.

1. Introduction

1a. Evidence of glaciation

The South West of England differs from most of the other coastal areas of the 'Irish Sea' in that, despite some late nineteenth century views to the contrary (Codrington 1898; Worth 1898) it has generally long been held to lie outside the limits of Pleistocene glaciations. Since 1960, however, Irish Sea ice has been shown to have impinged on the coast of North Somerset (Fig. 1) and the Atlantic coasts of Devon and Cornwall. Mitchell (1960) revived the ideas of Maw (1864)

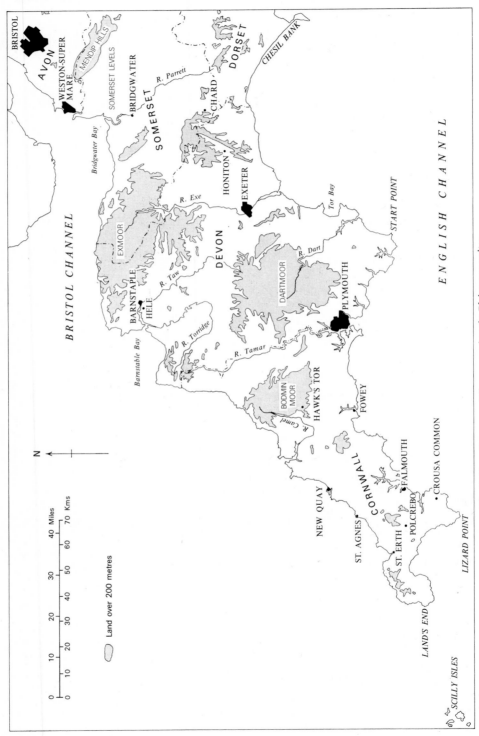

Fig. 1. Location map. Pliocene and early Pleistocene sites.

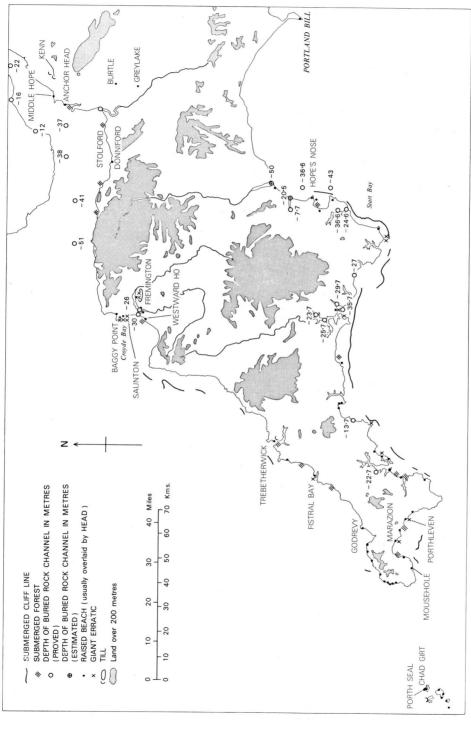

Fig. 2. Location map. Late Pleistocene and Holocene sites.

in dealing with the deposits at Fremington (Fig. 2) in North Devon. These are now seen as the most significant glacial clays and gravels in the peninsula (Kidson and Wood 1974). Mitchell and Orme (1967) presented evidence of till and ice margin deposits along the northern coasts of the Scilly Isles. Further evidence of ice penetration, at least into the coastal embayments of the area, has been demonstrated at Kenn in Somerset by Hawkins and Kellaway (1971). Much more speculatively, it has been suggested that ice penetrated by way of the Somerset Levels into the heart of central Southern England (Kellaway 1971) and even (Kellaway *et al.* 1975) that the English Channel itself has been subjected to one or more Quaternary glaciations.

1b. Low sea-levels

If the evidence for the role of ice in coastal development during the Quaternary is as yet not so convincing as to ensure general agreement, the reality of variation in sea-level is self-evident. Submerged forests, found well below low tide levels occur all round the coast (Fig. 2) and were recognised as evidence of former low sea-levels in the seventeenth century (Beale 1666) and, in places, were described in detail early in the nineteenth century (Horner 1815). Recent work (Clarke 1970; Kidson and Heyworth 1973) has shown their relationship to the post-glacial (Flandrian) transgression.

The buried rock channels of many rivers, ranging from the Severn to the Dart (Jones 1882; Codrington 1898; Worth 1898) were seen to indicate even greater variations in the relative levels of land and sea. It is now established that all the rivers of the peninsula, including those whose buried channels were not revealed by nineteenth century engineering works for railway bridges and tunnels, also have deep infilled channels (McFarlane 1955; Hawkins 1962; Durrance 1969, 1971; Clarke 1970). It is becoming increasingly clear that one of the stages of low sea-level to which these channels refer, probably of last glaciation (Devensian) age, occurred at 43–45 metres below British Ordnance Datum (=O.D.=Mean Sea Level Newlyn). Submerged cliffs, recently discovered (Cooper 1948; Hails 1975; Kelland 1975) give clear evidence of a stand of sea-level of considerable duration at this depth.

1c. High sea-levels

There is evidence too of higher interglacial sea-levels but little agreement on the Interglacials to which particular marine deposits are to be referred. Even levels attained by the sea in the last Interglacial (Ipswichian) are the subject of debate. A typical and characteristic cliff section, recognised and discussed by de la Beche (1839) and debated at length by Prestwich (1892) consists of shore platform, overlain by raised beach, in turn overlain by solifluction (head) and slope deposits. The notch at the back of the platform is often found at or about 7·6 metres above O.D., but varies in height with exposure. The platform may well have originated very early in the Pleistocene and been retrimmed by successive interglacial or interstadial seas. The raised beach is related to the Hoxnian Interglacial by some authors (Mitchell 1960, 1972; Stephens 1966, 1974) or to the Ipswichian by others (Zeuner 1959; Kidson 1971; Kidson and Wood 1974). Interpretations turn on whether or not a second, higher beach is discerned within the overlying head sequence and specifically on the relationship of the main raised beach to the Fremington Till, generally held to be Wolstonian in age. Much of the uncertainty stems from the fact that no mammalian faunas or artifacts have been found related to the coastal sections and that, unlike at Selsey on the English Channel coast (West and Sparks 1960) dateable organic deposits showing the Ipswichian age of the beach, have nowhere been found in the South West.

If the events of the later Pleistocene are still unclear, those of the Early Pleistocene are highly speculative. In the early years of the present century the officers of the Geological Survey recognised high-level surfaces of marine erosion backed by abandoned cliffs which they ascribed to the later Tertiary period. Among the most important of these was the so-called 430 feet (131 metres) 'Pliocene' platform (Reid and Flett 1907). Subsequently, similar surfaces indicating sea-levels up to 370 metres above that of the present were described (Balchin 1946, 1952) and dated as late Tertiary (post-Alpine). Included in these high-level surfaces in South West England was the continuation of another 'Pliocene Platform' first described by Wooldridge and Linton (1937) in South East England and related to a sea-level of about 200 metres above O.D. The revision of the Pliocene-Pleistocene Boundary by the International Geological Congress in London in 1948 (Boswell 1952) means that this surface, subsequently suggested as possibly equivalent to the Calabrian of the Mediterranean (Wooldridge 1950), must be referred to as 'earliest Pleistocene'. In South West England this in turn calls into question the age of all the lower features including the so-called 430 feet surface to which the only undoubted marine deposits found at high level have been related. It is the paucity of supporting depositional evidence at both 'low' and 'high' levels which leaves the Pleistocene chronology of South West England in such an imprecise state and which makes untenable the correlations with Mediterranean stages suggested by some authors (Bradshaw 1961; Clarke 1961–2, 1963). It may well be that such supporting evidence has been removed by the intensity of post-depositional erosion. Exposure to Atlantic gales has resulted, within the lifetime of present workers, in the elimination of many raised beach fragments recorded in the literature. It may be that despite the general freedom of the peninsula from glaciation, periglacial processes have been of such intensity that they too have contributed to the relative lack of supporting depositional evidence.

2. Pliocene and Early Pleistocene

2a. Deposits

i. The St. Erth Beds. At St. Erth in South Cornwall (Fig. 1) occur deposits which have been described as marine Pliocene (Reid 1890; Reid and Flett 1907) and related to the 430 feet (131 metres) 'Pliocene platform'. They consist of sands and clays overlain by 'head'. The clays have yielded a considerable marine fauna with strong Mediterranean affinities and include many extinct species. Reid (1890) compared the fauna with that of the Lenham Beds in South East England, which he also dated as Pliocene. He regarded it as dominated by deeper water species with a suggested depth of 50 fathoms (91 metres). Thus, despite the fact that the fossiliferous clays are found at only about 100 feet (30 metres) above O.D., he postulated that they were laid down by the sea responsible for cutting the 'Pliocene platform' at 430 feet (131 metres). He made no reference to suggestions by earlier authors that the clay was a 'boulder clay with marine shells'.

Mitchell (1965) re-examined the St. Erth Beds, which he took to be Pleistocene, and suggested that the marine clays and sands could have been deposited in the

Cromer Warm Period. He revived earlier references to boulder clay and suggested that lenses of this clay could have been thrust into the marine clays during disturbance by Lowestoft (Anglian) ice. He discounted Reid's correlation of the deposits with the 430 feet (131 metres) surface and suggested that, if the marine clay is in primary position, the sea-level to which it is related lies at 185 feet (56 metres) above O.D. Subsequently, he (1973) conducted detailed new investigations. On the basis of re-examinations of the mollusca, foraminifera, ostracoda and plants in the marine clay, he concluded that the deposits are Pliocene and should be placed no higher than the Boytonian unit of the Coralline Crag of East Anglia and the 'marnes à Nassa' of Normandy. He described the fauna as indicating a water depth at the time of deposition of 10 metres which in turn leads to a minimum sea-level 45 metres above present O.D. The sands include both beach and dune sands resting on a sub-areal surface.

ii. *High-level sands and gravels.* At a number of places in West Cornwall high-level deposits mainly of sands and gravels, but occasionally including clays, have been described. The gravels are frequently dominantly of quartz pebbles. Hill and MacAlister (1906) discussed the Polcrebo gravels (Fig. 1) occurring at a height of 500 feet (152 metres) and concluded that, "as the deposit accords approximately in elevation with the old Pliocene shoreline, it is probably a relic of the marine accumulation that lined the floor of the Pliocene sea." Reid and Scrivenor (1906) described an "old shore deposit" lying at a height of between 420 and 350 feet (128 and 107 metres) above O.D. surrounding St. Agnes Beacon (191 metres) rising like an island from the 'Pliocene platform' which according to Reid and Scrivenor is, "obviously an ancient sea floor or plane of marine denudation". Flett and Hill (1912) described gravels at Crousa Common at a height of 360 feet (110 metres) O.D. and concluded that they are marine Pliocene. None of these high-level deposits is fossiliferous and only their relationship to the 'Pliocene platform', to which Reid (1890) also referred the St. Erth Beds, links them with these dateable deposits. If, as some authors believe (see below), the platform is of Pleistocene age, the high-level deposits of Polcrebo, St. Agnes and Crousa Common must also be Pleistocene and younger than the St. Erth Beds now (Mitchell 1973) described as Late Pliocene. Milner (1922) examined all these deposits petrologically and concluded that, with the exception of kyanite and staurolite, the minerals demonstrated a local provenance. He suggested that the kyanite indicated marine currents from the north-east and the staurolite similar currents from Brittany. He accepted the general thesis of Pliocene age and marine origin put forward by Reid but pointed to the discrepant height of the Polcrebo gravels and attributed them to, "Quaternary erosion". Hendriks (1923) regarded the gravels as of fluvial origin resorted by the waves of the 'Pliocene' sea. By contrast Palmer (1930) ascribed the similar plateau gravels of the Bristol district (above 82 metres) to, "Early Pleistocene or possibly Late Pleistocene time". The last word has clearly not yet been written about the high-level gravels.

Mitchell (1960) referred to the gravels at Hele near Barnstaple, and similar deposits in the Scilly Isles, which he correlated with the St. Erth Beds. The Hele gravels which lie at a height of 185 feet (56 metres) led Mitchell to suggest aggradation to this level. He argued that neither ice nor sea-level have attained this height since their emplacement. Kidson and Wood (1974) have demonstrated that the Hele gravels are not marine but glacial outwash deposits. They argue that they are not Pliocene or early Pleistocene but much later, being part of the glacial sequence which includes the Fremington Till.

2b. High-level erosion surfaces

High-level erosion surfaces have been described in South West England at heights up to those of the summits of Exmoor, Dartmoor and Bodmin Moor. Many of these have been regarded as of marine origin. Balchin (1952) described surfaces surrounding Exmoor related to sea-levels at 373, 282, 251, 206, 130 and 85 metres. All were described as unwarped and therefore of post-Alpine age. Balchin attributed all of them to the later part of the Tertiary. Earlier he had discussed (1946) similar 'marine cut' surfaces in North Cornwall referring the higher ones to the Miocene and those related to sea-levels of 130 metres and below to the Pliocene. Wooldridge (1950), in reviewing the work of Balchin and others, concluded that over much of western Britain, "a Pliocene transgression, probably equivalent to the Calabrian stage of the Mediterranean, left evidence of planation up to a height of 1000 feet (305 metres) or even 1200 feet (366 metres), and that successive platforms at lower levels mark stages of retreat". However, in a study of the River Exe, Kidson (1962) failed to find evidence supporting sea-levels above 210 metres and would in common with many other workers, extend to South West England the view expressed by Wooldridge in relation to the southeast that, "I can see no evidence at all of marine planation above 700 feet" (213 metres). Kidson identified a strand line in the Exe basin which he correlated with the "Calabrian stage" suggested by Wooldridge. He showed that all the morphological evidence in the valleys of the Exe and its tributaries indicated adjustments to a falling base level and inferred that all the stages in this retreat could be accommodated in the Pleistocene. If this is indeed the case, the 'Pliocene bench' of Cornwall described by Reid (1890) must take its place in the story of the Early or Middle Pleistocene. Bradshaw (1961), reviewed the morphological evidence in the area of mid-Devon in the hinterland of Barnstaple and arrived at similar conclusions. Clarke (1963) working in the Camel Estuary area, identified cliff notches suggesting former sea-levels at 55, 38, 15, 7·6 and 4·6 metres. He regarded these as representing sea-levels in Pleistocene interglacials. This extended his earlier paper (1961–2) in which he correlated these, and others, with Deperet's Mediterranean levels. Weller (1960) described an early Pleistocene (Red Crag) shoreline at 205 metres in Bodmin Moor and (1961) in east Cornwall. He regarded the 131 metres ('Pliocene platform') as later Pleistocene. Everard (1960) found evidence in West Cornwall of a transgressive Pleistocene sea to a similar level and of still stands related to emergence during the later Pleistocene at levels of 162, 122, 105, 91 and 73 metres.

All the morphological evidence described above is totally unsatisfying since it lacks confirmation from deposits. Even the feature which is morphologically most convincing, the strandline at 206–210 metres, believed to mark the limit of the transgressing Early Pleistocene sea, has not gone unchallenged. Simpson (1964) dismissed it as an exhumed Upper Cretaceous shoreline. There seems little prospect that greater precision or certainty can be introduced into the debate unless new depositional evidence comes to light.

2c. Low-level shore platforms

If doubts exist in relation to the ages and origins of the high-level surfaces, it is not surprising that uncertainties arise with reference to the shore platforms of undoubted marine origin closer to contemporary sea-level. There is general agreement that they were initiated in the Pleistocene but here agreement ends. Many carry beach deposits of undoubted interglacial age and the platforms

S

must clearly be older than these. Since, however, the main beach deposits themselves have been attributed to the Ipswichian by some workers and to the Hoxnian by others, the age of the subjacent platform remains doubtful. Some (Zeuner 1959; Orme 1960, 1964) distinguished a number of distinct levels but others (Stephens and Synge 1966) argued that in some areas (Barnstaple Bay) multiple platforms are by no means certain. However, Wright (1967) probably reflects the generally accepted view when he states that after eliminating the effects of varying geological structure and lithology and of varying exposure, several distinct levels exist and, "these multiple levels reflect changes in sea-level". Some confusion arises in the literature from the use of heights to describe individual platforms. Confusion is worse confounded by referring them to different datum planes. The so-called '25 foot' (7·6 metres) platform, which is easily the most clearly marked and most widely present, and which is the equivalent of that on which the *Patella* beach in South Wales rests, covers a height range from 6–9·5 metres above O.D. Arkell (1943) referred to it as the '10 foot' (3 metres) platform simply because he used the 'present high water mark' as his datum. The platform heights which occur most frequently fall into four groups. Referred to O.D. the cliff notches at the upper limit of the platform lie at:

(a) -6 metres to 0 metre
(b) $+3·7$ metres to $+5·5$ metres
(c) $+6$ metres to $+9·5$ metres
(d) $+18$ metres to $+20$ metres

Table I summarises the wide range of views of the ages and heights of these features. Many follow Zeuner (1959) who distinguished three sea-levels at $+18$, $+7·5$ and $+3·6$ metres and attributed the higher two to the Ipswichian and the lowest to a Devensian interstadial. The major divergence of opinion appears to be between those who, like Zeuner (1959), place the cutting of the platforms in the Ipswichian and those who, like Mitchell (1960), argue for a much earlier, possibly Cromerian, Pleistocene age. Erosional features cannot easily be dated. Apsimon and Donovan (1960) may be justified in attributing the Howe Rock platform, south of Weston-super-Mare (Fig. 1) to the Paudorf (Devensian) Interstadial because it passes beneath the Devensian 'Lower Breccia' of Brean Down. It could be older. It is possible that the sea has returned during most of the Pleistocene warm periods to a level within the range of heights shown in Table I. Just as the present sea is retrimming pre-existing strandline features, so interglacial seas may have retrimmed the shore platforms produced in earlier interglacials. If this is so, the low-level shore platforms should perhaps be looked on as a polycyclic series. If this were the case, however, difficulties arise in relation to the Pleistocene ages, suggested above, for some of the high-level erosion surfaces. The resolution of these problems must await new dating techniques.

3. Later Pleistocene

3a. Introduction

Coastal sections in South West England, while apparently simple, have led to diametrically opposed views of both their stratigraphy and chronology. In this context three sites are of crucial importance: (i) Barnstaple Bay, (ii) Trebetherwick Point in the estuary of the River Camel, and (iii) the Scilly Isles (Figs 1 and 2). In discussing these sites in turn, it is as helpful to understand the areas of agreement

Table 1. Shore Platforms in South West England. Some recent views on heights and ages

Author	−6 m to 0 m	+3.7 m to +5.5 m	+6 m to +9.5 m	+18 m to +20 m
1. Arkell (1943*) Trebetherwick (Cornwall)		Hoxnian and Ipswichian — +4.6 to +6.1 m →		
2. Dewey (1948) a. Mousehole (Cornwall) b. Saunton (Devon)		+4.6 m		+20 m
3. Zeuner (1959) a. Hope's Nose to Hall Sands; Croyde-Saunton, Devon b. Hope's Nose to Plymouth Hoe, Devon c. Mousehole, Cornwall Portland, Dorset		Epi-Monastirian (Devensian Interstadial) +3.6 m	Late Monastirian (Ipswichian) +7.5 m	Main Monastirian +18 m (Ipswichian)
4. Mitchell (1960) Irish Sea, N. France, E and SE England		+1.5 m to +7.6 m (+5.5 m mean) Cromerian →		
5. Orme (1960, 1962†) South Devon		Devensian Interstadial +4.3 m	Late Ipswichian +7.3 m	Ipswichian +20 m
6. ApSimon & Donovan (1960) Brean Down, Somerset	Devensian −6 m to 0 m Interstadial			
7. Donovan (1962) Bristol Channel		+3 m (Swallow Cliff) Devensian Interstadial		
8. Bird (1963) Dodman, Cornwall		+3.7 m to +4.6 m Interglacial		
9. Clarke (1961) Camel Estuary, Cornwall		+4.6 m (2 m above H.W.M.) Interglacial (last stillstand)	+7.6 m Late Monastirian (4.6 m above H.W.M.) (Ipswichian)	+20 m Main Monastirian (Ipswichian)
10. Stephens & Synge (1960) Stephens (1966) Barnstaple Bay	0 to +15.2 m. Only possibly more than one platform Early Pleistocene, possibly Cromerian ↑			
11. Wright (1967) Teignmouth to Lizard‡		+3.7 m No reliable date	No reliable date	
12. Mitchell & Orme (1967) Scilly Isles		+1.8 m to +5.5 m Composite, possibly Hoxnian/ Ipswichian		
13. James (1968) South Cornwall		+4.9 m Epi-Monastirian (Devensian Interstadial)	+7.6 m Late Monastirian (Ipswichian)	
14. Kidson (1971) Barnstaple Bay	−1.5 m	+4.9 m to +5.5 m Composite from Cromerian to Flandrian →	+8.5 m	

* Arkell (1943) — This surface carries the raised beach which Arkell described as the '10-foot', 'Pre-glacial' or 'Patella' beach.
† Orme (1962) — suggests an Ipswichian retrimming of Hoxnian or even older coastal zone.   ‡ See also Everard *et al.* (1964).
Heights have been converted into metres O.D. and ages into modern British terminology.

between the so called 'Irish' and 'Anglo-Welsh' schools as to be aware of their differences of view. Both accept that, during the Devensian, the ice front lay well to the north with the result that the South West was subjected only to periglacial processes. Both accept that Wolstonian ice pressed into the Bristol Channel and the coastal embayments such as Barnstaple Bay. The southern limit of Wolstonian ice may well have been as far north as the Fremington—Ballycroneen (Southern Ireland) line, as suggested by Mitchell (1960), but his later revision (1972) extending the penetration of ice of this glaciation to the Scilly Isles finds ready acceptance. Few workers in the South West would, however, share recent views (Kellaway *et al.* 1975) on Saalian glaciation of the English Channel (for correlation between Wolstonian of Britain and Saalian of Europe see Mitchell *et al.* (1973).) Most accept the northern coasts of the Scilly Isles as the southern maximum of Wolstonian ice and remain agnostic on Anglian or earlier glaciations.

The differing approaches of the two schools of thought have been brought into sharpest focus in their interpretations of the 'raised' beaches and the overlying solifluction deposits. Prestwich (1892) and his contemporaries seem to have had little doubt that a simple tri-partite stratigraphical relationship existed consisting of shore platform, raised beach and overlying 'head' or rubble drift. Zeuner (1959) had no doubt that the old beaches of the South West had never been over-ridden by ice and equated them with the Tyrrhenian of the Mediterranean (=Ipswichian). His views were shared by Guilcher (1969), who regarded them as the equivalent of the Normannien of Western France, and by Kidson (1970, 1971). On the other hand the 'Irish school' (Mitchell 1960, 1972; Stephens 1966; Stephens and Synge 1966; Mitchell and Orme 1967) sought to establish that there are two beaches in South West England and to equate them with those in Southern Ireland which they regarded as of Hoxnian as well as of Ipswichian age. They assigned the most prominent of the old beaches of the South West of England to the Hoxnian Interglacial. This rested on their belief that these beach deposits pass beneath the till of accepted Wolstonian age at Fremington in North Devon. Recent work (Kidson 1971; Edmonds 1972; Kidson and Wood 1974) has shown that this is not so. Mitchell and Orme (1967) and Synge in Stephens and Synge (1966) also interpreted sections in the Scilly Isles as showing two beaches inter-bedded with glacial and periglacial deposits; a view not widely accepted.

3b. Barnstaple Bay

In North Devon raised beach deposits overlain by head are found resting on one or more shore platforms at Saunton and Croyde (Plate 1b) on the northern flank of Barnstaple Bay (Fig. 3d) and at Westward Ho! (Plate 2) on the southern. At the head of the estuary at Fremington Quay a similar sequence is found. The deposits at Saunton and Croyde incorporate erratic material including some of the 'giant' erratics which are to be seen (Fig. 2) around the coast of the South West Peninsula from Baggy Point in North Devon to Start Point on the English Channel coast of the county (Prestwich 1892; Worth 1898; Harker 1896; Ussher 1904; Reid and Scrivenor 1906; Reid 1907; Flett and Hill 1912). The glacial episode responsible for the emplacement of these erratic blocks clearly antedates the raised beach. Plate 1a shows the famous 'pink granite' erratic at Saunton firmly trapped between the shore platform and the base of the beach. Prestwich

Plate 1
a Saunton. 'Pink Granite' erratic below Ipswichian beach.
b Saunton. Ipswichian raised beach, resting on shore platform, overlain by Devensian Head.

Plate 1

a

b

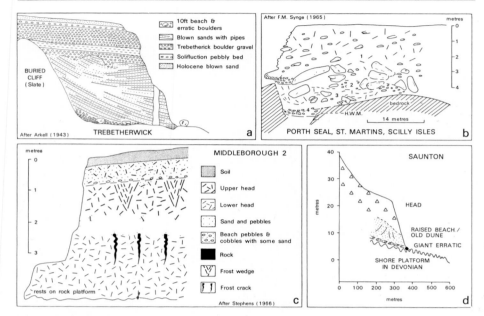

Fig. 3. Coastal sections.
    (a)   Trebetherwick Point.
    (b)   Scilly Isles.
    (c)   Croyde Bay.
    (d)   Saunton.

(1892) discussed the mollusca of all the raised beaches of the Peninsula, including those of Barnstaple Bay, and concluded that they, "agree therefore pretty closely with the molluscan fauna now living in the British seas". There can thus be little doubt that the deposits are interglacial. The simplest interpretation of the stratigraphy as demonstrated by the coastal sections would be:

Head	Devensian
Raised beach	Ipswichian
Erratic blocks	Wolstonian
Shore platform	Earlier Pleistocene

If this simple interpretation is well founded, the Fremington Boulder Clay (Plate 3a) should occupy the same horizon as the Erratic Blocks. However, Mitchell (1960) revived Maw's (1864) section through the Fremington boulder clay (Fig. 4a) and stated, "the Fremington beach runs below the boulder clay. and if the boulder clay is of Gipping (Wolstonian) age, the Fremington and Saunton beaches cannot lie in the last inter-glacial period". This was contrary to the view expressed by Zeuner (1959). Stephens (1961, 1966) re-examined the Barnstaple Bay sections and claimed to trace the Fremington beach to the area of the Fremington Till where it passed beneath the till. He divided the 'head' overlying all the raised beach sections into an upper cryoturbated head and a lower head

Plate 2
Westward Ho! Ipswichian raised beach.
a In profile.
b from lower shore platform.

Plate 2

a

b

also cryoturbated and incorporating 'weathered Irish Sea till' separated by a 'weathering sand layer' (Fig. 3c, Middleborough). His correlation table ran as follows:

	Saunton and Croyde	*Fremington*
Devensian	Stoney wash. Frost wedges. Upper head and sand.	Frost disturbances. Solifluction 'earth'.
Ipswichian	Weathering sand layer. No marine deposits known.	Weathering and erosion.
Wolstonian	Weathered Irish Sea till. Lower head.	Hele gravels. Fremington boulder clay.
Hoxnian	Sandrock (Partly dune sand) Raised beach shingle with erratics.	Raised beach shingle.
Anglian	Erratics including Saunton Gneissose granite.	
Early Pleistocene	Wave cut platforms below 15 metres O.D.	

In his second review of the Pleistocene history of the Irish Sea, Mitchell (1972) maintained his view of the stratigraphy and chronology of the Fremington sequence. Kidson (1971), however, drew attention to the fact that erratics of similar provenance to those on the coast passed into the area of the Fremington Till and were, in fact, incorporated within it (Dewey 1910; Taylor 1956, 1958; Arber 1964). He argued that the erratics and the till are deposits of the same (Wolstonian) glacial episode, that the Saunton and Westward Ho! raised beaches are younger (Ipswichian), and that the Fremington raised beach does not pass below the Fremington Till. Kidson and Wood (1974) showed (Fig. 4b) that the gravels beneath the Fremington Till are not beach but outwash gravels incorporated, with the Hele gravels, in a glacial suite of which the till and the erratics are integral parts. They supported the views of Zeuner (1959) and preferred the simple chronology. They regarded the raised beach deposits as Ipswichian and the overlying head, whatever facies variations it contains, as all of Devensian age.

3c. Trebetherwick Point

Arkell (1943) described a series of coastal sections, of which Figure 3a is a composite, at the mouth of the River Camel in Cornwall. He correlated the raised beach resting on the shore platform with the Saunton beach which he assigned to the Middle Acheulian Interglacial (=Hoxnian) but subsequently (1945) changed his views and placed it in the last (Ipswichian) Interglacial. The main interest in the Trebetherwick section is the Trebetherwick Boulder Gravel. Arkell regarded it as (a) younger than the main head. and (b) either a river gravel or a beach but in any event indicating a sea-level close to the level of the surface of the

Plate 3
a Brannam's clay pit in Fremington Till.
b Westward Ho! Soliflucted and cryoturbated raised beach gravel below O.D.

Plate 3

a

b

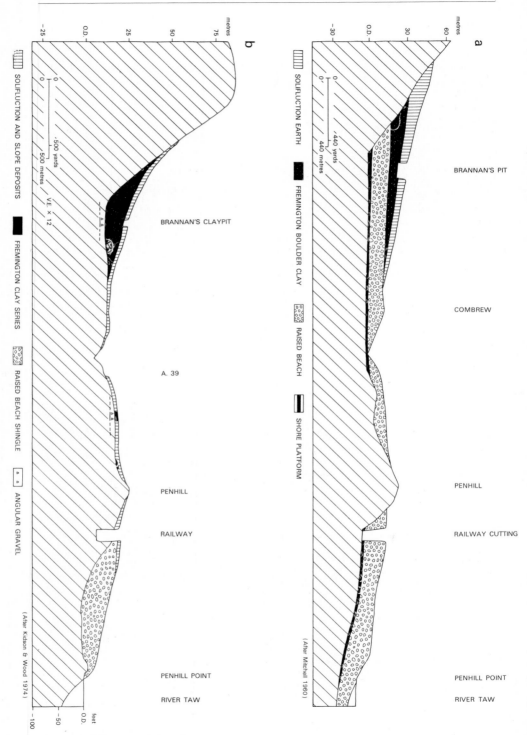

Fig. 4. Fremington stratigraphy
(a) after Mitchell.   (b) after Kidson and Wood.

gravel at about 17 metres above O.D. Mitchell and Orme (1967) described it as, "either outwash gravel or a raised beach". Mitchell regarded the overlying bed described by Arkell as a, "solifluction pebbly bed", as a till of weathered facies which he correlated with the Wolstonian tills of Fremington and the Scilly Isles. By implication any beach lying beneath the 'solifluction pebbly bed' must be Hoxnian or older. However, as noted by Arkell (1943), the Boulder Gravel and the 'solifluction pebbly bed' pass laterally into undoubted head. They could themselves be soliflucted river gravels. In the opinion of the present author, the Trebetherwick sections are comparable with those at Saunton. They show a shore platform overlain by a raised beach incorporating erratic blocks which is in turn overlain by head deposits of multiple facies, including the boulder gravel and the 'solifluction pebbly bed'. The 'erratic' content of these two beds was shown by George (in Arkell 1943) to be quite different from the Irish Sea drift and by Davison and Stuart to include rock types of Cornish origin, supporting the idea of solifluction from within the catchment of the River Camel. Even if this were not so it would be necessary, in an area where Devensian solifluction has clearly been significant, to show that any alleged till was in primary position.

3d. Scilly Isles

Barrow (1906) described in the Scilly Isles a coastal section reminiscent of those found all round the coast of the South West Peninsula. The raised beach is composed, according to Barrow, of a conglomerate made up of small uniform 'boulders' of granite 7–10 cm in diameter. He assumed it to have been formerly much more extensive. In his view, "its true position is seen on White Island where it rests on the bare granite and underlies the main Head, a glacial deposit in turn reposing on the latter". The surface of the Head forms a 'terrace feature' which in common with similar features elsewhere would now be described as a solifluction terrace. Of special significance is his view that, "the Head, in special localities, can be divided into two parts, an upper and a lower . . . separated by a curious glacial deposit". The glacial deposit is usually made up of foreign pebbles, with a large preponderance of flints but including greensand cherts, soft brown sandstones, basalts and metamorphic rocks. In only one place, on St. Martins, is the glacial deposit a tough clay. Barrow's interpretation is of special interest in view of what is written in section 3a above, on glacial limits and in section 3f below on sea-levels. "That these stones have been brought into their present position by ice admits of little doubt . . . (but) . . . it is quite clear that they must have been carried by floe ice". He observed that all the evidence points to, "powerful currents from . . . the north and west".

Mitchell and Orme (1967) re-examined the Pleistocene deposits of the Scilly Isles. They argued for a Gipping (Wolstonian) age for the glacial material which they regarded as indicating that Wolstonian ice moving from the north reached its southern limit when it pressed against the northern shores of the islands. They discerned an older (Chad Grit) raised beach beneath a lower head, in turn overlain by the glacial deposit and a younger (Porth Seal) beach covered by an upper head. However, nowhere were they able to describe a complete section showing all these features. Subsequently Synge (in Stephens 1970) described just such a complete section from Porth Seal (Fig. 3b). They suggested that the lower beach may be of Hoxnian and the upper of Ipswichian age. They compared the Scilly stratigraphy with that at Trebetherwick in which they equated the 'solifluction pebbly bed' described by Arkell (1943) with the Scilly Till.

If the interpretation of the thin and scattered deposits of the Scilly Isles as described by Mitchell and Orme and by Synge is accepted, they would represent two beaches separated by a glacial phase. Nowhere else in South West England is there clear evidence of such events. It is here suggested that the evidence from the Scilly Isles is equivocal and that there is an alternative interpretation which is at least as convincing. This accepts the view put forward by Barrow (1906) of a single beach resting directly on the shore platform and overlain by an upper and a main head. The possibility exists that the single beach is Ipswichian and the two heads are both distinct facies of Devensian solifluction deposits incorporating Wolstonian till caught up in these downslope movements. Barrow noted that parts of the beach itself were incorporated in the head and had clearly moved downslope. Any pre-existing till would inevitably be similarly affected. It is therefore suggested that the till is not in primary position and that the simple chronology suggested for Barnstaple Bay could apply equally to the deposits of the Scilly Isles.

3e. The Burtle Beds of Somerset

The most widespread interglacial marine deposits in South West England are to be found in the Somerset Levels. They take their name, the Burtle Beds, from the village of Catcott Burtle northeast of Bridgwater (see Figs 2 and 5). Ussher (1914) considered the possibility that the sands and gravels are of similar if earlier age to the submerged forests, but leaned towards a date comparable to that of the raised beaches. The fullest account of the sands and gravels of the Burtle Beds was given by Bulleid and Jackson (1937, 1941). They concluded, largely on the basis of the presence of washed-in terrestrial tests such as that of *Corbicula fluminalis* (Müll) and of antlers of the extinct deer *Dama dama* (*Cervus browni* (Dawkins)) that the Burtle Beds are interglacial and can be correlated with the gravels at Crayford in the Thames valley or with those at Clacton in Essex. This implied either a Hoxnian or an Ipswichian date. Welch (1955) suggested that the gravels at Kenn (Fig. 2) can be correlated with the Burtle Beds, but Hawkins and Kellaway (1971) reinterpreted them as till. ApSimon and Donovan (1956) regarded the Burtle Beds as belonging to the last interglacial (Ipswichian). They correlated with them the marine deposits lying below head at Weston-in-Gordano which they specifically felt could not be correlated with raised beach deposits at Middle Hope (Fig. 2), since the altitude of the Gordano marine deposits (11·2 to 13·6 metres O.D.) and of the Burtle Beds exceeds the sea-level necessary for the raised beach, and by implication of other raised beaches in the region. They referred the Gordano deposits to Zeuner's 'Main Monastirian' sea-level at 18 metres. In their view this correlation on an altitudinal basis is justified by their belief that the, "Bristol region was not affected by depression and recovery due to loading by Pleistocene ice-sheets, nor is there evidence of recent tectonic movement". The present author concurs with this view and would refer the Burtle Beds to a higher Ipswichian sea than that he believes to be responsible for most of the raised beaches of South West England.

Kellaway (1971) argued that the Burtle sands and gravels are not marine but glacial outwash. Despite evidence from Kidson and Haynes (1972) that the Burtle Beds show a rich marine microfauna, indicating a progressive estuarine/marine transgression, Hawkins and Kellaway (1973) re-asserted the glacial origins of the beds and challenged the interpretation by Kidson and Haynes that the Burtle Beds include a lower clay and silt member beneath the sands and gravels.

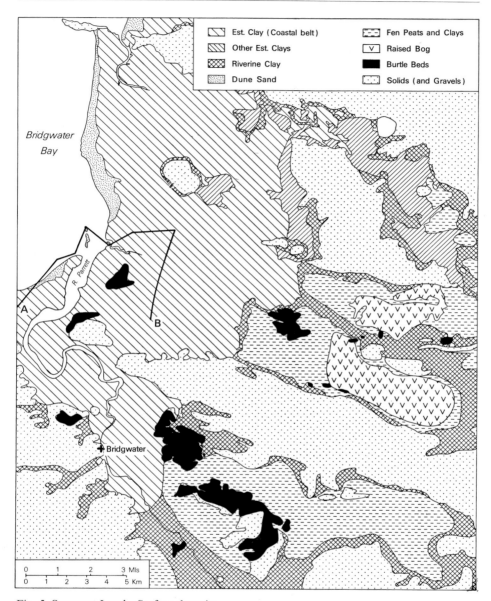

Fig. 5. Somerset Levels. Surface deposits.

Subsequently Kidson *et al.* (1974) gave a conclusive demonstration that the Burtle Beds, including sand and gravel overlying silt and clay, are *in situ* marine deposits. While restating the interglacial, probably Ipswichian age of the beds, they showed that they have been reworked at the margins during the Flandrian. A more thorough investigation of a specially excavated section (Fig. 6), including

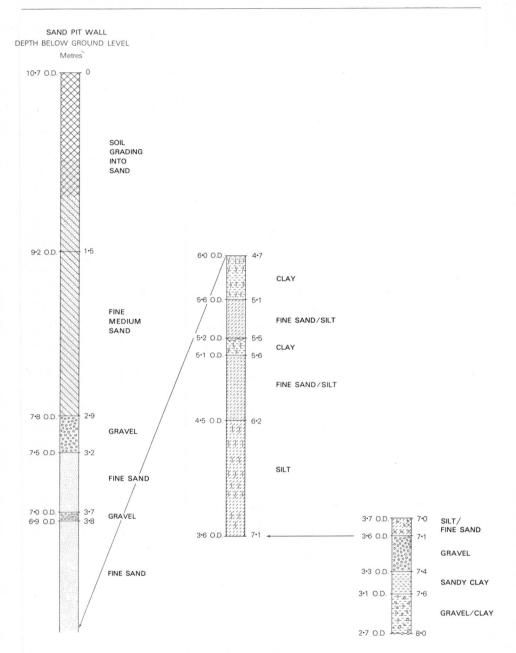

SAND PIT WALL
DEPTH BELOW GROUND LEVEL
Metres

10·7 O.D. — 0

SOIL
GRADING
INTO
SAND

9·2 O.D. — 1·5

FINE
MEDIUM
SAND

7·8 O.D. — 2·9   GRAVEL

7·5 O.D. — 3·2

FINE SAND

7·0 O.D. — 3·7   GRAVEL
6·9 O.D. — 3·8

FINE SAND

6·0 O.D. — 4·7   CLAY

5·6 O.D. — 5·1   FINE SAND/SILT

5·2 O.D. — 5·5   CLAY
5·1 O.D. — 5·6

FINE SAND/SILT

4·5 O.D. — 6·2

SILT

3·6 O.D. — 7·1

3·7 O.D. — 7·0   SILT/
FINE SAND
3·6 O.D. — 7·1   GRAVEL

3·3 O.D. — 7·4   SANDY CLAY

3·1 O.D. — 7·6   GRAVEL/CLAY

2·7 O.D. — 8·0

Fig. 6. Burtle Beds. Greylake Sandpit.

a new examination of the mollusca, foraminifera and ostracoda, is now under way and will be published in the near future. This clearly demonstrates the interglacial estuarine/marine transgressive nature of the Burtle Beds.

3f. Ipswichian sea-levels

It is implicit in what has already been written that no convincing section is to be found in South West England where two undoubted raised beaches are super-imposed indicating different former sea-levels. Not only does this apply to sea-levels in different interglacials, but also to levels attained within the last interglacial. This is not surprising on an exposed coast open to Atlantic gales where the waves of the present sea are rapidly destroying raised beach fragments which have survived until the present. However, in relatively sheltered estuarine areas, the possibility arises that evidence of differing sea-levels may remain. Arkell (1943) regarded his Boulder Gravel, with an upper surface at 17 metres above O.D., at Trebetherwick in the Camel Estuary as related to a higher sea-level than that of the underlying raised beach. Barrow (1906) argued that ice floes transported the foreign stones in the glacial deposit between the upper and lower heads on the Scilly Isles, and, by implication, suggested a higher sea-level than that responsible for the raised beach at a lower elevation. In the Somerset Levels the Burtle Beds were clearly deposited by a sea reaching heights of at least 15–18 metres above O.D. This compares with a sea-level in the 5–10 metre range suggested by the raised beach at Middle Hope (=Woodspring Hill in Ussher 1914; Palmer 1931). However, the raised beach described by Prestwich (1892) at Anchor Head, Weston-super-Mare (=Old Pier in Stephens 1973) indicates a higher sea-level, possibly within the Burtle range. It is tempting to relate these two levels to Zeuner's 'Main Monastirian' and 'Late Monastirian', i.e. both within the Ipswichian. However, it is premature to make such a correlation despite the fact that the evidence appears to support the possibility of two high sea-levels within the last inter-glacial. There are indications that the 'Burtle sea' may have been later in the interglacial than the lower sea-level at Middle Hope. While there is clear evidence in at least two localities that the Burtle Beds were reworked by the transgressing Flandrian sea, at heights of about 3·5 metres above O.D. (Kidson *et al.* 1974), no such evidence has been found at the 'Late Monastirian' level. Indeed, there are signs (Kidson and Haynes 1972) that the Burtle Beds have been subjected only to subaerial and periglacial processes between the time of their deposition and the later Flandrian. It has been suggested (Donovan 1962) that the Middle Hope (=Swallow Cliff) sea should be referred to a Devensian stadial but this can be discounted for reasons given in section 4c (p.281).

3g. Cave deposits

There is unfortunately no clear link between the coastal sections in South West England and the numerous cave deposits, with their rich mammalian faunas, found in the limestone caverns of the Mendips and the Devon limestone areas south east of Dartmoor. Many of the Mendip caves give Devensian sequences but the most important exposure is found in a quarry at Westbury-sub-Mendip which has an important Middle Pleistocene assemblage. Bishop (1974) has described a rich temperate mammalian fauna not later than Anglian in age and an earlier sparser fauna, in sands and gravels indicating a water table of the order of 240 metres above O.D., which is no later than Cromerian. These deposits are difficult if not impossible to link at present with the coastal Pleistocene.

The Tornewton Cave at Torbryan, near Buckfastleigh, on the fringes of Dartmoor (Sutcliffe and Zeuner 1958; Sutcliffe 1969) is the only British cave where an Ipswichian interglacial deposit can be directly observed lying between deposits of two cold periods. Sutcliffe (1975, 1976) has compared the deposits at Tornewton, on the basis of their mammalian faunas, with the gravels of the terraces of the Thames. He suggests that the palaeobotanical evidence from the terraces may have led to the error of interpreting the Ipswichian as a single temperate phase. He indicates that the cave faunas, as well as the mammalian faunas of the terraces, supports the Zeuner argument for two temperate phases (and two high sea-levels) within the interglacial. The evidence of the larger mammals is reinforced by that of the rodents. While the cave stratigraphy thus accords with the idea of two Ipswichian high sea-levels, it does not help to resolve the difficulty of which sea-level came first, the higher, as suggested by Zeuner, or the lower, as inferred above. Recent work on the north coast of the Bristol Channel (Sutcliffe and Bowen 1973) also lends support to the notion of more than one high sea-level in the last interglacial. In Minchin Hole, one of a series of sea caves in the Gower Peninsula, the '*Patella* beach' of pebbles, cobbles and boulders rests on and against an 'inner' sand beach and is in turn overlain by the so called 'Neritoides beach'. However, problems of wave surge in sea caves complicate the situation and the interpretation of these complex deposits awaits further detailed work.

4. The Devensian Glacial Period

Throughout the last glaciation, the South West was ice free. Periglacial processes of great severity were in operation as demonstrated by the thick deposits of 'head' which blanket not only the coastal cliffs but many inland slopes (Waters 1965). The block fields or 'clitter' surrounding the granite 'Tors' of Dartmoor bear witness to the efficacy of solifluction processes. Sea-levels reached as low as, and possibly lower than, 100 metres below O.D. and all the rivers of the area, in adjusting to these lower base levels, continued the process, probably initiated in earlier glacial episodes, of cutting deep valleys which at the present coastline can be found at depths below O.D. ranging from 7·7 to 51 metres (Fig. 2). Many of these buried rock valleys were located beneath the sediments of the modern estuaries during the phase of railway building in the nineteenth century. Others have been mapped more recently. Those below the postglacial sediments of the Somerset Levels are shown in Figure 8. (References to the literature are given in section 1b above.)

4a. 'Head' deposits

Throughout South and South West England, coastal sections are almost invariably capped by slope deposits to which Prestwich (1892) gave the name 'Head' or 'Rubble Drift'. Occasionally recent blown sand overlies the Head. In all cases the Head has demonstrably arrived in its cliff top location from the slopes above and behind the coast as a result of solifluction processes. Its surface adopts a terrace form, frequently reflecting the slope of the underlying shore platform. Fabrics invariably show preferred orientations related to the slopes. Head deposits reflect the facies of the 'country rock' and usually contain no erratics. Occasionally lenses of till and thin layers of loess interdigitate with or are interbedded with the Head. The till frequently leads to difficulties of interpretation.

Nowhere can it be shown to be in primary position and it has probably all been involved in the phases of periglacial activity responsible for Head production and emplacement. Prestwich and his contemporaries appear to have had no doubt that the head resulted from the last glaciation. However, Barrow (1906) pointed out in the Scilly Isles that the deposits are separated into a Lower or Main Head and an Upper Head. This division is characteristic throughout much of South West England. Such facies variation is not, of course, inconsistent with a Devensian age. A series of differing 'Heads' perhaps incorporating earlier (Wolstonian) tills and river gravels or even raised beach deposits can be readily visualised as resulting from climatic variation within the Devensian in which numerous cold and warm stades are known to have occurred. Gilbertson and Hawkins (1974) described a sequence of periglacial deposits at Clevedon, Somerset, overlying an Ipswichian beach. The periglacial deposits were believed to represent such Devensian climatic fluctuations. However, if the main raised beach deposits which underlie the Head are dated as Hoxnian, a problem is immediately created. Since Devensian ice is known not to have invaded the peninsula it must be assumed that some Ipswichian raised beaches survived that glacial period. Mitchell (1960) discerned no such deposits in the South West and later (1971) assigned to the Ipswichian only the sands and gravels of the Burtle Beds of Somerset and the deposits at Honiton in East Devon, discovered during motorway building, bearing a fauna almost exclusively consisting of *Hippopotamus*. Stephens (1966) and Stephens and Synge (1966) discerned a 'weathering sand horizon' or 'sand and pebbles' between the Upper and Lower Head in Croyde Bay. This they assigned to the Ipswichian. Stephens recognised 'weathered Irish Sea till' within the Lower Head and ascribed it to the Wolstonian glaciation, leaving only the Upper Head as evidence of the Devensian cold phase. Figure 3c shows this division of the Head at one section at Middleborough House, Croyde Bay. It has, however, been argued in section 3b above that the raised beach is not Hoxnian but Ipswichian. If this is indeed the case, Stephens' interpretation becomes untenable and the true stratigraphic picture is that demonstrated at Saunton (Fig. 3d) where the (Ipswichian) raised beach is overlain by a single Head sequence. Mottershead (1971) discussed the Head deposits on the South coast of Devon where upper and lower facies also occur. Both were interpreted as of Devensian age. The Upper Head included silty material which Mottershead regarded as loess. At Doniford, on the coast of West Somerset, Gilbertson and Mottershead (1975) examined a Head sequence of some complexity (Fig. 7). They ascribed the whole to the Devensian despite the variegated facies and could find no evidence of glacial activity. This would appear to be the correct interpretation throughout the South West, though in this section river gravels, moved downslope by periglacial activity, may dominate the profile.

On the coasts of South and West Britain there is evidence that some raised beach deposits have similarly been moved downslope. In the present intertidal zone occur deposits of beach pebbles, cobbles and boulders which have been subjected to solifluction processes. They are most frequently to be seen, when beach levels are low, close to O.D., and they often form the basement on which peat and submerged forest deposits of middle Flandrian date rest as in Bridgwater and Barnstaple Bays. Plate 3b shows such a deposit at Westward Ho! in Barnstaple Bay. It represents beach boulders and cobbles soliflucted in the Devensian from the Ipswichian beach, which still survives as a raised beach (Plate 2). The vertical stones, sometimes shattered, bear witness to frost action.

T

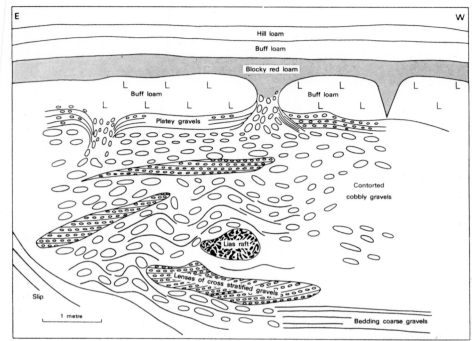

After Gilbertson & Mottershead (1975)

Fig. 7. Donniford. Section in Devensian Head.

4b. Loess

Barrow (1906) discussed at length a very fine material capping the 'normal' or main Head in the Scilly Isles. Here it is described locally as 'clay' or 'ram' or 'iron cement'. It is widespread on many islands of the group. Barrow recognised that the 'iron cement' is aeolian in character and petrologically similar to the matrix of the 'glacial deposit'. On White Island he described a section as follows:

1. Upper or Recent Head
2. Iron-cement bed or glacial deposit
3. Main or Lower Head
4. Old Beach
5. Sloping surface of the granite

He compared the 'iron-cement' with the 'Limon' on the Brittany coast north of Morlaix both in terms of particle size and of its curious property of 'setting' when dry as well as in its stratigraphical position. Barrow regarded the Upper Head as 'formed in post glacial times' and the limon or loess or 'iron-cement' as post-dating the 'Main Head' of glacial (=last glaciation=Devensian) age. It was not, however, until Combe *et al.* (1956) showed that parts of the Lizard peninsula were covered with a fine silty material, the mineralogy of which differed from that of the underlying bedrock, that it was realised that loess is more widespread in South West England. Subsequently, Findlay (1965) demonstrated that thin weathered aeolian silty drifts occur extensively in the Mendip Hills and Harrod *et al.* (1973) described a similar cover on the East Devon Plateau, to the South East of Dartmoor and at other scattered localities throughout the county. Harrod *et al.* (1973) suggested that the 'aeolian silt' was deposited in the last glaciation (Devensian) and that it may be similar in age to the Upper Head of the South Devon coast (Mottershead 1971).

4c. Devensian sea-levels

i. *Low sea-levels.* There is considerable evidence in the South West of low Devensian sea-levels. Stride (1962) drew attention to submarine features including ridges, benches and 'littoral deposits' between the Scilly Isles and the south coast of Ireland in 60 and 70 fathoms (110 and 128 metres) of water. He compared them with near shore deposits at similar depths off western Mexico for which a date of 19300 ± 300 has been reported. While such long range comparisons have obvious dangers, a late Pleistocene age for littoral deposits which have clearly been overtaken by the rising waters of a transgressing sea is a probability. It may be that the buried rock channels of the rivers of the Peninsula discussed in section 1c above are graded to base levels of the order suggested by Stride. However, none of them has been examined in detail to depths greater than about 45 metres below O.D. In these upper parts the valleys are infilled with Flandrian (Pollen Zone IV and upwards) sediments as shown by Clarke (1970) for the Exe and Kidson and Heyworth (1973) for the rivers of the Somerset Levels. However, Anderson (1968) showed that the buried rock channels of many of the South Wales rivers draining to the Bristol Channel are partly infilled with boulder clay and fluvioglacial deposits. He suggested that some of the glacial deposits are of Würm (Devensian) and others of Riss (Wolstonian) age. Codrington (1898) and Worth (1898) claimed that the rock channels of many of the rivers of South Devon are partly infilled with till. Such till, if it exists, could only date from the Wolstonian or earlier glaciations. Much of this 'evidence' is speculative, not least as far as age is concerned. Nevertheless, it serves to reinforce what appears to be a common sense presumption, namely that the buried rock valleys were cut during low sea-levels not only in the Devensian but also in earlier glacial periods. Durrance (1969) argues for two distinct phases of channel cutting in the estuary of the River Exe. The channels are perhaps best seen as policyclic or polyphase in age and origin.

Recent evidence has shown that, whatever were the lowest levels to which the Devensian sea fell, it paused at about 43 metres below present Ordnance Datum for a considerable period. Hails (1975) and Kelland (1975) have shown the existence of a submerged cliffline at this level, together with buried channels grading into it in Start Bay (Fig. 2). This confirms comparable findings by Cooper (1948) off Plymouth. Clarke (1970) traced the buried channel of the Exe to a similar submerged cliff off Torbay. McFarlane (1955) had earlier, quite independently, extrapolated the long profiles of the buried channels of the River Erme in South Devon and of the Rivers Taw and Torridge in North Devon and related them to a base level of about 45 metres below O.D. The evidence for a stillstand of considerable duration at a level 43 to 45 metres below present O.D. can thus be described as convincing.

However, Donovan and Stride (1975) in discussing buried cliff-like features around Devon and Cornwall with bases at 44 metres, 54 metres and 64 metres below Ordnance datum, which they interpret as degraded coastlines, ascribe them to the Late Tertiary. Such an age is possible but questionable. In the light of the discussion above, at least the "44 metre" cliff shown in Figure 2 must be regarded as marking a Devensian sea-level, even if it originated earlier.

ii. *High sea-levels*. Donovan (1962) argued that erosion features and terrace deposits of the River Severn and the Bristol Channel indicate a sea-level not more than a few metres above or below that of the present in the period 24000 to 45000 years age. He equated the Swallow Cliff platform and beach, to the north of Weston-super-Mare, with the main terrace of the Severn and thus gave it an Upton Warren Interstadial date (38000 to 42000 B.P.). He correlated it with Zeuner's (1959) Epi-Monastirian sea at 3 metres above present O.D. Similarly, he equated the Howe Rock platform at Brean Down, south of Weston-super-Mare, with the Worcester Terrace of the Severn to which he attributed a sea-level 7 metres below that of the present and assigned it to the Paudorf Interstadial. It may be that Donovan's correlations of the Bristol Channel shoreline features with fluvial terraces in the English Midlands will prove to have been justified. There is, however, little reliable dateable evidence in the locality to sustain them. Wood (In Callow and Hassall 1969) reported ^{14}C dates on shells from the Middle Hope (Swallow Cliff) raised beach ranging from $33240 ^{+760} _{-700}$ B.P. (NPL 126A) to $38990 ^{+1690} _{-1390}$ B.P. (NPL 126B). He regarded this as confirmation of Donovan's findings and correlations. However, ^{14}C dates on shells from the Burtle Beds in Somerset and the raised beach at Saunton in Devon (Kidson 1971), neither of which can be younger than Ipswichian, fall within a similar range of dates. It is now accepted that ^{14}C dates on shell can only give dates which are too young. They appear to be almost valueless. There is thus no direct coastal evidence of high Devensian interstadial sea-levels though there is a strong presumption that such higher levels were attained in one or more warm interstadial period. Local evidence of one climatic amelioration, correlated with the Alleröd, at Hawks Tor on Bodmin Moor (Fig. 1), is given by Conolly *et al.* (1950).

5. The Holocene Period

Godwin (1943, 1948) and Godwin *et al.* (1958) demonstrated that in the Somerset Levels, as elsewhere in Britain, the postglacial period has been dominated by the recovery of the sea from the very low levels attained during the Devensian. Kidson and Heyworth (1973) have extended Godwin's work in Bridgwater Bay and Clarke (1970) and Hails *et al.* (1975) and others to the English Channel. Much of the continental shelf, including probably the whole of the Bristol Channel area, and parts of the Western Approaches, had been dry land at the maxima of the last glaciation. The transgressing Flandrian sea submerged this land surface, reworked much of the superficial deposits, including glacial detritus and periglacial material, resting upon it and pushed much of the sand, pebble and cobble sizes of sediment into the present nearshore zone and on to contemporary beaches. The most obvious manifestations of the transgression are the submerged forests which have been recorded (Fig. 2) at the following localities: Sharpness, in the Severn Estuary; near the mouth of the River Axe (Somerset); Stolford; Minehead; Porlock; Saunton; Westward Ho!; Dunbar Sands; Perranporth; Lower St. Columb Porth; the Hayle Estuary; Prah Sands; Mawgan Porth; Marazion; Porthleven; Porth Mellin; Falmouth; Millendraeth Bay (Near Looe) and Tor Bay. At Westward Ho! in Devon, the discovery of Mesolithic occupation sites has resulted in detailed descriptions of the submerged forest beds (Rogers 1946; Churchill 1965). Not all the submerged forests are still visible; some appear to have been lost by coastal

Table 2. Radiocarbon dates from coastal sites in southwest England (excluding Somerset Levels and Bridgwater Bay)

Location	Laboratory No.	14C date B.P.	Height/O.D.
Prah Sands, Cornwall	IGS C14/71 (St 3694)	1805 ± 100	+2·0
Porth Mellin, Cornwall	IGS C14/72 (St 3692)	2345 ± 100	−0·3
Avonmouth	IGS C14/27 (St 3257)	3110 ± 100	+3·96
Topsham, Devon	Birm 534 W3	3300 ± 120	−3·06
Newton Abbot, Devon	SRR 163	3332 ± 70	c −1·0
Larrigan (near Penzance)	BM 29	3656 ± 150	c −1·6
Kingston Seymour, Avon	I 4846	3690 ± 110	+1·3
Topsham, Devon	Birm 533 W2	3910 ± 130	−3·66
Westward Ho! Devon	IGS C14/42 (St 3402)	4995 ± 105	2 metres below HWM (c +1·1 O.D.)
Kingston Seymour, Avon	I 4844	5600 ± 110	+0·15
Westward Ho! Devon	Q672	6585 ± 130	−2·1
Teignmouth, Devon (offshore)	NPL 86	8580 ± 800	−23·8

erosion. Others are normally covered by sand or mud and are only exposed at times of low beach levels. In some of the forests, large trunks and stumps, in growth positions, of oak and pine are found. Often the trees can be seen to be rooted in thin soils on the weathered bedrock surfaces or in soliflucted deposits, including soliflucted beach gravels. The forests, with their associated beds of peat, occur down to and below the levels of the lowest tides. Clarke (1970) recorded a peat bed at 24 metres below O.D. off Teignmouth in South Devon (NPL 86 in Table 2) and mud with a high tree pollen content of Pollen Zone IV–V at 43 metres below O.D. In the Somerset Levels peats resting directly on bedrock have been found in boreholes at depths of more than 21 metres below O.D., below

Table 3. Radiocarbon dates from the Somerset Levels and Bridgwater Bay

Site/Borehole	Lab. No.	Age in 14C years B.P.	Present height (Metres O.D.)	Ht. before compaction (Metres O.D.)	Ht. rel. to contemp. water-table (Metres)	Ht. of contemp. water-table (Metres O.D.)
1 Viper's Platform	Q311	2410 ±110 2460 ±110	+3·8	Unchanged	0·0	+3·8
2 Shapwick Heath Trackway	Q39	2470 ±110	+3·8	Unchanged	0·0	+3·8
3 Westhay Track	Q308	2800 ±110	+3·6	Unchanged	0·0	+3·6
4 Meare Heath Track	Q52	2840 ±110	+3·8	Unchanged	0·0	+3·8
5 5A	NPL146	3460 ±90	+0·6	+1·4	−1·2	+2·6
6 Penzoy Farm	HAR 545	3900 ±80	+3·5	Unchanged	+1·0	+2·5
7 Honeycat Track	Q429	4215 ±130	+2·4	Unchanged	0·0	+2·4
8 Hill Farm, Catcott Burtle	HAR 463	4280 ±70	+3·5	Unchanged	+1·0	+2·5
9 Hill Farm, Catcott Burtle	BIRM 529	4340 ±120	+3·5	Unchanged	+1·0	+2·5
10 Blakeway Farm Track	Q460	4460 ±130	+2·4	Unchanged	0·0	+2·4
11 Honeygore Track	Q431	4750 ±130	+2·4	Unchanged	0·0	+2·4
12 6V	I 3395	4790 ±120	+2·3	+2·5	+0·5	+2·0
13 6F	I 3396	5250 ±140	+0·8	+1·4	+0·5	+0·9

Table 3 — continued

Site/Borehole	Lab. No.	Age in 14C years B.P.	Present height (Metres O.D.)	Ht. before compaction (Metres O.D.)	Ht. rel. to contemp. water-table (Metres)	Ht. of contemp. water-table (Metres O.D.)
14 6F	I 3397	5330 ±120	−0·5	Unchanged	−0·5	0·0
15 5B	NPL147	5380 ±95	−1·0	−0·2	−0·5	+0·3
16 Tealham Moor	Q120	5412 ±130	0·0	+0·5	0·0	+0·5
17 Shapwick Heath	Q423	5510 ±120	+0·3	+0·5	0·0	+0·5
18 Tealham Moor	Q126	5620 ±120	0·0	+0·5	0·0	+0·5
19 5C	NPL148	6230 ±95	−2·0	−1·8	0·0	−1·8
20 Burnham-on-Sea	Q134	6262 ±130	−4·6	−4·0	−1·0	−3·0
21 6P	I 2689	6890 ±120	−5·95	−5·0	+0·2	−5·2
22 6J	I 2688	7060 ±160	−7·3	−6·5	0·0	−6·5
23 8B	I 3713	7320 ±120	−8·45	−8·4	−0·7	−7·7
24 6P	I 2690	7360 ±140	−8·95	−8·2	−0·2	−8·0
25 212/3	I 4403	8360 ±140	−21·3	−20·0	−0·5	−19·5
26 211/34	I 4402	8480 ±140	−19·5	−19·0	−0·5	−18·5

marine sands and clays (Kidson and Heyworth 1973) (I. 4403 in Table 3). Austen (1851) recorded stumps of trees, rooted in terrestrial sediments and overlain by marine beds, at 20 metres below O.D. at Pentuan, Cornwall.

There can be no doubt that these submerged and buried terrestrial deposits covering a vertical range of at least 47 metres represent stages in the drowning of the land surface by the transgressing Holocene (Flandrian) sea. The rise of sea-level began very early in the postglacial and took place initially so rapidly that by 6000 B.P. only about 6 metres of rise remained to be achieved at a decelerating pace. Figure 10, which is based on the radiocarbon dates given in Table 3, shows the rate of rise in the area of the Somerset Levels compared to similar Holocene sea-level rise in Southern Sweden (Mörner 1969) and The Netherlands (Jelgersma 1966). The Somerset curve has been corrected for gravitational compaction (as in Table 3) but no allowance has been made for isostatic rebound since the area is believed (see Section 6a below) not to have suffered such rebound at least in the postglacial period. The ^{14}C dates are unadjusted.

As sea-level rose, waterlogging developed in the coastal areas since the ground water table was also, of course, affected. This caused or accelerated peat accumulation and was followed shortly by submergence. In the present nearshore and offshore areas all that has subsequently occurred has been a deepening of the sea above the submerged beds. In the river valleys, however, all of which had been cut down, many to at least 30 metres below O.D., in what are now the estuary areas, the rise in sea-level has been accompanied by the deposition of sands, silts and more or less brackish clays. In these areas this deposition has maintained the land surface at or close to sea-level throughout the Holocene. At times the rate of sedimentation has been more rapid than sea-level rise; at others the reverse has been the case. Where sedimentation was rapid, vegetation re-established itself on the near horizontal surfaces and peat formation resumed. A succession of alternating peats and clays was thus built up. In sediments from the early part of the Holocene, when sea-level rise was rapid, peats, apart from a basal peat bed, are rare and thin. As sea-level rise slowed however, peats formed more extensively and thicker beds developed. A typical section therefore consists of a well developed basal peat bed (the buried forest layer) which blankets the sub-Holocene surface, followed by thick clays, silts and sands with an occasional thin peat and then by intercalated peats and freshwater clays of more or less equal thickness. In some areas a very thick upper peat forms the present surface. This top peat has accumulated during the late Holocene period of very slow sea-level rise. Because the rate of peat growth in these areas has exceeded the rate of sea-level rise, they have remained free from inundation by the sea for up to 5000 years.

This pattern can best be seen in the Somerset Levels. Here the deeply entrenched river valleys (Fig. 8) have been infilled in the last 9000 years. The coastal submerged forest exposures can be traced inland, and at an increasing depth, where the equivalent bed forms the basal peat resting on the solid floors of the infilled valleys (Fig. 9). Above this bed is the characteristic section described above. Over much of the inland areas of the Somerset Levels a peat bed, 5 metres or more in thickness, forms the topmost layer. In its upper part this is of raised bog origin and lies above the general water table (Godwin 1948, 1955a, 1955b, 1956). Apart from these raised bog areas and the coastal dunes, all the other Holocene sediments infilling the Somerset Levels were plainly laid down close to 'sea-level'. The present surface height is generally below high water mark of Spring Tides but above that of Neap Tides. This appears to have been the general relationship throughout the

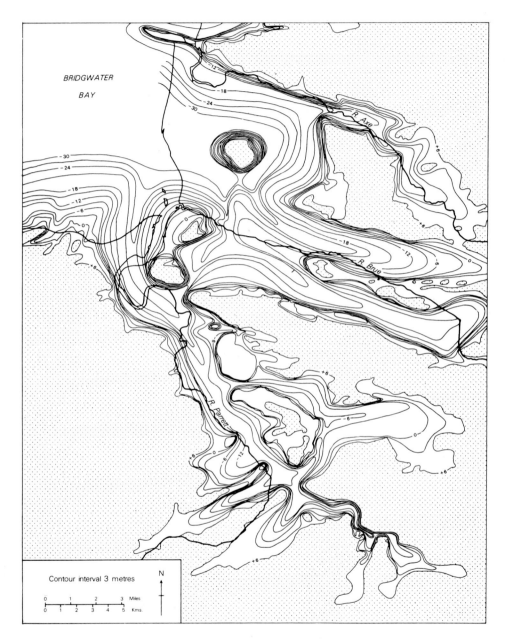

BRIDGWATER

BAY

R. Axe

R. Brue

R. Parrett

Contour interval 3 metres

N

0   1   2   3   Miles

0   1   2   3   4   5   Kms.

Fig. 8. Somerset Levels. Sub-Holocene surface.

Holocene as sedimentation and sea-level rise proceeded in step (Kidson and Heyworth 1973). In the areas of fen and carr peats, where the relationship between sea-level, the ground water table and peat growth is very close, this correlation is very clear. Many of the flooding episodes recognised by Godwin (1960) in the levels, and which stimulated the construction of the prehistoric wooden trackways across the bogs, can be shown to relate to the contemporary sea-level, although climatic factors may have triggered disturbance of the equilibrium.

The sea-level curve for the Somerset Levels, presented in Figure 10, shows a smooth almost exponential rise in sea-level. There appears to be no evidence in the Somerset Levels, or in South West England generally, of minor regressions and renewed transgressions. Intercalating peats and clays which elsewhere have been interpreted as marking such oscillations in sea-level advance have been here regarded as responses to minor variations in the relative rates of sedimentation and sea-level rise (Kidson and Heyworth 1973). Godwin (1943, 1956) interpreted the thicker marine clays on the coastal fringes of the Somerset Levels in terms of a renewed transgression in Romano-British times. Kidson and Heyworth do not accept this interpretation and suggest that the cumulative effects of occasional high sea-levels, caused by exceptional storms, storm surges and long period tides during the later stages of the transgression, account for all the evidence put forward by Godwin.

Figures 11 and 12 are attempts to show changes in the coastal margins of the Somerset Levels at selected times during the Holocene. The various coastal outlines reflect not only the steady advance of the sea but also the varying response to it. Although the coastline of 4000 B.P. (Fig. 12) lay much further inland than that of today, sea-level was then slightly lower than now. At this time the waters of the transgressing Flandrian sea reworked the margins of some of the Burtle Bed 'islands' as the dates from Penzoy Farm and Hill Farm, Catcott Burtle (Table 3) serve to show. The coastline of today has been pushed further seaward as a result of embanking and drainage over the last 2000 years. There is no evidence in the Somerset Levels, or in the rest of South West England, of Holocene sea-levels higher than that of the present day.

6. Discussion and Conclusions

6a. Isostatic and tectonic movements

Churchill (1966) examined the relationship to modern sea-level of peat beds formed at sea-level 6500 years ago. He deduced a tilt from northwest to south east, hinged along a line from the mouth of the Bristol Channel to the area of north east Yorkshire, giving uplift of 9–12 metres in North West England and depressions of 3–6 metres in the Netherlands. The information available from South West England does not support these conclusions. No evidence of isostatic rebound has so far been found in the Bristol Channel area (ApSimon and Donovan 1956; Kidson and Heyworth 1973). Detailed work, as yet unpublished, in Cardigan Bay (Kidson and others) shows that dates from peat beds in West Wales fit almost perfectly on the sea-level curve from Bridgwater Bay (Fig. 10), indicating

Fig. 9. Somerset Levels. Section line A–B on Figure 5.

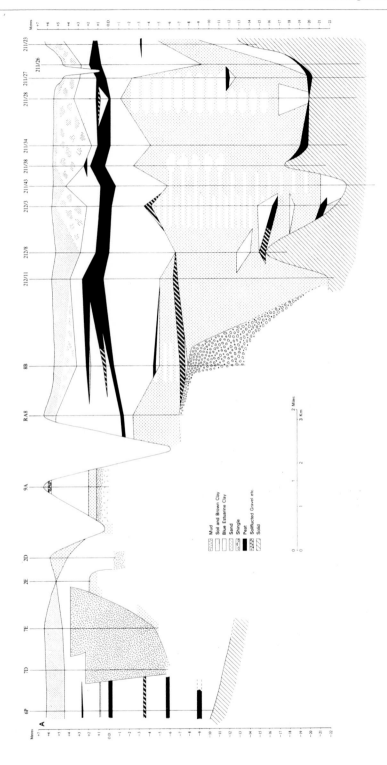

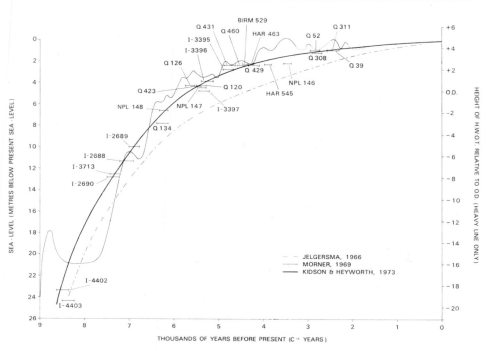

Fig. 10. Somerset Levels. Holocene sea-level curve.

no differential movement between the two areas during the Holocene. In the British Isles isostatic adjustment centres in North West Scotland and diminishes with distance from that centre. It is believed that much of South Wales and South West England has remained unaffected by isostatic rebound at least during the Holocene. It is known that South East England is affected by tectonic warping resulting from geosynclinal subsidence in the Flemish Bight area of the North Sea. Woodridge and Linton (1937) showed that this hinged about an axis passing north-south through Eastern England; the so-called 'Hitchin Line'. Again South West England appears to be sufficiently far removed from these tectonic disturbances to have remained unaffected. Indeed, the uniformity in height of erosion features such as the so-called 'Calabrian surface' and the low-level shore platforms suggests that Central South and South West England have remained essentially stable for much of the Quaternary.

6b. Sea-levels

Many questions remain unanswered concerning sea-level movements during the Pleistocene. High-level erosion surfaces in the South West have been interpreted by many authors as indicating a rapid secular fall in sea-level during the Pleistocene. Glacio-eustatic fluctuations about this falling base level would indicate low sea-levels during earlier glaciations higher than high sea-levels in later interglacials. If this were indeed the case, such relatively high glacial sea-levels would make possible the emplacement from stranded ice bergs of the giant erratics which are found close to present High Water Mark, all round the coast of the

Table 4. Correlation of Quaternary coastal features in southwest England

Period	Deposits	Erosion Features	Locations	Sea Level
Holocene	Raised bogs. Coastal clays. Submerged forests. Sands, gravels, silts, clays, peats infilling upper parts of buried rock channels of all rivers. Much of sediment of present beaches. Basal peat bed	Trimming of shore platforms.	Somerset Levels. As in Figure 2. Widespread Bridgwater Bay.	Rising as in Figure 10.
Devensian	Loess. Upper and Lower Heads including soliflucted tills (of Wolstonian age?) raised beach gravels and river gravels. Boulder clay in buried river channels.	Buried rock channels of rivers.	Scilly Isles ('iron cement') Lizard Peninsula, East and South Devon, Mendips. Widespread including Scilly Isles, Trebetherwick, Westward Hol, Croyde Bay, Donniford, Bridgwater Bay. All major rivers. Severn and its northern tributaries.	Circa 100 metres below O.D. At glacial maxima. Stage at 42 to 45 metres below O.D. Possible higher stages during interstadials.
Ipswichian	Burtle Beds (Late) with Anchor Head Raised Beach? Raised beaches (Earlier).	Trimming of upper parts of shore platforms.	Somerset Levels Trebetherwick?, Scilly Isles? English Channel coast. Widespread. As in Figure 2.	18–20 metres above O.D. 6–9½ metres above O.D.
Wolstonian	Till and outwash sands and gravels. Some till in buried channels of some (northerly) rivers. Erratics including 'giant' erratics.	Buried rock channels of rivers?	Fremington, Gele, Kenn, Scilly Isles. Widespread. Widespread from Croyde Bay and Saunton to Start Point as in Figure 2.	Low, relative to preceding interglacial but possibly higher than Ipswichian.
Earlier Pleistocene	High level gravels?	Shore platforms:	Bristol area. Crousa Common, Polcrebo, St. Agnes.	Fluctuations but perhaps in the context of a secular fall from late Pliocene levels in excess of 210 m above O.D.
Earliest Pleistocene	Some giant erratics?	High level erosion surfaces below 210 m? "Calabrian" surface?	Widespread.	

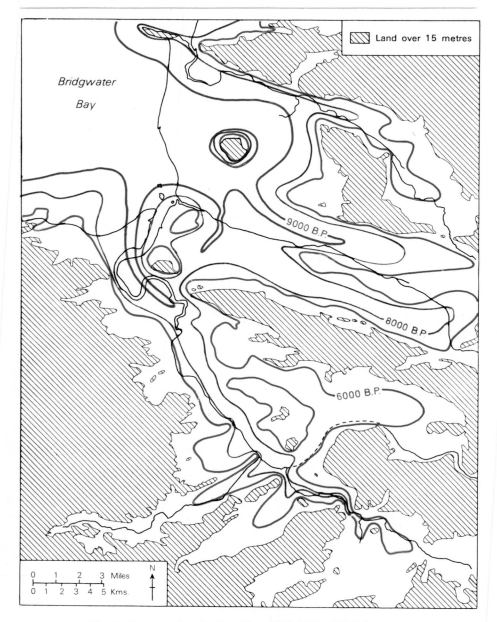

Bridgwater
Bay

Land over 15 metres

9000 B.P.

8000 B.P.

6000 B.P.

0   1   2   3 Miles
0 1 2 3 4 5 Kms.

N

Fig. 11. Somerset Levels. Coastlines 9000, 8000, 6000 B.P.

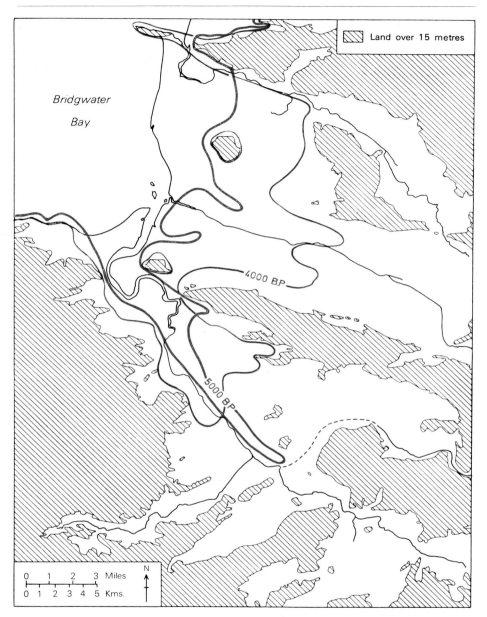

Fig 12. Somerset Levels. Coastlines 5000, 4000, B.P

peninsula (section 3b above) well beyond the accepted glacial limits. In the absence of a secular fall of sea-level the 'giant' erratics present a major problem. The ice rafting method of transport could then only be invoked in the context of time lags between ice advance and retreat and fall and rise of sea-level. An alternative explanation of glacial transport during an early Pleistocene glaciation with more southerly limits than any yet accepted appears improbable since no erratics have been found at levels higher than the highest raised beaches.

However convenient for the explanation of the distribution of giant erratics, the secular fall of sea-level is not supported by the evidence of shore platforms and the raised beaches in South West England. These suggest that, at least during the later Pleistocene, sea-level returned again and again to something close to that of the present. The shore platforms are certainly older than the Ipswichian deposits which rest upon them and the 'giant erratics' can be no younger than the Wolstonian. Many authors have inferred that the platforms originated in the early Pleistocene. It is, of course, possible to reconcile the conflicting evidence of the high-level surfaces and the low-level shore platforms by suggesting a rapid secular lowering of sea-level during the early Pleistocene followed by the maintenance, apart from glacio-eustatic oscillations, of a level close to that of today in the later Pleistocene. This must, however, remain entirely speculative at this stage since no hard evidence to support it is available.

6c. Conclusions

Clear evidence in South West England of the events of the Quaternary is confined to the later Pleistocene and the Holocene. Wolstonian ice, moving down the Irish Sea passed up the Bristol Channel and impinged on the fringes of the coastal lowlands. It entered Barnstaple Bay where the Fremington Till sequence represents the most significant glacial deposit of the peninsula. It may have reached the northern shores of the Scilly Isles. The train of giant erratics which encircles the peninsula, close to present sea-level, probably stems from this glaciation but the possibility remains that some may have originated earlier in the Pleistocene. Evidence of high sea-levels in the Ipswichian interglacial abounds in the form of raised beach and nearshore deposits. The most widespread evidence, in the form of raised beaches, refers to a sea-level 6–9·5 metres above that of the present. This probably dates from the earlier part of the Interglacial. Less extensive evidence suggest a higher sea-level in the later part of the interglacial at 18–20 metres higher than present O.D. The Burtle Beds in the Somerset Levels are the major legacy from this transgression. In the Devensian the ice limits lay well to the north of the area but intense solifluction processes produced head deposits of considerable thickness. These deposits are of very variable facies but two distinctive phases are recognised giving an upper and a lower head. Incorporated in the head are lenses of soliflufted till, river gravels and beach deposits. Above lie thin deposits of blown sand in the coastal areas and a veneer of loess in some inland sites. The post-glacial is represented by peat deposits, submerged forests and silts, sands and clays deposited by the Flandrian sea transgressing from the low levels of the Devensian. At the height of this glaciation, sea-levels may have been lower than 100 metres below O.D. but a Devensian still stand at 43–45 metres below O.D. is clearly established. The Flandrian transgression began rapidly but decelerated after about 6000 B.P. It was continuous, without minor regressions, and never attained levels higher than that of the present day.

Evidence of pre-Wolstonian events is difficult to interpret. No Hoxnian beaches are recognised in the area. The evidence of high-level erosion surfaces is equivocal and the high level gravels resting upon them add little to our knowledge of the early Pleistocene. The low-level shore platforms are so extensive and well developed that the inference that they originated in the Middle or Early Pleistocene is hard to resist. Early Pleistocene sea-levels remain debateable. A suggested stratigraphy and chronology is given in Table 4 on page 291.

References

ANDERSON, J. G. C. 1968. The concealed rock surface and overlying deposits of the Severn Valley and Estuary from Upton to Neath. *Proc. S. Wales Inst. Engrs* **83**, 27–47.

APSIMON, A. M. and DONOVAN, D. T. 1956. Marine Pleistocene deposits in the Vale of Gordano, Somerset. *Proc. Univ. Brist. Spelaeol. Soc.* **7**, 130–136.

—— and —— 1960. The stratigraphy and archaeology of the late-glacial and post-glacial deposits of Brean Down, Somerset. *Proc. Univ. Brist. Spelaeol. Soc.* **9**, 67–136.

ARBER, M. A. 1964. Erratic boulders within the Fremington clay of North Devon. *Geol. Mag.* **101**, 282–283.

ARKELL, W. J. 1943. The Pleistocene rocks of Trebetherwick Point, North Cornwall: their interpretation and correlation. *Proc. Geol. Assoc.* **54**, 141–170.

—— 1945. Three Oxfordshire Palaeoliths and their significance for Pleistocene correlation. *Proc. prehist. Soc. New Ser.* **11**, 20–32.

AUSTEN, R. A. C. 1851. Superficial accumulations of the Coasts of the English Channel. *Q. J. geol. Soc. Lond.* **7**, 118–136.

BALCHIN, W. G. V. 1946. The geomorphology of the North Cornish coast. *Trans. R. geol. Soc. Corn.* **17**, 317–344.

—— 1952. The erosion surfaces of Exmoor and adjacent areas. *Geogr. J.* **118**, 453–476.

BARROW, G. 1906. Geology of the Isles of Scilly. *Mem. geol. Surv. U.K.* 15–34.

BEALE, Dr. 1666. Some promiscuous observations made in Somersetshire. *Phil. Trans. R. Soc.* **18**, 323.

BIRD, E. C. F. 1963. Coastal landforms of the Dodman district. *Proc. Ussher Soc.* **1**, 56–57.

BISHOP, M. J. 1974. Preliminary report on the Middle Pleistocene mammal bearing deposits of Westbury-sub-Mendip, Somerset. *Proc. Univ. Brist. Spelaeol. Soc.* **13**, 301–318.

BOSWELL, P. G. H. 1952. The Plio-Pleistocene boundary in the east of England. *Proc. geol. Assoc. London* **63**, 301–312.

BRADSHAW, M. J. 1961. *Aspects of the geomorphology of North West Devon.* M.A. Thesis, Univ. of London.

BULLEID, A. and JACKSON, J. W. 1937. The Burtle Sand Beds of Somerset. *Proc. Somerset archaeol nat. Hist. Soc.* **83**, 171–192.

—— 1941. Further notes on the Burtle Sand Beds of Somerset. *Proc. Somerset archaeol nat. Hist. Soc.* **87**, 111–116.

CALLOW, W. J. and HASSALL, G. I. 1969. National Physical Laboratory Radiocarbon Measurement VI. *Radiocarbon* **11**, 130–136.

CHURCHILL, D. M. 1965. The Kitchen Midden site at Westward Ho!, Devon, England: ecology, age and relation to changes in land and sea levels. *Proc. prehist. Soc.* **31**, 74–84.

—— 1966. The displacement of deposits formed at sea-level, 6,500 years ago in Southern Britain. *Quaternaria* **7**, 239–257.

CLARKE, B. B. 1961–2. The superficial deposits of the Camel estuary and suggested stages in its Pleistocene history. *Trans. R. geol. Soc. Corn.* **19**, 259–279.

—— 1963. Erosional and depositional features as evidence of former Pleistocene and Holocene strandlines. *Proc. Ussher Soc.* **1**, 57–59.

CLARKE, R. H. 1970. Quaternary sediments off South East Devon. *Q. J. geol. Soc. Lond.* **125**, 277–318.

CODRINGTON, T. 1898. On some submerged rock-valleys in South Wales, Devon and Cornwall. *Q. J. geol. Soc. Lond.* **54**, 251–278.

COMBE, D. E., FROST, L. G., LE BAS, M. and WATTERS, W. 1956. The nature and origin of the soils over the Cornish Serpentine. *J. Ecol.* **44**, 605–615.

CONOLLY, A. P., GODWIN, H. and MEGAW, E. M. 1950. Studies in the post-glacial history of British vegetation. XI—Late-glacial deposits in Cornwall. *Phil. Trans. R. Soc. B* **234**, 397–469.

COOPER, L. H. M. 1948. A submerged cliff line near Plymouth. *Nature* **161**, 280.

DE LA BECHE, H. T. 1839. *Geology of Cornwall, Devon and West Somerset*. London 395–434.

DEWEY, H. 1910. Notes on some igneous rocks from North Devon. *Proc. Geol. Assoc.* **21**, 429–434.

—— 1948. *British regional geology. South West England*. H.M.S.O., London 63–67.

DONOVAN, D. T. 1962. Sea levels of the last glaciation. *Bull. geol. Soc. Am.* **73**, 1297–1298.

—— and STRIDE, A. H. 1975. Three drowned coastlines of probable late Tertiary age around Devon and Cornwall. *Marine Geology* **19**, M35–M40.

DURRANCE, E. M. 1969. The buried channels of the Exe. *Geol. Mag.* **106**, 174–189.

—— 1971. The buried channel of the Teign Estuary. *Proc. Ussher Soc.* **2**, 299–306.

EDMONDS, E. A. 1972. The Pleistocene history of the Barnstaple area. *Rep. No. 72/2 Inst. geol Sci.* 12 pp.

EVERARD, C. E. 1960. Valley and structural trends in West Cornwall. *Abstr. Proc. Third Conf. Geologists and Geomorphologists in South West England*. Penzance 20–22.

——, LAWRENCE, R. H., WITHERICK, M. E. and WRIGHT, L. W. 1964. Raised beaches and marine geomorphology. *In*, K. F. G. Hoskins and G. J. Shrimpton (Editors). Present views of some aspects of the geology of Cornwall and Devon. *Roy. geol. Soc. Corn.*, Penzance 283–310.

FINDLAY, D. C. 1965. The soils of the Mendip district of Somerset. *Mem. Soil Surv. Gt Br.* Harpenden.

FLETT, J. S. and HILL, J. B. 1912. Geology of the Lizard Meneage. *Mem. geol. Surv. U.K.* 229–237.

GILBERTSON, D. D. and HAWKINS, A. B. 1974. Upper Pleistocene deposits and landforms at Holly Lane, Clevedon, Somerset. *Proc. Univ. Brist. Spelaeol. Soc.* **13**, 349–360.

—— and MOTTERSHEAD, D. N. 1975. The Quaternary deposits at Doniford, West Somerset. *Fld Stud.* **4**, 117–129.

GODWIN, H. 1943. Coastal peat beds of the British Isles and North Sea. *J. Ecol.* **31**, 199–247.

—— 1948. Studies of the post-glacial history of British vegetation. X—correlation between Climate, Forest Composition, Prehistoric Agriculture and Peat Stratigraphy in Sub-Boreal and Sub-Atlantic Peats of the Somerset Levels. *Phil. Trans. R. Soc. B* **233**, 275–286.

—— 1955a. Studies of the post-glacial history of British vegetation. XIII—The Meare Pool region of the Somerset Levels. *Phil. Trans. R. Soc. B* **239**. 161–190.

—— 1955b. Botanical and geological history of the Somerset Levels. *Advmt Sci.* **12**, 319–322.

—— 1956. *The history of the British flora*. Cambridge University Press, 384 pp.

—— 1960. Prehistoric trackways of the Somerset Levels: their construction, age and relation to climatic change. *Proc. prehist. Soc.* **26**, 1–36.

——, SUGGATE, R. P. and WILLIS, E. H. 1958. Radiocarbon dating and the eustatic rise in ocean level. *Nature, Lond.* **181**, 1518–1519.

GUILCHER, A. 1969. Pleistocene and Holocene sea level changes. *Earth Science Rev.* **5**, 69–97.

HAILS, J. R. 1975. Submarine geology, sediment distribution and Quaternary history of Start Bay, Devon. *Q. J. geol. Soc. Lond.* **131**, 1–5, 19–35.

HARKER, A. 1896. Report on a block of stone trawled in the English Channel and two erratics near the Prawle, South Devon. *Trans. Devon Assoc. Advmt. Sci.* **28**, 531–532.

HARROD, T. R., CATT, J. A. and WEIR, A. H. 1973. Loess in Devon. *Proc. Ussher Soc.* **2**, 554–564.

HAWKINS, A. B. 1962. The buried channel of the Bristol Avon. *Geol. Mag.* **99**, 369–374.

—— and KELLAWAY, G. A. 1971. Field Meeting at Bristol and Bath with special reference to new evidence of glaciation. *Proc. Geol. Assoc.* **82**, 267–291.

—— and —— 1973. 'Burtle Clay' of Somerset. *Nature, Lond.* **243**, 216–217.

HENDRIKS, E. M. L. 1923. The physiography of southwest Cornwall, the distribution of chalk flints, and the origin of the gravels of Crousa Common. *Geol. Mag.* **60**, 21–31.

HILL, J. B., and MACALISTER, D.A. 1906. Falmouth, Truro and the mining district of Camborne and Redruth. *Mem. geol. Surv. U.K.* 88–103.

HORNER, L. 1815. Sketch of the geology of the South Western part of Somersetshire. *Geol. Trans. I.G.C. 1950* **1**, 338–84.

JAMES, H. C. L. 1968. Aspects of the raised beaches of South Cornwall. *Proc. Ussher Soc.* **2**, 55–56.

JONES, E. D. 1882. Description of a section across the River Severn based upon borings and excavations made for the Severn Tunnel. *Proc. Geol. Assoc.* **7**, 339–351.

JELGERSMA, S. 1966. Sea-level changes during the last 10,000 years. *Proc. Symp. on World Climate 8000–O.B.C.* 54–71.

KELLAND, N. C. 1975. Submarine geology of Start Bay determined by continuous seismic profiling and core sampling. *Q. J. geol. Soc. Lond.* **131**, 7–17.

KELLAWAY, G. A. 1971. Glaciation and the Stones of Stonehenge. *Nature, Lond.* **233**, 30–35.

——, REDDING, J. H., SHEPHARD-THORN, E. R. and DESTOMBES, J. -P. 1975. The Quaternary history of the English Channel. *Phil Trans. R. Soc. A* **279**, 189–218.

KIDSON, C. 1962. The denudation chronology of the River Exe. *Trans. Inst. Brit. Geogr.* **31**, 43–66.

—— 1970. The Burtle Beds of Somerset. *Proc. Ussher Soc.* **2**, 189–191.

KIDSON, C. 1971. The Quaternary history of the coasts of South West England with special reference to the Bristol Channel coast. *In*, K. J. Gregory and W. L. D. Ravenhill (Editors). Exeter Essays in Geography, University of Exeter Press. 1–22.

—— and HAYNES, J. R. 1972. Glaciation in the Somerset Levels. The Evidence of the Burtle Beds. *Nature, Lond.* **239**, 390–392.

——, —— and HEYWORTH, A. 1974. The Burtle Beds of Somerset—glacial or marine? *Nature, Lond.* **251**, 211–213.

—— and HEYWORTH, A. 1973. Flandrian sea level rise in the Bristol Channel. *Proc. Ussher Soc.* **2**, 565–584.

—— 1976. The Quaternary deposits of the Somerset Levels. *Q .J. Eng. Geol. In Litt.*

—— and WOOD, R. 1974. The Pleistocene stratigraphy of Barnstaple Bay. *Proc. Geol. Assoc.* **85**, 223–237.

MAW, G. 1864. On a supposed deposit of Boulder clay in North Devon. *Q. J. geol. Soc. Lond.* **20**, 445–451.

MCFARLANE, P. B. 1955. Survey of two drowned river valleys in Devon. *Geol. Mag.* **92**, 419–429.

MILNER, A. B. 1922. The nature and origin of the Pliocene deposits of the country of Cornwall and their bearing on the Pliocene geography of the South West of England. *Q. J. geol. Soc. Lond.* **78**, 348–377.

MITCHELL, G. F. 1960. The Pleistocene history of the Irish Sea. *Advmt Sci.* **17**, 313–325.

—— 1965. The St. Erth Beds—and alternative explanation. *Proc. Geol. Assoc.* **76**, 345–366.

—— 1972. The Pleistocene history of the Irish Sea: second approximation. *Sci. Proc. R. Dubl.* **4**, 181–199.

—— 1973. The late Pliocene marine formation at St. Erth, Cornwall. *Phil. Trans. R. Soc.* **B 266**, 1–37.

—— and ORME, A. R. 1967. The Pleistocene deposits of the Isles of Scilly. *Q. J. geol. Soc. Lond.* **123**, 59–92.

——, PENNY, L. F., SHOTTON, F. W. and WEST, R. G. 1973. A correlation of Quaternary deposits in the British Isles. *Geol. Soc. Lond. Special Report No.* 4, 99 pp.

MÖRNER, N.-A. 1969. Eustatic and climatic changes during the last 15,000 years. *Geol. en Mijnbouw.* **48**, 389–399.

MOTTERSHEAD, D. N. 1971. Coastal head deposits between Start Point and Hope Cove, Devon. *Fld Stud.* **3**, 433–453.

ORME, A. R. 1960. The raised beaches and strand lines of South Devon. *Fld Stud.* **1**, 1–22.

—— 1962. Abandoned and composite sea cliffs in Britain and Ireland. *Ir. Geogr.* **4**, 279–291.

—— 1964. The geomorphology of southern Dartmoor and the adjacent area. *Rep. Trans. Devon Associat. Advmt Sci.* Dartmoor Essays, 31–72.

PALMER, L. S. 1931. On the Pleistocene succession of the Bristol district. *Proc. Geol. Assoc.* **42**, 345–361.

PRESTWICH, J. 1892. The raised beaches and 'Head' or rubble-drift of the South of England: their relation to the valley drifts and to the glacial period and on a late post-glacial submergence. *Q. J. geol. Soc. Lond.* **48**, 263–342.

REID, C. 1890. The Pliocene deposits of Great Britain. *Mem. geol. Surv. U.K.*

—— 1907. Geology of the country around Mevagissey. *Mem. geol. Surv. U.K.* 55–63.

—— and FLETT, J. S. 1907. Geology of the Land's End. *Mem. geol. Surv. U.K.* 68–84.

—— and SCRIVENOR, J. B. 1906. Geology of the Country near Newquay. *Mem. geol. Surv. U.K.* 62–71.

ROGERS, E. H. 1946. The raised beach, submerged forest and Kitchen Midden of Westward Ho! and the submerged stone row of Yelland. *Rep. Trans. Devon Ass. Advmt Sci.* **3**, 109–135.

SIMPSON, S. 1964. The supposed 690 ft marine platform in Devon. *Proc. Ussher Soc.* **1**, 89–91.

STEPHENS, N. 1961. Re-examination of some Pleistocene sections in Devon and Cornwall. *Abstrs Proc. Fourth Conf. Geol. and Geomorphologists in South West of England,* 22–23.

—— 1966. Some Pleistocene deposits in North Devon. *Biul. peryglac.* **15**, 103–114.

—— 1970. The west country and southern Ireland. *In*, C. A. Lewis (Editor) *The glaciations of Wales and adjoining regions.* Longmans, London.

—— 1973. Southwest England. *In*, Mitchell *et al.* A correlation of Quaternary deposits in the British Isles. *Geol. Soc. Lond. Special Report No.* 4.

—— and SYNGE, F. M. 1966. 'Pleistocene Shorelines.' *In*, G. Dury (Editor) *Essays in Geomorphology.* London, 1–66.

STRIDE, A. H. 1962. Low Quaternary sea-levels. *Proc. Ussher Soc.* **1**, 6–7.

SUTCLIFFE, A. J. 1969. Pleistocene faunas of Devon. *In*, F. Barlow (Editor). *Exeter and its region.* University of Exeter.

—— 1975. A hazard in the interpretation of glacial-interglacial sequences. *Quaternary Newsletter* **17**, 1–5.

—— 1976. The British glacial-interglacial sequence: a reply. *Quaternary Newsletter* **18**, 1–7.

—— and BOWEN, D. Q. 1973. Preliminary report of excavations in Minchin Hole, April-May 1973. *News. William Pengelly Cave Studies Trust* **21**, 12–25.

—— and ZEUNER, F. E. 1958. Excavations in the Torbryan Caves, Devonshire. *Proc. Devon archaeol Expl. Soc.* **5**, 127–145.

TAYLOR, C. W. 1956. Erratics of the Saunton and Fremington areas. *Rep. Trans. Devon Ass. Advmt Sci.* **88,** 52–64.
—— 1958. The Saunton Pink Granite Erratic. *Rep. Trans. Devon Ass. Advmt Sci.* **90,** 179–186.
USSHER, W. A. E. 1904. Geology of Kingsbridge and Salcombe. *Mem. geol. Surv. U.K.*
—— 1914. A geological sketch of Brean Down with special reference to the marsh deposits of Somerset. *Proc. Somerset archaeol nat. Hist. Soc.* **60,** 17–40.
WATERS, R. S. 1965. The geomorphological significance of Pleistocene frost action in South West England. *In,* J. B. Whittow and P. D. Wood (Editors). *Essays in Geography for Austin Miller.* Reading, 39–57.
WELCH, F. B. A. 1955. Note on gravels at Kenn, Somerset. *Proc. Univ. Brist. Spelaeol. Soc.* **7,** 137.
WELLER, M. R. 1960. The erosion surfaces of Bodmin Moor. *Trans. R. geol. Soc. Corn.* **19,** 234–242.
—— 1961. The palaeography of the 430 ft shoreline stage in East Cornwall. *Abstr. Proc. Fourth Conf. Geol. and Geomorphologists in South West England,* 23–24.
WEST, R. C. and SPARKS, B. W. 1960. Coastal interglacial deposits of the English Channel. *Phil. Trans. R. Soc.* B **243,** 95–133.
WOOLDRIDGE, S. W. 1950. The upland plains of Britain: their origin and geographical significance. *Advmt Sci.* **7,** 162–175.
—— and LINTON, D. L. 1937. Structure, surface and drainage of South East England. *Trans. Inst. Brit. Geogrs* **10,** 1–124.
WORTH, R. H. 1898. Evidences of glaciation in Devon. *Rep. Trans. Devon Ass. Advmt Sci.* **30,** 378–390.
WRIGHT, L. W. 1967. Some characteristics of the shore platforms of the English Channel coast and the northern part of the North Island, New Zealand. *Z. Geomorph.* NF Band **11,** 36–46.
ZEUNER, F. E. 1959. *The Pleistocene period* (2nd edition). London, 276–307.

C. Kidson, The University College of Wales, Department of Geography, Llandinam Building, Penglais, Aberystwyth, Dyfed, Wales SY23 3DB.

The Quaternary evolution of the south coast of England

D. N. Mottershead

This paper considers the evidence available of Quaternary events affecting the south coast of England. Evidence of major changes in climate and sea-level is discussed, in addition to the enigmatic presence of far-travelled erratic material in a region in which there is no supporting evidence of direct glaciation. Gaps in present knowledge are indentified, and fruitful areas for possible future research suggested. A chronology is suggested which reflects the consensus interpretation of Quaternary events in the area.

1. Introduction

This paper is concerned with that part of the English coast between Land's End (SW 342 252) and Dungeness (TR 095 170); a straight line distance of nearly 500 km. This coastline transects geological outcrops ranging from Palaeozoic to Tertiary in age and in doing so impinges upon several geological provinces (Fig. 1).

In Cornwall and South Devon the Palaeozoic hard rock massif creates a resistant coastline, cliffed and highly indented. From Exmouth eastwards the coastline is formed of gently dipping rocks of Cretaceous and Jurassic age. These strata offer no great resistance to marine erosion, and consequently form an erosional cliffline which sweeps round Lyme Bay. Within the Hampshire Basin unconsolidated Tertiary strata form the bulk of the coastlines. These rocks will erode readily, as for example in Bournemouth Bay. Much of the Tertiary coastline, however, occurs in sheltered areas around the Solent, and this minimises coastal recession. Passing eastwards beyond the Hampshire Basin is the low coastline of

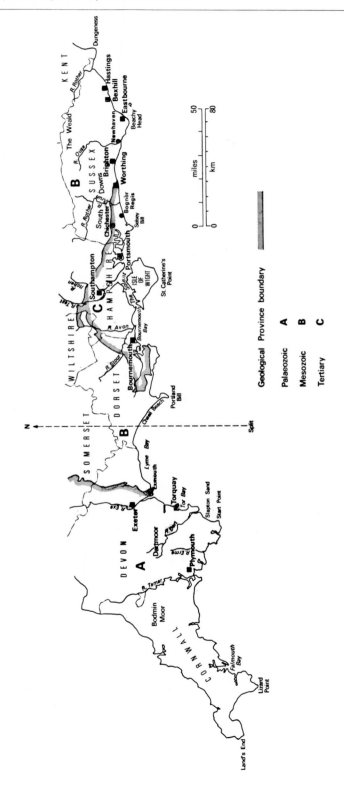

Fig. 1. Location Map.

the Sussex Coastal Plain. The western part of this is in the lee of the Isle of Wight and thereby protected to some extent from the effect of westerly gales. East of Brighton, a cliffed coastline appears again as first the Chalk, and then resistant members of the lower Cretaceous succession outcrop. This abundance of erosional coastlines means that features of Quaternary age close to sea-level are preserved only in selected localities, where protected by the outcrop of a resistant lithology, as at Portland Bill (SY 655 696) and throughout Cornwall and South Devon, or where sheltered from forces of marine erosion, as is the northeast coast of the Isle of Wight.

Zones of coastal deposition occur both in sheltered estuaries, and in areas where material is concentrated by longshore drift. The coastline is punctuated at intervals by estuaries, of which the most important belong to the Tamar, Dart, Exe, Solent, Arun and Ouse. Coastal accumulations form the three great shingle structures of the south coast, the bars of Slapton and Chesil and the cuspate foreland of Dungeness. Additionally, shingle spit accumulations have created sheltered harbours, where quiet water deposition has taken place throughout the latter part of Quaternary time, as in Poole Harbour, and Portsmouth and Langstone harbours.

2. Quaternary features

Along these varied types of coastline different lines of evidence exist for Quaternary events. These are as follows and will be treated in turn:—
  (a) High-level surfaces of denudation.
  (b) Raised shore platform and clifflines.
  (c) Erratic boulders.
  (d) Raised beaches.
  (e) Interglacial estuarine organic deposits.
  (f) Solifluction and related deposits.
  (g) Submerged channels, platforms and clifflines.
  (h) Postglacial submerged sediments and organic deposits.

2a. High-level surfaces of denudation

For a period of some 30 years, the views of Wooldridge and Linton, first put forward in 1938, have held sway regarding the high-level surfaces in southeast England. These authors postulated the existence of a surface of marine erosion at 600–650 ft (183–198 m) OD. The dissected surface they describe is widespread throughout the southeast and within the area of current interest occurs along the South Downs and around the margins of the Hampshire Basin. In certain localities in the North Downs, this bench is backed by a steeper slope interpreted as a degraded cliff. In places, again mainly in the North Downs, this bench carries pockets of sand and shingle with marine faunal remains of earliest Quaternary (Red Crag) age. To this morphological and sedimentological evidence, Wooldridge and Linton relate the evidence of drainage patterns. The central Wealden area, from which this 183–198 m beach is absent, carries a drainage pattern well adjusted to geological structure. Around the Wealden margins, where the bench is allegedly present, marked discordance of drainage occurs, as for example with the rivers Arun, Adur and Ouse, which break completely through the South Downs escarpment in draining to the English Channel coast.

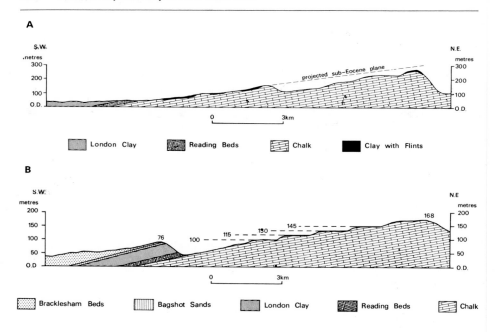

Fig. 2. Alternative interpretations of the early Quaternary evolution of the South Downs. 'A' indicates solutional lowering of the Chalk beneath the stripped sub Eocene plane, 'B' shows polycyclic evolution.

On the basis of these features, Wooldridge and Linton conclude that there was a marine transgression at the beginning of the Quaternary attaining a maximum altitude of 198 m OD. This transgression submerged most of the Wealden margins, but left the central Weald upstanding as an island. The Hampshire Basin was likewise submerged up to the level of the Wiltshire and Dorset Downs. As the sea-level fell, extended consequent streams formed, discordant to the geological structure beneath, accounting for the present drainage patterns. The subsequent fall in sea-level was supposedly intermittent, with stillstands creating benches of marine erosion at various levels below 198 m (Brown 1960).

This polycyclic view of the early Quaternary evolution has been followed in the coastal areas of Sussex by Sparks (1949) and Small and Fisher (1970). In Hampshire, Everard (1956) adopts a similar approach in mapping erosional flats and postulating a sequence of falling sea-levels with associated strandlines (Fig. 2). These same approaches have been followed in southwest England by a variety of authors (Brunsden 1963; Brunsden *et al.* 1964). These authors locate the early Quaternary shoreline more precisely at 690 ft (210 m), at which altitude well-marked flats occur backed by a degraded cliffline. Brunsden has pointed to the drainage relations of the Dart, which below 210 m is markedly discordant in the same way as the Sussex rivers. Direct sedimentary evidence of an early Quaternary high sea-level in the southwest is lacking, although the existence of quartz gravels at Crousa Down (Fig. 3) at an altitude of 105–110 m OD has been taken to indicate evidence of a marine transgression (Hendriks 1923; Flett and Hill 1946). There is,

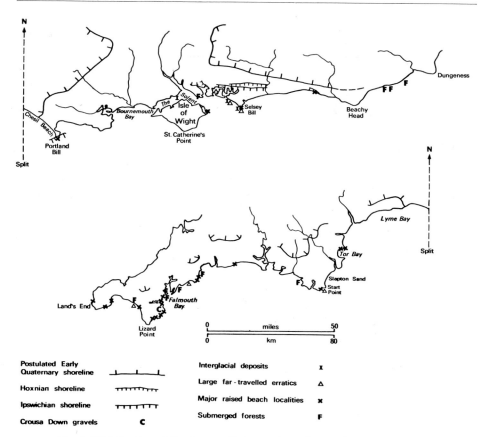

Fig. 3. Distribution of Quaternary features along the south coast.

however, little secure evidence as to their age. Nevertheless, several authors have found supporting evidence in morphology and drainage/structure relationships.

This view of early Quaternary events has been questioned as a result of recent work on the chalklands of the South Downs. As far as the south coast is concerned it is unfortunate that the marine deposits of earliest Quaternary age are almost entirely limited to the confines of the London Basin. Thus direct sedimentary evidence of the Waltonian transgression is absent from our area. Secondly, D. K. C. Jones (1974) has re-examined the discordance of drainage in southeast England and concludes that it can be explained by antecedence, and does not require superimposition from an overlying cover of marine strata associated with a high sea-level. And thirdly, soil studies by Hodgson, Catt and Weir (1967) of the backslope of the South Downs, suggest that the Clay-with-flints present there is residual material derived from a cover of Reading Beds formerly extending beyond the present outcrop and up the backslope of the Downs. They point out that the surface of the Downs is very close to a projected sub-Eocene surface indicated by the present contact between the Eocene rocks and the Chalk (Fig. 2). This line of argument is extended by Hodgson, Rayner and Catt (1974) who analyse the statistical distribution of the altitudes of erosional flats mapped by Sparks, and conclude that they are illusory rather than real.

In this part of the south coast, then, the evidence on which the polycyclic thesis is based, has recently been subject to reappraisal. The alternative view put forward by the workers cited above is that the chalklands have evolved during the late Tertiary and throughout the Quaternary by progressive stripping of the overlying Eocene strata, leaving only a residual mantle of Clay-with-flints. Where this has been stripped from the higher parts of the Downs solutional erosion was able to attack the Chalk surface directly and consequently to lower it. Thus the morphology of the Chalk backslope is interpreted as being entirely a function of subaerial processes.

This alternative view of early Quaternary events is, by its very nature, only applicable to the eastern part of the south coast. In the southwest the almost complete absence of Tertiary strata renders it inapplicable. Direct sedimentary evidence of the early Quaternary transgression is largely limited to the London Basin, and lacking in both east and west sections of the south coast, and current interpretation rests on morphological evidence alone. Perhaps this deficiency will be remedied as further detailed soil mapping continues.

Though doubt has been cast upon the reality of polycyclic surfaces on the chalklands of south central and southeast England, there is stronger evidence for such features at lower altitudes in the same region, developed upon rocks of Tertiary age. Thus Everard (1954a, 1956) describes the widespread gravel-covered terraces of the Hampshire Basin ranging in altitude from c. 130 m down to below present sea-level. These altitudes correspond with those in the Weymouth lowland where Sparks (1952) has also found evidence of polycyclic landscape evolution. In South Devon several authors have identified terrace levels and planation surfaces at these intermediate altitudes (Orme 1960; Kidson 1962; Brunsden 1963). The morphological evidence of a falling base level, punctuated by intermittent stillstands allowing terraces and limited planation surfaces to develop, is thus well established.

2b. Raised platforms and clifflines

Raised shore platforms at altitudes of 35 m OD and below are widespread along the south coast (Fig. 3). They were recognised as such and described by writers as early as Borlase (1758) and de la Beche (1839). They have subsequently been described from a variety of localities by a large number of authors. Recently Everard *et al.* (1964), Stephens and Synge (1966), Stephens (1970), Kidson (1971), James (1976), and Mottershead (1976b) have provided summaries of earlier work.

The raised shore platforms often form the basal member of Quaternary sequences within the region and as such are of considerable stratigraphic importance. There are, however, considerable problems associated with the interpretation and dating of these features. These problems are concerned with identifying the number of platforms present and the maximum height to which each extends as indicated by the wave-cut notch. This latter feature is, unfortunately, visible in only a few localities although James (*op. cit.*) describes a method for estimating its altitude. One problem which has clouded the interpretation of the wave cut platforms is the persistent use of the term '25-foot platform' in the literature, to which isolated fragments were referred in the absence of altitudes derived from precise levelling.

A closer examination of existing literature suggests a consensus which supports a concentration of shore platforms at four main levels. It should be emphasised that these altitudes relate to the platform height and neglect any depth of marine deposits upon them.

(i) The highest such feature at 30–35 m OD is to be found only in West Sussex and in the vicinity of Portsmouth. It is strongly developed, extending almost continuously over a distance of more than 25 km and intermittently beyond that. It has a maximum width of up to 4 km northwest of Chichester. Though usually mantled by deposits it has recently been exposed in section at Boxgrove Common (SU 925 083), where it can be seen rising gently northwards and truncating the dip of the underlying Chalk along an almost plane surface. It is perhaps surprising that a feature so well developed in one area should apparently be absent from the rest of the south coast.

(ii) There appears to be evidence of scattered fragments of platform c. 20 m OD. Of these the most outstanding, and yet isolated, example is that at Portland Bill. This platform is apparently matched by a fragment of similar altitude at Ports Down, Hampshire (SU 601 061), and at Mousehole, Cornwall (SW 159 277), Plymouth Hoe, Devon (SX 477 536), and several other fragments in southwest England as indicated by James (1976).

(iii) Strong evidence throughout the region exists of a platform extending to altitudes of 7–12 m OD. This is particularly extensive in West Sussex where it has been shown by Hodgson (1964) to attain a width from front to back of 13 km and to rise to at least 5·7 m OD. The long-known exposure of the wave-cut notch and fossil cliff associated with this feature at Black Rock, Brighton (TQ 335 033) shows it to attain a maximum altitude of 9 m OD (Smith 1936). In southwest England, Zeuner (1959), Orme (1960), and Wright (1967), all refer to a platform rising to between 7 and 8 m OD, although James prefers an altitude of 12–13 m OD especially on the exposed coasts near Land's End.

(iv) A still lower platform, washed by the present sea, and rising to between 4 and 5 m OD is present at many localities in southwest England and has been described by many of the previously mentioned authors. The notch and fossil cliff associated with this level is displayed at Pendower (SW 909 381) at 4·4 m OD.

The dating of these platforms presents problems. Two lines of argument have been followed in the past. Zeuner attempted long distance correlations with the Mediterranean sequence on the basis of altitude alone. An alternative approach has been to use the stratigraphy of the overlying deposits as an inductive tool. There is no good reason to suppose, however, that the platforms immediately predate the deposits lying on them, which range from giant erratics to beach material and Head deposits. These problems will be examined in later discussion.

2c. Erratic boulders

Boulders and stones of far-travelled material have long been known from the south coast, and provide evidence which poses problems even at the present day. Figure 3 indicates some of the more important localities at which these erratics are found. They range in size from the Giants Rock, at Porthleven, a block of microcline gneiss of some 50 tonnes, to material of gravel calibre. It is the large blocks which pose the most intriguing problem.

The giant erratic material is derived principally from three sources. Of local origin are the Devonian igneous rocks of Plymouth Hoe, the quartzites of the Dorset coast and the Jurassic, Cretaceous and Tertiary (greywether sarsen) erratics of the Hampshire and Sussex coast east of the Isle of Wight. Of more distant origin are the granitic and gneissic rocks. These appear to belong to two suites, those of southwest England deriving from sources to the north, whilst those of the Sussex coast can be matched with outcrops in Brittany, the Channel Isles and Cotentin. (Kellaway *et al.* 1975).

These erratics are widespread in distribution geographically but of limited altitudinal extent. Care must be taken in the interpretation of these features, for many erratics are found as part of human constructions, and are clearly not *in situ*. In addition it is possible that others were released as ballast from wrecks. Only those found in a stratigraphic context are discussed here, for it is these alone which are able to provide uncontestable evidence as to their origin. Many of the giant erratics are to be found on the present shore platform, whilst others are to be found on low-level rock platforms. In West Sussex Hodgson (1964) states that the majority are above present sea-level at altitudes up to a few metres above OD. Calkin (1934) records the presence of erratic pebbles from altitudes of 24–27 m in West Sussex.

It is difficult to conceive any rational explanation of the presence of the giant erratics which does not involve the action of ice. The majority of authors has favoured the mechanism of ice rafting by bergs, whilst the presence of an ice sheet in the English Channel has even been postulated, though strong arguments have been raised against this (Kidson and Bowen 1976). Neither of these possibilities is without its complications. There is little or no solid supporting evidence for the presence of an ice sheet in the English Channel. On the other hand, the presence of ice bergs in the Channel is perhaps more plausible, but raises the problem of how the Breton erratics reached southern England, unless pack ice formed along the French coast picking up the erratics before fragmenting and releasing erratic-bearing bergs into the English Channel. There is also the unsolved problem of why the erratics are so strongly concentrated in West Sussex, whilst being very thinly scattered along the coast of southwest England.

The effect of marine processes in the distribution of erratic material must also be considered. It is likely that foreign stones in contemporary beaches may have been trawled up from the Channel floor by the Flandrian transgression. It is equally possible that the erratic stones on the 30–35 m platform in West Sussex were carried up and emplaced by an earlier higher transgressive sea. These arguments apply to foreign material of finer calibre, whose altitudinal distribution is not necessarily directly related to the altitude attained by any glacial agency. Similarly it is possible that erratic stones including the giant ones, have been reduced in altitude by marine erosion, during the course of which cliffs have receded, and overlying material including erratics, has fallen to the foreshore and become incorporated into beach deposits. This may explain erratic blocks which are contained entirely within beach deposits a few metres above OD in Sussex (Hodgson 1964).

Any chronological consideration of the erratics must therefore take into account the original mode of transport of the material to the coast of southern England. In addition it must evaluate the present location of the material and whether it is *in situ* or subsequently disturbed and redeposited. In this connection it should be mentioned that Reid (1892) was of the opinion that the erratics were the earliest of the superficial materials on the Sussex coast and occur redistributed throughout all younger beds.

Finally, if floating ice was the agent by which the erratics came to the south coast, problems are raised of the sea-levels which obtained when this took place. The arguments pertinent to these problems will be taken up in a later section.

2d. The raised beaches

Raised beaches occur with a widespread distribution along the south coast (Fig. 3). It is unfortunate that the altitude and depth of these deposits are not always recorded precisely or in a standardised manner by the authors who have studied them. Some authors refer altitudes to OD and others to local high water mark, with the result that it is difficult to compare different sites in a precise manner. Furthermore, some authors quote only the altitude of the base of the beach thus rendering it difficult to suggest a maximum altitude for the transgression responsible for it. These difficulties point up the necessity for further work and in particular for precise height information on the raised beaches. Notwithstanding these problems, the raised beaches will be considered in four altitudinal groups.

(i) *The 30–35 m Raised Beach.* Sands and shingle of this deposit (previously referred to as the Hundred Foot, Goodwood, or Slindon Raised Beach) are widespread in east Hampshire and west Sussex, and have been widely recorded (e.g. Prestwich 1892; Reid 1892; Calkin 1934; Dalrymple 1957; ApSimon and Shackley 1976). Many of the exposures were of a temporary nature and are no longer visible today. The deposits attain a depth of up to 4 m, and ApSimon and Shackley show that they rise to 38–40 m OD. The sands are generally decalcified, and only at one locality, Waterbeach (SU 895 084), are organic remains recorded (Reid 1903). Here evidence of a temperate marine fauna is found. At Boxgrove (SU 923 080) the sands can be seen resting on a plane erosional bench cutting across the dip of the underlying Chalk, which is itself bored by marine organisms on its surface.

ApSimon and Shackley describe artifacts found in the upper part of the sediments. They are generally slightly abraded and apparently contemporaneous with the beach material. Uniformly of Acheulian type, they suggest strongly a Hoxnian age for this beach.

Kellaway *et al.* (1975) have recently reinterpreted this sediment as glacial outwash in conflict with all previous authors. This interpretation poses problems associated with the contemporaneity of temperate fauna and human artifacts, the erosional plane on which the sediments lie, the marine borings in the underlying Chalk, and the absence of supporting evidence of glaciation in the area at this altitude. Accordingly the marine interpretation is accepted in this paper.

(ii) *The 15–20 m Raised Beach.* Isolated fragments of beaches within this altitudinal range occur over a distance of over 300 km between the Foreland (SZ 658 876) and Penlee. Well developed beaches occur at the Foreland (White 1921) and Portland Bill (Prestwich 1875; Arkell 1947) both attaining an altitude of c. 20 m OD. Beaches at Ports Down (ApSimon and Shackley 1976), Plymouth (Masson-Phillips 1959) and Penlee (Reid and Flett 1907) occur at altitudes of 17–18 m OD. A survey of the literature underlines the need to standardise the altitudinal information on the above features, for the task of comparing them is confused by the variety of units and reference heights employed by different authors. Thus any comparison of altitudes at the present time can only be approximate. A substantial faunal list has been described for the Portland beach in which *Littorina neritoides* L., *Littorina littorea* (L.), *Patella vulgata* (L), and *Rissoa sp.* Frèminville are common. This fauna at the present day is distributed around the shores of northern Britain and Baden Powell (1930) suggests that it is indicative of temperatures at Portland 3–4°C lower than present. Erratic igneous stones are recorded in the Portland beach, whilst that at Plymouth contains locally derived erratic material.

(iii) *The Low Raised Beaches.* The greatest extent of raised beach deposits at below 15 m occurs on the lower coastal plain of the Hampshire/Sussex border.

Figure 4 shows its relationship in this area with the 30–35 m Raised Beach and platform. Here Hodgson (1964) has studied the extent and contents of this beach extensively. Its upper surface rises from 2·8 m OD to a maximum of 14 m OD, and it consists of sands and shingle, usually decalcified, though chalk pebbles occur where decalcification has not taken place. Erratic igneous boulders are present within the beach. Hodgson also recorded the presence of *Modiolus modiolus* (L.) and *Littorina saxatilis* (Olivi), whilst Prestwich (1859) found *Mytilus edulis* L., *Cardium edule* L. and *Tellina balthica* Philippi in addition. This beach appears to be the correlative of that of Black Rock where its maximum elevation is 12 m OD. In the southwest beach deposits at low altitudes are widespread, but any organic material, which may originally have been present, has been leached away (James 1976). Two exceptions to this are the beaches at Hope's Nose (SX 950 637) and Thatcher Rock (SX 944 628) both in Torbay. The former rises to 10–11 m OD and its fauna, described by Ussher and Lloyd (1933), contains *Ostrea edulis* L., *Mytilus edulis* L. and *Patella vulgata* L. That of Thatcher Rock is described by Hunt (1888) and also indicates temperate conditions, though slightly cooler than the present day.

(iv) Raised beaches are also abundant on the lowest rock platforms, at altitudes up to 4 m OD, but no faunal records appear to be published for these.

All of the raised beaches for which faunal records exist are therefore seen to contain material indicating temperate conditions, and they are referable in general to interglacial conditions. The faunas are not sufficiently diagnostic for assignation to a particular interglacial, although the artefacts of the 30–35 m beach suggest a Hoxnian age. The difficulty of effecting correlations on a faunal basis has led most authors to attempt correlations on an altitudinal basis, but in the absence of precise data on elevations this proves unsatisfactory. Until such data are available it seems that we can go no further than the broad altitudinal groupings considered above.

2e. Interglacial estuarine deposits

At two foreshore localities on the south coast, Selsey and Stone, there exist organic deposits of interglacial age (Fig. 3). They were first described by Reid (1892, 1893) and subsequently re-examined in more detail by West and Sparks (1960). The Stone deposits have recently been analysed by Brown *et al.* (1975).

At Selsey, the organic deposits occur between −4 and −1 m OD on the fore-shore in depressions cut into the Bracklesham Beds. They appear to be overlain by the raised beach which outcrops in the nearby low cliff. A sequence of silt and clay deposits occurs containing erratic (quartzite, limestone, and igneous) stones at the base. On the basis of pollen analysis the deposit is dated as Ipswichian. A temperate mammalian fauna is present together with a molluscan fauna including *Corbicula fluminalis* (Müll). Plant macrofossil and molluscan analysis suggests that the lower part of the deposit is of freshwater origin, the upper part estuarine, representing a marine transgression at −1·76 m OD in zone Ip IIb. West and Sparks suggest that this same transgression was responsible for the raised beach shingle above.

At Stone, in contrast, the organic deposits lie between two alluvial gravels at levels slightly above OD. A sequence of peats and estuarine clays indicates alternating brackish and freshwater conditions, suggesting an intermittent rise in sea-level. West and Sparks accord the same date to these deposits also, although Brown *et al.* (1975) suggest that the latter part of Ip IIb a more appropriate age. Brown *et al.*

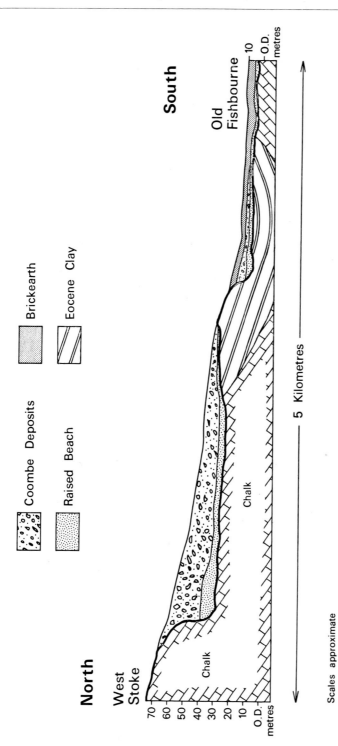

Fig. 4. Beaches and platforms of the Sussex Coastal Plain (after Hodgson 1964).

interpret the organic sediments as being of intertidal origin, and indicating an inter-mittently rising sea-level. Unlike Selsey, the overlying gravels are fluvial in origin and rest on a possibly eroded surface, thus rendering it impossible to demonstrate the height at which the transgression culminated.

Despite their distinctions, and their different stratigraphical relationships, they both indicate that during the mid-Ipswichian, sea-level was close to OD. Further-more this was a transgressive sea, which at Selsey probably rose to form the lower raised beach.

2f. Solifluction and related deposits

The most extensive Quaternary deposits indicative of cold conditions on the south coast of England are the solifluction deposits. These are widely preserved on the coastal slopes and platforms of the English Channel margin. They are well described in Geological Survey publications: Ussher (1904, 1907), Reid (1902), Reid and Flett (1907), White (1913, 1924) and Edmunds (1929). They have considerable regional variation, both in terms of facies and local terminology. In the southwest they are generally referred to as 'Head' deposits, whilst on the Chalk outcrop the local equivalent is often described as 'Coombe Rock' (Mantell 1822) or more recently, as Hodgson (1964) has preferred, by the term 'Coombe Deposit'.

These deposits often overlie raised beach deposits and raised coastal platforms. At the foot of a fossil cliffline they attain a depth of c. 25 m, as at Black Rock, Brighton and at several localities in Devon and Cornwall. Generally the deposit slopes seaward, thinning away from the old cliffline. These morphological relation-ships have been described by Mottershead (1971).

Head deposits are comprised of a *mélange* of materials derived locally from a position upslope. They are very poorly sorted and contain material ranging fre-quently from coarse frost-shattered angular boulders down to clay (Mottershead 1976a). There is considerable variation in appearance from one bedrock type to another. A particularly distinctive variation occurs on the Chalk outcrop where the Coombe Rock contains fractured flints and angular chalk fragments set in a chalky matrix. Sometimes it contains in addition derived Tertiary material. Occasionally the Head is crudely bedded, with layers dipping off the slope from which it is derived. This can be seen at Rickham Sand (SX 753 368) and Landing Cove (SX 776 354). Williams (1968) has estimated that between 3 and 10·5 m depth of material were removed by solifluction at sites in southern England. At certain localities in Sussex, Acheulian artifacts and temperate mammalian bones have been found (White 1924) but it is generally considered that these are derived fossils, and not contem-poraneous with the formation of the deposits. Thus they are not directly related to the age of the solifluction beds.

The main problems associated with the Head deposits are concerned with differentiation and dating. There is a general lack of datable organic materials within the head. Throughout the area no convincing example of a temperate deposit or weathering profile has been reported from within the Head. Conse-quently it has not been possible to assign the Head to more than one period of formation.

The coastal Head has been regarded by many authors as belonging to the period of the last glaciation, resting as it does on the temperate raised beach deposits, and generally not being overlain by any substantial depth of younger material. This poses problems in east Sussex, for the Coombe Deposit there overlies the two raised platforms and beaches, which arguably belong to two different interglacials. If this interpretation is correct, then the lower, younger platform would be overlain by one Coombe Deposit, whilst the upper one may reasonably be expected to carry the products of the two succeeding cold phases. As yet however, no work has distinguished in the field between two periods of Coombe Deposit formation although Dalrymple (1957) discriminates between the Wolstonian accumulation and the Devensian cryoturbation of this material at Slindon, West Sussex. It is possible that further detailed study of the sedimentology and weathering of the Coombe Deposit on the higher platform may elucidate this subject.

Interesting and significant variations occur in the upper parts of solifluction deposits. In southwest England there is a Main Head deposit often several metres in depth, and this is frequently overlain by a finer grained Upper head of less than 1 m depth. This deposit contains fewer stones, and more silt sized material than the Main Head beneath.

In southeast England the finer calibre of avilable sediment has led to the development of a sequence of chalky muds with incorporated silty material (Sparks 1949) of aeolian origin. One such deposit has been described from Cow Gap (TV 595 951) near Beachy Head, Sussex by Kerney (1963) another near Portsmouth by Shakesby (1973) and Gordon (1973). Here, as at many other sites on the Chalk, there exists a well-preserved terrestrial molluscan fauna, from which Kerney has elucidated a pattern of changing environmental conditions. Additionally two fossil soils are present which are attributed to the Alleröd and Bölling oscillations and are associated with a sharp increase in molluscan abundance. By implication the underlying Coombe Deposits are allotted to the Devensian. The volume of deposits overlying is interpreted by Kerney as representing a period of vigorous erosion in Zone III.

In both east and west, therefore, are common threads of evidence. The occurrence of silty material in both locations suggest a phase of dry conditions with abundant wind, followed by moist conditions during which solifluction and slope wash took place. Phases of soil formation are represented in the southeast, and in both regions cryoturbation structures sometimes affect these deposits.

2g. Submerged channels, platforms and clifflines

There exists a series of erosional features submerged below present sea-level, both within estuaries and in the offshore zone. This range of features comprises river channels cut in bedrock and associated terraces, both grading toward low sea-levels. In the offshore zone submarine platforms have been identified, as a result of recent surveys, fringing the present coastline and separated from each other by marked breaks of slope.

The existence of buried channels was recognised very early by Colenso (1832) who described a section in tin workings at Pentuan (SX 017 474) with a bedrock floor at −15·5 m OD. In 1898, Codrington noted the submerged rock valley of the Dart. More recent investigations have shown such buried channels to exist in several major rivers. Jones (1971) refers to the buried channel of the Ouse in Sussex, with borehole evidence of a rockhead altitude of −29·6 m OD at Newhaven. Seismic profiling has revealed a major buried channel in the eastern Solent (Dyer 1975). Everard (1954b) and Curry *et al.* (1968) have shown that this buried channel is associated with submerged gravel-covered terraces in Southampton Water.

Buried channels and terraces within the Teign and Exe valleys, as shown by borehole and seismic data, have been described by Durrance (1969, 1971,

1974). For the Exe he identifies two sets of buried channels and terraces, an older one cut in bedrock and infilled with gravels down to below −50 m OD, and a younger one cut into both bedrock and gravels grading to below −30 m OD. Clarke (1970) traces the extension of the buried channel offshore in Torbay to a level of at least −46 m OD. The Erme is shown by Macfarlane (1955) to have a bedrock profile cut to the same level, whilst the Dart (Green 1949) has a buried channel extending to a depth of at least 42 m.

Buried channels are therefore seen to be a common occurrence, and have been found associated with all the major river valleys examined to date. Dyer and Durrance, with sufficient evidence to construct long profiles, conclude that the Solent and the Exe systems are graded to −46 m and −52 m OD respectively.

Cooper (1948) drew attention to a submerged platform seaward of Plymouth. He described a submerged cliff 11 m high, with its base at −42·6 m OD. Hails (1975) describes a similar feature at the same depth in Start Bay. Kellaway *et al.* (1975) show this feature occurring again offshore of Portland Bill and extending up the eastern Channel from the Isle of Wight to beyond Beachy Head. Donovan and Stride (1975) claim the existence of two further submerged clifflines, with bases at 48–59 m below OD and 58–69 m OD. Although in some cases they may be coincident with major geological boundaries, in other areas, west of the Lizard for example, they are cut entirely in Palaeozoic rocks.

The evidence is strongest for the most-observed upper cliffline at −42 m OD the closest to shore. It may well be significant that the postulated base levels of the Solent, Exe, Dart, and Erme buried channels occur at the same height. This would suggest a major stillstand of sea-level, during which time river channels became graded to this altitude.

The age of these erosional features cannot be ascertained with any certainty although they must predate any overlying deposits.

2h. Postglacial submerged sediments and organic deposits

Submerged sediments are commonly found infilling buried channels and over-lying submerged platforms and terraces. Within estuaries and the many sheltered harbours of the south coast (e.g. Poole Harbour, Portsmouth Harbour) sediments have accumulated subsequent to the erosional episodes responsible for the features described earlier. Gravels, silts, peats and submerged forests are often found in these situations, and permit an interpretation of events during the most recent rise in sea-level. Peat layers and submerged forests which occur in these sedimentary successions were formed above the sea-level of their time, and also afford evidence of their age. Thus they are of great value in reconstructing the history of the sea-level rise during which they were formed.

The existence of submerged forests, in which tree stumps are found, often in the position of growth, and associated with leaf litter, roots and an underlying soil, has long been known. Borlase (1757) offers what appears to be the earliest observation of this feature on the south coast in the scientific literature, in describing a forest of oak, hazel and willow at least 3·6 m below high tide level. Colenso 1832 observed a submerged forest exposed in tin workings at −12·5 m OD. James (1847) and Meyer (1870) describe submerged forest layers in Portsmouth Harbour, exposed during construction of the Dockyard. The memoirs and maps of the Geological Survey indicate the widespread distribution of submerged forests at just below present sea-level. It has since become possible to date these deposits by pollen analysis, and more recently radiocarbon dating. Godwin and

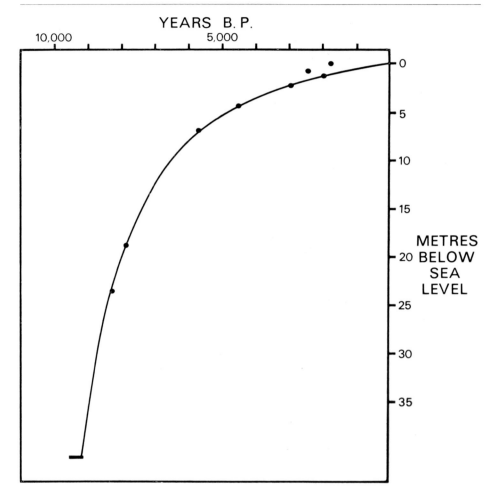

Fig. 5. Flandrian sea-level curve for Start Bay (after Morey 1976).

Godwin (1940) and Godwin (1945) provide examples of pollen dating of submerged peats in Southampton and Portsmouth Harbours respectively.

Figure 5 plots the altitude of organic layers in Start Bay for which dates are available. Morey (1976) has collated the data of other workers and presented it in a form which enables the Flandrian rise in sea-level to be observed. On account of the limited amount of data available, a smooth curve is obtained, lacking the oscillations described by Tooley (1974). A rapidly rising sea-level occurred during the early Flandrian, levelling off to its present position around 5500 BP.

A more detailed picture of the stratigraphy associated with this transgression for one locality, Southampton Water, is given by Hodson and West (1972) (Fig. 6).

Sea-level movements since about 5500 years ago are shown by two studies of

Poole Harbour (Gilbertson 1967; Byrne 1975). Both these authors identify a fall in sea-level of approximately 2 m between 5000 and 3800 BP. A widespread clay deposit within the harbour records a subsequent transgression rising to just above present sea-level by 2500 BP. It seems likely that future investigations in other sheltered harbours of the south coast will demonstrate the geographical extent of this event.

In addition to forming the deposits described above, the Flandrian transgression had a major influence on the form of the coastline of southern England. Major valleys and inlets became drowned by the rising sea-level to create the present highly indented coastline especially in the southwest and in Hampshire. The Isle of Wight became detached from the mainland. In addition, recent research has shown that major shingle structures were created at this time. Thus, Hails (1975) concludes that the shingle barrier of Slapton Sands on Torbay has been pushed ashore by the rising Flandrian sea-level, whilst Carr and Blackley (1973,1974) ascribe the shingle barrier of Chesil Beach, Dorset, to a similar origin.

3. Discussion

Whilst there are many points of agreement between the different authors, many unresolved problems exist concerning the Quaternary evolution of the south coast, and many points of disagreement have always existed with respect to the middle Quaternary. (See table 1, p. 316).

Fundamental questions are being raised about early Quaternary conditions. The view of Wooldridge and Linton of an early Quaternary sea-level at c. 200 m OD, accepted by many geomorphologists for areas throughout the length of the south coast has recently come under challenge. This view postulates that the surface morphology of coastal lands reflects a polycyclic evolution associated with an intermittently falling base-level. The alternative view of early Quaternary landscape evolution, developed upon the chalklands, is of a prolonged period of subaerial development. It is difficult to reconcile this latter mode of evolution with that suggested for the Palaeozoic region of the southwest.

Concerning the middle Quaternary, the principal questions revolve around the age and height of the raised marine erosion platforms, and the age of the deposits upon them. In this respect the erratic material is critical, especially the less mobile giant erratics. Occurring in isolation as they do, without the supporting presence of till, they pose great problems of stratigraphic interpretation. It is not possible to say conclusively whether they are in the position to which they were originally delivered to the south coast or whether they are reworked. The erratic gravel may have been repeatedly reworked by many processes, and consequently appears scattered throughout a variety of younger sediments. The giant erratics have either remained in their original position or moved downhill under gravity. Thus it is possible that their arrival either predates or postdates the platforms on which they occur. Erratics in a similar position on the Bristol Channel coast have been regarded as either Anglian in age (Edmonds 1972) or Wolstonian (Kidson and Wood 1974).

If the simplest assumption is made, that all the erratic material (both giant and gravel) was delivered to the south coast area during one phase, then the presence even of small erratics in the 30–35 m beach indicates that it predates this beach.

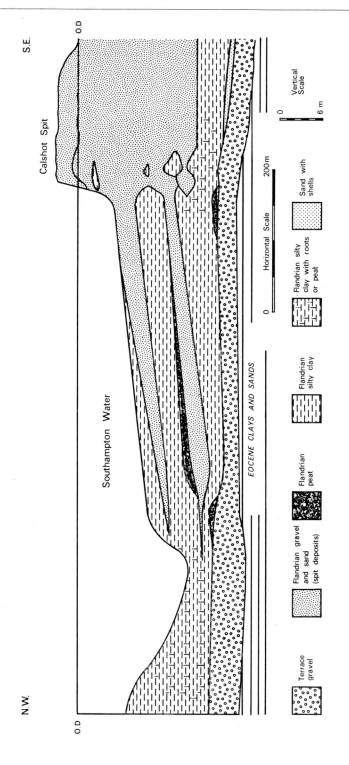

Fig. 6. Sequence of Flandrian deposits at Calshot, Southampton Water (after Hodson & West 1972).

Table 1. Summary of Quaternary features of the South Coast

Stage	Deposits	Erosion Features	Sea-level
FLANDRIAN	Coastal shingle accumulations Chalk dry valley infills Submerged forests and peats Marine silts and clays	Re-trimming of shore platform	Rising toward present level
DEVENSIAN	Upper Head Coombe Deposit/Main Head Cryoturbation of Wolstonian Deposit?	Submerged coastal rock platforms. Marine silts and clays Buried channels of major rivers	As low as −100 m, major stage at −42 − −35 m OD
IPSWICHIAN	Lower Raised Beaches Organic estuarine deposits	Lower platforms re-occupied	Rising to c. 10 m OD
WOLSTONIAN	Coombe Deposit ?	Dissection of Hoxnian platform	Low, relative to the adjacent interglacials
HOXNIAN	30–35 m Raised Beach	Upper Platform of Hampshire/Sussex	Rising to c 35 m OD
ANGLIAN	Erratic material	Low level shore platform	C. present sea-level
EARLIER QUATERNARY	Plateau Gravels of Hampshire Basin Crousa Gravels?	Widespread terrace levels of Hampshire Basin Bevelled spurs and interfluves, with superimposition of major rivers OR Solutional lowering of Chalk surface	Intermittently falling with periodic still-stands
EARLIEST QUATERNARY	Scattered littoral deposits of Red Crag type	Marine erosional bench with degraded cliff OR Solutional lowering of Chalk surface	183–210 m OD

Accepting the archaeological evidence as indicating a Hoxnian date for this beach, then the erratics may be assigned to a phase of low sea-level in the Anglian. Providing that the giant erratics on present shore platforms are close to their original position of deposition, this indicates that the lowest shore platform, washed by the present sea, may be as old as Anglian.

The age of higher shore platforms can only be related to the beaches upon them. The evidence for suggesting a Hoxnian age for the 30–35 m beach has already been described. The 15 m beach in Sussex is regarded by Sparks and West as Ipswichian in age, on account of its close association with the Selsey interglacial deposit, although the two features have not been observed in direct stratigraphic contact. This is, however, a dating with which most authors agree for the low-level beaches. A problem is therefore posed of where to fit into the sequence the beaches at 15–20 m. Though there is no solid evidence of their age, most authors refer these to the Ipswichian. Two interpretations of sea-level change for the last interglacial have been put forward. Zeuner (1959) regard the beaches as being younger at successively lower altitudes, putting the 30–35 m (Tyrrhenian) beach into the penultimate interglacial, and the 15–20 m (Monastirian) and 15 m (late Monastirian) beaches into the last interglacial. Recently Hollin (1970, 1974) has suggested that the Ipswichian contained two transgressions, an earlier one rising to the 15 m beach level, and a later surge to the 15–20 m level. It is evident that some positive method of dating these deposits is clearly required before these questions can be resolved.

The general patterns of events outlined above reflects the consensus of opinion of a large number of researchers. One conflicting interpretation has recently been put forward by Kellaway *et al.* (1975), who identify erratic materials at Selsey as a moraine and deposits as the higher platform in Sussex as fluvioglacial outwash. Many workers strongly doubt the field interpretation placed upon these features in that paper, and consequently the chronological interpretation put forward.

In Devensian and Flandrian times the pattern seems clearer and there is little disagreement between authors. The Devensian was characterised by massive solifluction, which strongly modified the form of the land surface and created much of the present day landscape. That this was a time of lower sea-level is evidenced by the marine erosion of many of the coastal Head deposits at the present day. Sea-level recovered to its present position during Flandrian time causing sediments to accumulate in sheltered waters.

4. Conclusion

It is evident, therefore, that major changes have taken place during the Quaternary along the south coast of England. In addition to major changes of climate and geomorphological process, there have also been changes in the coastline, both vertically and in plan. The greater resistance of the Palaeozoic rocks of the southwest restricted trimming by high interglacial seas to narrow benches. In contrast the more vulnerable Chalk and Tertiary beds of Sussex enabled widespread platforms to develop, which remain apparently protected from marine erosion by the presence of the Isle of Wight. In the intervening area, between the Exe and the Solent, recent marine erosion has removed all traces of former sea-levels with the sole exception of Portland Bill. Much of the present form of the coastline, with its drowned valleys and shingle accumulations, is therefore a direct consequence of the Flandrian transgression.

References

APSIMON, A. M. and SHACKLEY, M. L. 1976. Two new exposures of the Portsdown raised beach, near Fareham, Hampshire. *Proc. Hants Field and Arch. Soc.* (forthcoming).

ARKELL, W. J. 1947. The Geology of the country around Weymouth, Swanage, Corfe and Lulworth. *Mem. geol. Surv. G.B.*

BADEN POWELL, D. W. F. 1930. Notes on raised beach mollusca from the Isle of Portland. *Proc. malac. soc Lond.* **19,** 67–76.

BECHE, H. T. DE LA 1839. *Report on the geology of Cornwall, Devon and West Somerset.* London, H.M.S.O.

BORLASE, W. 1757. An account of some trees discovered underground on the shore at Mount's Bay in Cornwall. *Phil. Trans. R. Soc. Lond.* **50,** 51–53.

—— 1758. *The natural history of Cornwall.* Oxford. Privately published.

BROWN, E. H. 1960. The building of southern Britain. *Z. geomorph.* **NF4,** 264–274.

BROWN, R. C. *et al.* 1975. Stratigraphy and environmental significance of Pleistocene deposits at Stone, Hampshire. *Proc. geol. Ass.* **86,** 349–363.

BRUNSDEN, D. 1963. The denudation chronology of the River Dart. *Trans. Inst. Br. Geog.* **32,** 49–63.

—— KIDSON, C., ORME, A. R. and WATERS, R. S. 1964. Denudation chronology of parts of south-western England. *Field Studies* **2,** 115–132.

BYRNE, S. 1975. *Environmental changes in Poole Harbour during the Flandrian.* Unpublished B.A. dissertation, University of Durham.

CALKIN, J. B. 1934. Implements from the higher raised beaches of Sussex. *Proc. prehist. Soc.* **7,** 333–347.

CARR, A. P. and BLACKLEY, M. W. L. 1973. Investigations bearing on the age and development of Chesil Beach, Dorset, and the associated area. *Trans. Inst. Br. Geog.* **58,** 99–111.

—— and —— 1974. Ideas on the origin and development of Chesil Beach, Dorset. *Proc. Dorset. nat. Hist. and Arch. Soc.* **95,** 9–17.

CLARKE, R. H. 1970. Quaternary sediments off south-east Devon. *Q. J. geol. Soc.* **125,** 277–318.

CODRINGTON, T. 1898. On some submerged rock valleys in South Wales, Devon and Cornwall. *Q. J. geol. Soc. Lond.* **54,** 251–278.

COLENSO, J. W. 1832. A description of Happy Union tin streamwork at Pentuan. *Trans. R. geol. Soc. Corn.* **4,** 29–39.

COOPER, L. H. N. 1948. A submerged ancient cliff near Plymouth. *Nature Lond.* **161,** 280.

CURRY, D., HODSON, F. and WEST, I. M. 1968. The Eocene succession in the Fawley Transmisson Tunnel. *Proc. Geol. Ass.* **79,** 179–206.

DALRYMPLE, J. B. 1957. The Pleistocene deposits of Penfold's Pit, Slindon, Sussex, and their chronology, *Proc. Geol. Ass.* **68,** 294–303.

DONOVAN, D. T. and STRIDE, A. H. 1975. Three drowned coast lines of probable Late Tertiary age around Devon and Cornwall. *Mar. Geol.* **19,** 35–40.

DURRANCE, E. M. 1969. The buried channel of the Exe. *Geol. Mag.* **106,** 174–189.

—— 1971. The buried channel of the Teign estuary. *Proc. Ussher Soc.* **2,** 299–306.

—— 1974. Gradients of buried channels in Devon. *Proc. Ussher Soc.* **3,** 111–119.

DYER, K. R. 1975. The buried channels of the 'Solent River', southern England. *Proc. Geol. Ass.* **86,** 239–245.

EDMONDS, E. A. 1972. The Pleistocene history of the Barnstaple area. *Rep. No.* **72/2,** *Inst. geol. Sci.*

EDMUNDS, F. H. 1929. The Coombe Rock of the Hampshire and Sussex coast. *Summ. Prog. geol. Surv.* 63–68.

EVERARD, C. E. 1954a. The Solent River: a geomorphological study. *Trans. Inst. Br. Geog.* **20,** 41–58.

—— 1954b. Submerged gravel and peat in Southampton Water. *Pap. and Proc. Hants. Field Club* **18,** 263–285.

—— 1956. Erosion platforms on the borders of the Hampshire Basin. *Trans. Inst. Br. Geog.* **22,** 33–46.

——, LAWRENCE, R. H., WITHERICK, M. E. and WRIGHT, L. W. 1964. Raised beaches and marine geomorphology. *In,* Hosking, K. F. G. and Shrimpton, G. J. Present views of the geology of Cornwall and Devon. *R. geol. Soc. Corn.,* 283–310.

FLETT, J. S. and HILL, J. B. 1946. The Geology of Lizard and Meneage. *Mem. geol. Surv. G.B.*

GILBERTSON, D. D. 1967. *Post-glacial sea-levels in Poole Harbour, Dorset.* Unpublished B.A. thesis. University of Lancaster.

GODWIN, H. 1945. A submerged peat bed in Portsmouth Harbour. Data for the study of postglacial history. IX. *New Phytol.* **44**, 152–155.

—— and GODWIN, M. E. 1940. Submerged peat at Southampton: data for the study of post-glacial history. V. *New Phytol.* **39**, 303–307.

GORDON, P. J. 1973. *The molluscan fauna of the superficial deposits in Rake Bottom, Butser Hill.* Unpublished B.A. dissertation. Portsmouth Polytechnic.

GREEN, J. F. N. 1949. History of the River Dart, Devon. *Proc. Geol. Ass.* **60**, 105–124.

HAILS, J. R. 1975. Sediment distribution and Quaternary history. *J. geol. Soc., Lond.* **131**, 19–36.

HENDRIKS, E. M. L. 1923. The physiography of Southwest Cornwall, the distribution of Chalk flints and the origin of the gravels of Crousa Down. *Geol. Mag.* **60**, 21–32.

HODGSON, J. M. 1964. The low-level Pleistocene marine sands and gravels of the West Sussex Coastal Plain. *Proc. Geol. Ass.* **75**, 547–561.

——, CATT, J. A. and WEIR, A. M. 1967. The origin and development of Clay-with-flints and associated soil horizons on the South Downs. *J. Soil. Sci.* **18**, 85–102.

——, RAYNER, J. H. and CATT, J. A. 1974. The geomorphological significance of Clay-with-flints on the South Downs. *Trans. Inst. Br. Geog.* **61**, 119–131.

HODSON, F. and WEST, I. M. 1972. Holocene deposits of Fawley, Hampshire and the development of Southampton Water. *Proc. Geol. Ass.* **83**, 421–442.

HOLLIN, J. T. 1970. Antarctic ice surges. *Antarctic J. U.S.* **5**, 155–6.

—— 1974. Last interglacial sea-levels in the U.K. and eastern North America. *Quaternary Newsletter* **12**, 3–4.

HUNT, A. R. 1888. The raised beach of the Thatcher Rock; its Shells and their teaching. *Trans. Devon Assoc.* **21**, 225–253.

JAMES, H. 1847. On a section exposed by the excavation at the New Steam Basin in Portsmouth Dockyard. *Q. J. geol. Soc.* **3**, 249–251.

JAMES, H. C. L. 1976. Problems of dating raised beaches in south Cornwall. *Trans. R. geol. Soc. Corn.* (in press).

JONES, D. K. C. 1971. The Vale of the Brooks. *In,* Williams, R. B. G. (Editor). Guide to Sussex Excursions *Inst. of Br. Geog.* 43–46.

—— 1974. The influence of the Calabrian transgression on the drainage evolution of south-east England. *Inst. Brit. Geog. Spec. Pub.* **7**, 139–158.

KELLAWAY, G. A., REDDING, J. H., SHEPARD-THORN, E. R., and DESTOMBES, J-P. 1975. The Quaternary history of the English Channel. *Phil. Trans. R. Soc. Lond. A.* **279**, 189–218.

KERNEY, M. P. 1963. Late-glacial deposits on the chalk of southeast England. *Phil. Trans. Roy. Soc. B.* **246**, 203–254.

KIDSON, C. 1962. The denudation chronology of the River Exe. *Trans. Inst. Br. Geog.* **31**, 43–66.

—— 1971. The Quaternary history of the coasts of southwest England, with special reference to the Bristol Channel coast. *In,* Gregory, K. J. and Ravenhill, W. (Editors) *Exeter essays in geography* 1–22.

—— and WOOD, R. 1974. The Pleistocene stratigraphy of Barnstaple Bay. *Proc. Geol. Ass.* **85**, 223–238.

—— and BOWEN, D. Q. 1976. Some comments on the history of the English Channel. *Quaternary Newsletter* **18**, 8–10.

MACFARLANE, P. B. 1955. Survey of two drowned river valleys in Devon. *Geol. Mag.* **92**, 419–429.

MANTELL, G. A. 1822. *The Fossils of the South Downs.* London, Lupton Relfe.

MASSON-PHILLIPS, E. H. 1959. The raised beaches at Plymouth Hoe, South Devon. *In,* Zeuner, F. E. *The Pleistocene Period* 374–5.

MEYER, C. J. A. 1870. On the lower Tertiary deposits recently exposed at Portsmouth. *Q. J. geol. Soc.* **27**, 74–89.

MOREY, C. R. 1976. The natural history of Slapton Ley Nature Reserve IX. The morphology and history of the Lake basins. *Field Studies* **4**, 353–368.

MOTTERSHEAD, D. N. 1971. Coastal head deposits between Start Point and Hope Cove, Devon. *Field Studies* **3**, 433–453.

—— 1976a. Quantitative aspects of periglacial slope deposits in southwest England. *Biul. peryglac.* **25**, 35–57.

—— 1976b. The Quaternary history of the Portsmouth region. *In,* Mottershead, D. N. and Riley, R. C. (Editors). *Portsmouth Geographical Essays* Vol. II, 1–21.

ORME, A. R. 1960. The raised beaches and strandlines of south Devon. *Field Studies* **1**, 109–130.

PRESTWICH, J. 1859. On the westward extension of the old raised beach of Brighton. *Q. J. geol. Soc. Lond* **15**, 215–221.

—— 1875. Notes on the phenomena of the Quaternary period in the Isle of Portland and around Weymouth. *Q. J. geol. Soc.* **31**, 29–54.

—— 1892. The Raised Beaches and 'Head' or Rubble-Drift of the south of England. *Q. J. geol. Soc. Lond.* **48**, 253–343.

REID, C. 1892. The Pleistocene deposits of the Sussex coast and their equivalents in other districts. *Q. J. geol. Soc. Lond* **48,** 344–61.

—— 1893. A fossiliferous deposit at Stone, on the Hampshire coast. *Q. J. geol. Soc. Lond.* **49,** 325–9.

—— 1902. Geology of the country around Southampton. *Mem. geol. Surv. G.B.*

—— and FLETT, J. S. 1907. The Geology of the Land's End district. *Mem. geol. Surv. G.B.*

SHAKESBY, R. A. 1973. *A study of the valley deposits in Rake Bottom, Butser Hill.* Unpublished dissertation. Portsmouth Polytechnic.

SMALL, R. J. and FISHER, G. C. 1970. The origin of the secondary escarpment of the South Downs. *Trans. Inst. Br. Geog.* **49,** 97–107.

SMITH, B. 1936. Levels in the raised beach, Black Rock, Brighton. *Geol. Mag.* **73,** 423–426.

SPARKS, B. W. 1949. The denudation chronology of the dip-slope of the South Downs. *Proc. Geol. Ass.* **60,** 288–293.

—— 1952. Stages in the physical evolution of the Weymouth lowland. *Trans. Inst. Br. Geog.* **19,** 17–29.

STEPHENS, N. 1970. The West Country and Southern Ireland. *In,* Lewis, C. A. (Editor). *The Glaciations of Wales and adjacent regions.* London, Longmans 267–314.

—— and SYNGE, F. M. 1966. Pleistocene shorelines. *In,* Dury, G. H. (Editor) *Essays in geomorphology.* London, Heinemann 1–52.

TOOLEY, M. J. 1974. Sea-level changes during the last 9000 years in north-west England. *Geog. J.* **140,** 18–42.

USSHER, W. A. E. 1904. The Geology of the country around Kingsbridge and Salcombe. *Mem. geol. Surv. G.B.*

—— 1907. The Geology of the country around Plymouth and Liskeard. *Mem. geol. Surv. G.B.*

—— and LLOYD, W. 1933. The Geology of the country around Torquay. *Mem. geol. Surv. G.B.*

WEST, R. G. and SPARKS, B. W. 1960. Coastal interglacial deposits of the English Channel. *Phil. Trans. R. Soc.* **243,** 95–133.

WHITE, H. J. O. 1913. The Geology of the country around Fareham and Havant. *Mem. geol. Surv. G.B.*

—— 1921. The Geology of the Isle of Wight. *Mem. geol. Surv. G.B.*

—— 1924. The Geology of the country near Brighton and Worthing. *Mem. geol. Surv. G.B.*

WILLIAMS, R. B. G. 1968. Some estimates of periglacial erosion in southern and eastern England. *Biul. peryglac.* **17,** 311–335.

WOOLDRIDGE, S. W. and LINTON, D. L. 1938. The influence of the Pliocene transgression on the geomorphology of southeast England. *J. Geomorph.* **1,** 40–54.

WRIGHT, L. W. 1967. Some characteristics of the shorelines of the English Channel coast and the northern part of the North Island, New Zealand. *Z. geomorph.* **11,** 36–46. 36–46.

ZEUNER, F. E. 1959. *The Pleistocene Period: its Climate, Chronology and Faunal Successions.* London, Hutchinson and Co. (Publishers) Ltd.

D. N. Mottershead, Portsmouth Polytechnic, Department of Geography, Lion Terrace, Portsmouth PO1 3HE.

Acknowledgments

We are grateful to the following to reproduce here, in whole or in part or in revised form, the following figures: the Director, Institute of Geological Sciences, Fig. 1, opposite p. 14 and Fig. 3, p. 19; R. A. Garrard and Elsevier Scientific Publishing Co., Fig. 1, opposite p. 70 and Fig. 10, opposite p. 90; N. Stephens and the editor of *Irish Geography*, Fig. 1, p. 181, Fig. 2, p. 182, Fig. 3, p. 185 and Fig. 4, p. 186; N. Stephens and the Royal Society of London, Fig. 6B, p. 188; F. M. Synge, N. Stephens and the Royal Irish Academy, Fig. 6C, p. 188; F. Mitchell and N. Stephens, Fig. 7, p. 193; N. Stephens, A. E. P. Collins and the Royal Irish Academy, Fig. 7, p. 193; A. G. Smith and the Royal Irish Academy, Fig. 7, p. 193; A. E. P. Collins, F. Mitchell and N. Stephens, Fig. 8, p. 195; the Geologists' Association, Fig. 3a, p. 268; F. M. Synge and the Longman Group Ltd., Fig. 3b, p. 268; N. Stephens and the editor of *Biuletyn peryglacjalny*, Fig. 3c, p. 268; F. Mitchell and the British Association for the Advancement of Science, Fig. 4a, p. 272; C. Kidson and the Geologists' Association, Fig. 4b, p. 272; C. Kidson, A. Heyworth and the Geological Society of London, Fig. 5, p. 275, Fig. 8, p. 287, Fig. 9, p. 289, Fig. 11, p. 292, Fig. 12, p. 293; D. D. Gilbertson, D. N. Mottershead and the Field Studies' Council, Fig. 7, p. 280; J. M. Hodgson and the Geologists' Association, Fig. 4, p. 309; C. R. Morey and the Field Studies' Council, Fig. 5, p. 313; F. Hodson, I. M. West and the Geologists' Association, Fig. 6, p. 315.

Author index

Numbers in bold type refer to pages on which references are listed and those followed by an asterisk refer to text figures.

W

Place index

Abbot Moss 149
Aberaeron 242
Aberarth 242, 243
Aberayron 70
Aberdyfi 244
Abergele 251
Abermawr 241, 248
Aberogwen 250
Aberporth 242
Abervalley 250
Aberystwyth 85, 243
Allonby 139
Alston 121
Altcar 133
Alt Mouth 133, 134
Anglesey 14, 15, 20, 63, 249, 252
Annagassan 214, 216
Annalong 184
Annan 175
Antrim 7
Antrim Co. 179, 189, 196
Antrim Plateau 183
Applecross Peninsula 102
Ardglass 182, 183, 184, 188, 191
Ardnamurchan 103
Ardnamurchan Peninsula 102
Ards Peninsula 183, 186
Arklow 206
Arklow Rock 200, 206
Arnside Moss 136, 139
Arran 110
Askrigg Block 14
Avoca 206, 208
Ayrshire 114, 116, 117
Azores 95

Baggy Point 266
Baginbun Head 204
Balbriggan 211
Balcary Point 17, 27
Ballagan Point 184
Ballaquark 169
Ballateare 164
Ballure 165, 167, 168, 170
Ballycotton 75
Ballycroneen 3, 75, 218, 266
Ballyhalbert 194, 196
Ballyhillin 189
Ballykeel 183, 184, 188
Ballykeerogemore 205
Ballymadder 204
Ballymadder Point 204
Ballymartin 181
Ballymoyle Hill 206
Ballyquintin Point 183, 184, 188
Ballyteige Bay 203, 206
Ballytrent 203
Baltic Sea 1, 7, 8
Baltray 212
Banc-y-warren 248
Bangor 27
Bank End Moss 142
Bannow Bay 203, 208
Bann Valley 189
Bardsey Island 24, 90, 246, 252
Barmouth 252
Barmouth Bay 80, 85, 88
Barnstaple Bay 2, 75, 262, 264, 266–270, 279, 294
Barron Wood 123
Barrow 200

Subject index